Machine Learning in Educational Sciences

Myint Swe Khine
Editor

Machine Learning in Educational Sciences

Approaches, Applications and Advances

Springer

Editor
Myint Swe Khine
Curtin University
Bentley, WA, Australia

ISBN 978-981-99-9378-9 ISBN 978-981-99-9379-6 (eBook)
https://doi.org/10.1007/978-981-99-9379-6

This Springer imprint is published by the registered company Springer Nature Singapore Pte Ltd.
The registered company address is: 152 Beach Road, #21-01/04 Gateway East, Singapore 189721, Singapore

Paper in this product is recyclable.

Preface

Educational research involves collecting and analyzing data to improve teaching, learning, and administrative practices in formal and non-formal settings. Researchers and practitioners use a variety of approaches, including qualitative, quantitative, and mixed methods to identify effective teaching methods, develop new educational programs, evaluate the effectiveness of educational policies, and understand the factors that contribute to student success. Educational research is a complex and challenging field, but it is also very rewarding. It has the potential to make a real difference in the lives of students and educators around the world. Educational research can help us to better understand how students learn and how to create more effective learning environments and improve the quality of education.

Artificial intelligence (AI) has been making headlines news in recent years, with promises of better customer service with AI-powered chatbots, new drugs and treatments for diseases, and personalized learning experiences for students. AI is also rapidly transforming the education landscape and has the potential to revolutionize education. AI-powered tools are being used for immersive learning, real-time feedback, and automating tasks such as grading papers and managing student records. All of these can help students learn more effectively. However, these applications of AI would not be possible without machine learning. Machine learning is a key driver and a powerful tool that enables AI to learn and refine over time. Educational researchers increasingly use machine learning in their quest to improve the quality of education. This book is about approaches, applications, and advances in the use of AI and machine learning in educational sciences. It is hoped that the book will add to the growing body of literature on the applications of machine learning in educational sciences.

I would like to express my sincere gratitude to the contributors who shared their valuable experience in the use of machine learning in education sciences. I am also grateful to the anonymous reviewers who spent their valuable time reviewing the manuscripts and providing helpful suggestions. Their contributions have made this volume possible and I am deeply indebted to them.

Bentley, Australia Myint Swe Khine

Contents

Editor and Contributors

About the Editor

Prof. Myint Swe Khine currently teaches at the School of Education, Curtin University, Australia. He has more than 30 years of experience in teacher education. He received Master's degrees from the University of Southern California, USA, University of Surrey, UK, and the University of Leicester, UK, and a Doctoral degree from Curtin University, Australia. He worked at the National Institute of Education, Nanyang Technological University, Singapore, and was a Professor at Emirates College for Advanced Education in the United Arab Emirates. He has wide-ranging research interests in teacher education, science education, learning sciences, psychometrics, measurement, assessment, and evaluation. He is a member of the Editorial Advisory Board of several international academic journals. Throughout his career, he has published over 40 edited books. The most recent volumes include Methodology for Multilevel Modelling in Education Research: Concepts and Applications (Springer, 2022), and Rhizomatic Metaphor: Legacy of Deleuze and Guattari in Education and Learning (Springer, 2023).

Contributors

Balqis Albreiki is a Ph.D. Candidate in Computer Science at the United Arab Emirates University. Her research interests mainly focus on developing intelligent machine learning models that can be applied to solve specific educational challenges, such as identifying students at risk of failure. She has published numerous journal and conference papers, showcasing her ability to conduct high-quality research that contributes to the field of Computer Science. In recognition of her outstanding achievements in research and innovation, Balqis was awarded the prestigious Chancellor's Innovation Award at UAE University in 2021.

Pallavi Asthana is currently working as a full-time Assistant Professor in the Department of Electronics and Communication Engineering at Amity University, Uttar Pradesh, India. She has done her M.Tech. in Digital Systems and is pursuing Ph.D. in Computer Science and Engineering. Her area of interest includes Instrumentation, machine learning, Control Systems, etc. Currently, she is working in the field of Engineering Education and developing the tools for the quantitative measurement of course learning outcomes. She has published more than 25 research papers in reputable conferences and journals. She has also written 6 book chapters. Pallavi Asthana is a member of the IEEE and IETE, UK. She has been a key contributor to organizing various international conferences and workshops at Amity University. She is also a member of the Internal Quality Assurance Cell at Amity University.

László Bognár is an engineer and mathematician and has been teaching various engineering and mathematics subjects at the University of Dunaújváros in Hungary for more than 20 years. He was the Rector of the university for 8 years. Besides teaching, he is intensively involved in industrial projects as a consultant. His special fields are quality statistics, statistical process control, multivariate statistical analysis, and stochastic processes. He has been elected President of the Fejér County Chamber of Engineers and has received several awards. Today, his focus is on the theory and practice of learning analytics in education. With his team, he has developed unique machine learning models to track student activity within a course and alert students who are at risk of dropping out.

Richard Brooker is a data scientist with 11 years of experience across multiple domains, including marketing, logistics, recommendation engines, computer vision, natural language processing, analytics, data mining and enrichment, and causal analysis. He holds a master's degree in mathematics from Warwick University and has worked for Channel 4, the BBC, Uber, and WPP. Richard is currently the lead data scientist at Alef Education and was a technical reviewer on the book "The Reinforcement Learning Workshop," published by Packt Publishing. His research interests include generative AI and applications of machine learning to education.

Eduardo C. Cascallar received a Ph.D. degree from the University of Texas, Austin. Special Guest Professor at KU Leuven University and Managing Director for Assessment Group International (Belgium/USA). International consultant for the World Bank, OECD, Asian Development Bank, and UNESCO. He was the Principal Research Scientist at the American Institutes for Research. He served as Director of Assessment at the City University of New York and as Coordinator for Testing and Research at the Center for the Advancement of Language Learning in Washington, for the US government, and was in the faculty of Georgetown University and the University of California, Los Angeles, and he was Research Scientist at Educational Testing Service. He has worked in predictive methodologies in educational, medical, and national security applications.

Ying Cheng Ph.D., is a Professor in the Department of Psychology at the University of Notre Dame and Director of the LAMBS Lab. She is also the Associate Director of the Lucy Family Institute for Data and Society.

Yiran Cui Doctor of Education, is an Associate Researcher at the College of International Education, Shandong University. Her research focuses on Blended Learning, Educational Informatization, and the Research on Quantitative Analysis Based on International Large-Scale Educational Data.

Chih-Pu Dai is an Assistant Professor of Learning Design and Technology at the University of Hawai'i at Mānoa. Chih-Pu earned his Ph.D. in Instructional Systems and Learning Technologies from Florida State University. His research interests focus on enhancing and supporting immersive learning experiences in advanced learning technologies, such as game-based learning environments, virtual reality, and mixed reality, through the applications of artificial intelligence and machine learning. His research has been published in leading journals such as Computers & Education, Educational Technology Research and Development, and the British Journal of Educational Technology. Chih-Pu is an active member of professional organizations such as the International Society of Learning Sciences (ISLS), and the American Educational Research Association (AERA).

Madelynn Denner is currently pursuing a Master's in Applied and Computational Math and Statistics from the University of Notre Dame. Madelynn is a graduate research assistant in the Learning Analytics and Measurement in Behavioral Sciences (LAMBS) Lab at the University of Notre Dame.

Manohar Dhulipala has a bachelor's degree in Applied Mathematics and a minor in Computer Science from the University of the Pacific. His interests include machine learning, cryptography, and graph theory.

Christine DiStefano is a Professor of Educational Research and Measurement at the University of South Carolina. She teaches courses in test theory, educational measurement, and classroom assessment. Her research interests include analysis of survey data within the structural modeling framework, Rasch analysis, and investigations of validity and test performance to assess children's social-emotional/behavioral health.

Sebastian Dziallas is an Assistant Professor of Computer Science at the University of the Pacific. He previously served as a founding faculty member at Fulbright University in Vietnam. He received a Ph.D. in Computer Science from the University of Kent and a B.S. in Engineering from Olin College of Engineering. His research interests in computing education research include using narrative methodologies to understand student experiences and exploring novel work-based learning approaches.

Kiran Fahd completed her Ph.D. degree in Business Systems and Analytics entitled "An Innovative Big Data Analytics Method for Decision Makers in the Higher Education Sector" at the University of Newcastle, NSW, Australia. Dr. Fahd has published several journal and conference articles on big data analytics and machine learning studies.

Janet Hanson is currently a faculty in the Department of Education Leadership at the University of North Texas at Dallas. She previously worked as an Associate Professor at Azusa Pacific University in the Educational Leadership Department and Teacher

Education Department where she wrote and taught masters level research courses on topics such as leadership for diverse schools, research and leadership, community engagement, ethnography, etc., Hanson was an active committee member of the Institutional Review Board at APU for six years. She is a published author in peer-reviewed journals and has a book titled, Manage Your Mindset: Maximize Your Power of Personal Choice published by Rowman and Littlefield.

Bramah Hazela is working at the Department of Computer Science & Engineering as an Assistant Professor at Amity School of Engineering & Technology Lucknow, Amity University Uttar Pradesh Lucknow Campus, India since July 2012. He earned a Doctoral degree in Engineering in 2018. He has 20 years of rich academic experience. He has published 40 research papers and many book chapters in Scopus-indexed journals. He has presented papers at international conferences sponsored by Springer, IEEE, and ACM. He also authored four books and served as Session Chair at various International Conferences. He is also a member of the IET UK, IEEE, ACM, and CSI societies. He has served as an invited faculty member at various engineering colleges. He has also served as an academic member of state universities and state-funded autonomous engineering institutes. His current research interests are formal language theory, Natural Language Processing, blockchain, data science, and assessment of learning outcomes.

Alain Hennebelle is a Teaching Professional in the School of Computing and Information Systems at The University of Melbourne and Deakin University in Melbourne, Australia. He earned his Master's in Computer Science from France at Ecole des Mines de St-Etienne and University of St-Etienne, and a micro-master in Statistics and Data Sciences from Massachusetts Institute of Technology (MIT), USA. He is a Chief Architect with extensive experience in Artificial Intelligence, Machine Learning, data science, software engineering, and Development Operations (DevOps), working in international companies, such as Sun Microsystems (Oracle), and Bull (France) in cooperation with IBM. His research interests include Machine Learning algorithms and statistics, cloud computing, and smart city applications.

Brittany Ho is a third-year student at the University of the Pacific, where she is pursuing a major in computer science. Her interests include applying various Natural Language Processing (NLP) techniques and machine learning models to data emanating from diverse fields.

Maxwell R. Hong received a Ph.D. in Quantitative Psychology from the University of Notre Dame. Dr. Hong has worked as a Research Assistant in the LAMBS Lab at the University of Notre Dame.

Leila Ismail is the Founding Director of the Intelligent Distributed Computing and Systems (INDUCE) Lab in the Department of Computer Science and Software Engineering of the College of Information Technology of the United Arab Emirates University, and a Visiting Professor in the School of Computing and Information Systems, Faculty of Engineering and Information Technology at The University of Melbourne, Cloud Computing and Distributed Systems (CLOUDS) Lab. She has vast industrial and academic experience in Artificial Intelligence and Machine Learning publishing her work in top-ranked venues, regularly invited as keynote speaker throughout the world. She is a recipient of several prestigious awards. Leila is the Editor of the Information Innovation Technology in Smart Cities, published by Springer Nature in 2018.

Fabian Kieser is a Research Assistant at the University of Education in Heidelberg, Germany. He studied mathematics and physics for the teaching profession at grammar schools at the University of Heidelberg, Germany. His research interest is in physics problem-solving in higher education and school. He is also interested in how current methods of natural language processing can be used for different teaching and learning scenarios.

Tanya Linden is a Senior Teaching Fellow at the University of Melbourne. Her research interests are in the application of information technology in various areas of education, enterprise systems, and promoting STEM careers to female students. She is well-published in all these areas and regularly presents her work throughout North America, Europe, and Australasia. Her publication "Scrum-Based Learning Environment: Fostering Self-Regulated Learning" in the Journal of Information Systems Education received the Best Paper Award 2018. Her latest grants support projects addressing the underrepresentation of females in STEM. Tanya is an Associate Editor for the international Springer journal Education and Information Technologies.

Yangyang Luo received a B.S. degree in software engineering from Chongqing University, Chongqing, China, in 2011 and an M.S. degree in software engineering from Chongqing University, Chongqing, China, in 2014, and a Ph.D. degree in education from Tsinghua University, Beijing, China in 2021. He is currently working as Lecturer at the Institute of Higher Education at Lanzhou University, Lanzhou, China. His main research interest centers on digital learning environments and educational data mining. He is also interested in learning analytics field and methodologies of artificial intelligence application in education.

Gafur Mammadov is a third-year computer science student at the University of the Pacific. His research interests are machine learning and data science. He interned at Nvidia from May 2023 to December 2023 working at the intersection of high-performance computing and machine learning.

Shah J. Miah is a Professor of Business Analytics, Program Director, Bachelor of Business Analytics, and Head of the Business Analytics, Economics & Politics at Newcastle Business School, The University of Newcastle, New South Wales, Australia. Since receiving his Ph.D. degree from Griffith University in the area of Business Decision Support Systems, his research interests have expanded to Artificial Intelligence (AI) oriented information systems (IS).

Xin Miao is the lead researcher at Alef Education. She holds a master's degree in International Education Policy from Harvard University and a graduate certificate in International Development from Johns Hopkins University SAIS. She has bottom-up education practice experience in China's highly professionalized K12 education system and has been working in R&D in the EdTech industry in the UAE, focusing on education research, policy analysis, evidence-based R&D in the private sector to support K12 education transformation in this region and globally.

Sumita Mishra is a full-time Assistant Professor in the Department of Electronics and Communication Engineering at Amity University, Uttar Pradesh, India. She has received the M.Tech. degree in optical communication from the Shri Govindram Seksaria Institute of Technology and Science, Indore, India. She received a Ph.D. degree from Dr. Ram Manohar Lohia Avadh University, Faizabad, India. Dr. Mishra is a member of IET (UK), IAENG, and has been a Senior Member of IEEE. She has several publications in international journals and prestigious conferences. She is a reviewer and editorial board member of many highly reputed journals. Dr. Mishra had key responsibilities in organizing several International Conferences and various international events. Her current research interests include deep learning, visible light communication, and the Internet of Things. She is also involved in international collaborations and is guiding several Ph.D. students in this area.

Samantha Monroe is a Junior Education Researcher at Alef Education. She holds a Master's degree in International Education Policy from Harvard University and an undergraduate degree in Psychology and Special Education from Marist College. She has interned for Educate Lanka and the Center for Global Development, has worked as a research consultant for the KHDA in Dubai, and received a Fulbright grant as an English Teaching Assistant in Malaysia for 2018. Her research interests include international education research, policy, and analysis, particularly for inclusive education initiatives.

Mariel F. Musso received Ph.D. in Psychology (National University of San Luis, Argentina) and Ph.D. in Educational Sciences (KU Leuven University, Belgium). Research Scientist at the Developmental Cognitive Neuroscience Lab at the Mind Brain and Behavior Research Center, University of Granada. She is an Associate Researcher at National Council for Scientific and Technical Research (CONICET) and Professor at UADE, Argentina. She is an international consultant in many national assessment projects for Assessment Group International and the World Bank, in countries of Southeast Asia, Latin America, the Middle East, and Europe. She has published many papers in journals and several book chapters in her main areas of interest.

Teresa M. Ober is an Associate Research Scientist in the Center for K-12 Teaching, Learning, and Assessment at Educational Testing Service (ETS).

Geophrey Odero is a Ph.D. candidate (Educational Research and Measurement) and an instructor of Statistics at the University of South Carolina. He teaches courses in Elementary Statistics, Statistical Reasoning, and Elementary Statistics for Biological and Health Sciences. His research interests include Item Response Theory, Scale Construction, Multivariate Statistics, Educational and Psychological Measurement, Structural Equation Modeling, Survey Design, Machine Learning, Latent Class Clustering/Cluster Analysis, and Computing Education.

Vivek Pallipuram is an Associate Professor of Computer Engineering at the University of the Pacific, USA. He earned his Ph.D. in Computer Engineering from Clemson University in 2013 and completed post-doctoral training at the University of Delaware in 2015. His research interests include statistical modeling, machine learning, cloud computing, and high-performance computing. Over the past 5 years, he extensively published machine learning-based solutions for cloud computing and education. His work appears in prominent venues including IEEE Cluster, IEEE Cloud, IEEE Network Computing Applications (NCA), Journal of Supercomputing, and Concurrency and Computation: Practice and Experience, among others.

Bo Pei is a postdoctoral researcher in the Lucy Family Institute for Data and Society and ND Learning at the University of Notre Dame.

Zahiriddin Rustamov is a Data Science graduate from the University of Malaya, Malaysia, having completed his Master's degree in 2022. He holds a Bachelor's degree in Information Technology, specializing in Software Systems Development, obtained in 2021, and a diploma in Computer Science and Management Mathematics from Tunku Abdul Rahman University of Management and Technology, completed in 2019. Currently, Zahiriddin is pursuing a Ph.D. in computing at the United Arab Emirates University, focusing on artificial intelligence research, particularly in the areas of computer vision, machine learning, and deep learning.

Jaloliddin Rustamov holds a B.S. degree in Information Technology, specializing in Software Systems Development, from Tunku Abdul Rahman University of Management and Technology, Kuala Lumpur, Malaysia, earned in 2021. He also obtained an M.S. degree in Data Science from the University of Malaya, Malaysia, in 2022. His research interests encompass artificial intelligence, machine learning, deep learning, and data science.

Angela Starrett is an Assistant Professor of Educational Research and Measurement at the University of South Carolina. She teaches courses in statistics, structural equation modeling, and classification methods. Her research interests include applying advanced methodologies to rural education research, mixture models with complex data, and investigations of validity when adapting interventions in rural settings.

Peter Wulff studied physics education at Free University Berlin, Germany, did his in-service teacher training there, and then moved to IPN Kiel, Germany to do his Ph.D. Ever since his graduate studies, he was interested in integrating qualitative and quantitative methods in science education. During his postdoc years, he utilized machine learning and natural language processing to analyze pre-service physics teachers' written reflections. Being appointed assistant professor (tenure-track) at Heidelberg University of Education, Germany, he now also applies machine learning to analyze problem-solving in science and physics.

Zizhong Xiao is a current graduate student at the University of Maryland, College Park (UMD). He earned his B.A. Psychology at the University of California, Riverside, where he earned an NSF Graduate Fellowship. Before coming to UMD, he taught English in Malaysia for one year on a Fulbright student scholarship and obtained a M.S. Research Psych. at Azusa Pacific University. He is interested in examining how interpersonal relationships and everyday interactions strengthen people's happiness and well-being.

Xiangyu Xu is a Ph.D. student in Applied and Computational Math and Statistics at the University of Notre Dame. Among other research experiences, he has worked as a graduate research assistant in the Learning Analytics and Measurement in Behavioral Sciences (LAMBS) Lab at the University of Notre Dame.

Jin Eun Yoo received a B.A. degree from Seoul National University and two M.S. degrees in educational psychology and applied statistics and a Ph.D. degree in measurement, statistics, and research methodology from Purdue University, West Lafayette, IN, USA. She was a Psychometrician with Pearson, Austin, TX, USA, and a Research Scholar with the Department of Computer Science, San Francisco State University. She has been a Professor at the Korea National University of Education. Her research interests include machine learning techniques in social science and methodological issues of educational statistics. She is a member of the American Educational Research Association and the National Council on Measurement in Education. She currently serves as an Associate Editor for Frontiers in Psychology: Quantitative Psychology and Measurement and Innovation and Education.

Xiang Yu is a Ph.D. student in Applied and Computational Math and Statistics at the University of Notre Dame. Xu is a former graduate research assistant in the LAMBS Lab at the University of Notre Dame.

Chong Ho (Alex) Yu holds a Ph.D. in Educational Psychology with a focus on Measurement, Statistics, and Methodological Studies, and also a Ph.D. in Philosophy with a specialization in History and Philosophy of Science (Arizona State University). He is a former professor and Director of Data Analytics at Azusa Pacific University. Currently, he is a Professor and the Program Director of Data Science at the Department of Mathematics, at Hawaii Pacific University. He is a three-time winner of the SAS faculty scholarship (2016, 2017 SAS Global Forum, and 2017 Western Users of SAS Software Conference). In addition, he also won the Distinguished SAS Educator Award in 2021. He authored numerous books on education and data science. His most recent books on these topics are Data Mining and Exploration: From Traditional Statistics to Modern Data Science (2022, CRC Press), and Creating Changes in Science and Math Education: Lessons from Hong Kong (2020, Springer, coauthor: H. S. Lee).

Nazar Zaki is a highly accomplished Professor of Computer Science with a specialization in AI and machine learning. He received his Ph.D. from Universiti Teknologi Malaysia (UTM). Dr. Zaki is the founder of the Big Data Analytics Center, an institution that strives to make a lasting impact through innovative data analytics research and services. His research interests are focused on developing intelligent algorithms to address issues in diverse fields, including biology, healthcare, and education. Dr. Zaki has published more than 120 scientific papers in respected journals and conferences, and his work has been recognized through various scholarship awards.

Introduction

Exploring the Potential of Machine Learning in Educational Research

Myint Swe Khine

1 Introduction

Since Arthur Lee Samuel, a pioneer in the field of computer gaming and artificial intelligence, coined the term "machine learning" in his paper published in 1959, machine learning as a field of study has been growing exponentially and wide-ranging applications are developed in the areas of healthcare, business, industries, security operations, and many more. Artificial intelligence and machine learning are two closely related terms that are often used interchangeably. But in essence, Artificial Intelligence (AI) is a broad term that refers to the ability of machines to perform tasks that are typically associated with human intelligence, such as learning, reasoning, and self-correction, Machine Learning (ML) is a subset of AI that allows machines to learn from data without being explicitly programmed [1]. The following table summarizes the differences between artificial intelligence and machine learning (Table 1).

Despite extensive use in other fields of study, the application of machine learning methods in educational research is still in its infancy. As computing power grows, improved algorithms and abundant data are available, researchers are looking into how machine learning can improve education and training processes. Machine learning is making it possible to analyze study patterns, personalize instruction, adapt assessments, and predict student performance with high accuracy and precision. In recent years, applications of machine learning in educational sciences are gradually emerging in the literature.

This book provides comprehensive coverage of machine learning in educational sciences. The book is organized into five parts. While Part I introduces an overview of the book, Part II deals with the foundations of machine learning. In Part III, studies related to predicting student performance with machine learning are presented. The

M. S. Khine (✉)
Curtin University, Bentley, WA, Australia
e-mail: m.khine@curtin.edu.au

M. S. Khine (ed.), *Machine Learning in Educational Sciences*,
https://doi.org/10.1007/978-981-99-9379-6_1

Feature	Artificial Intelligence (AI)	Machine Learning (ML)
Scope	Broad field encompassing various methods to achieve human-like intelligence	Subset of AI that uses algorithms to learn from data
Goal	Create intelligent machines	Empower machines to learn and improve by themselves
Methods	Rule-based systems, logic, expert systems, machine learning, etc.	Primarily uses algorithms trained on data
Examples	Self-driving cars, chess-playing programs, chatbots	Spam filters, facial recognition, recommendation systems

Table 1 Differences between artificial intelligence and machine learning. *Source* [1]

applications of machine learning techniques in assessment processes are described in Part IV of the book, and Part V covers using machine learning in educational research. The chapters in this book attempt to consolidate and document recent explorations in theory building, conducting systematic, rigorous, and multidisciplinary research on machine learning in education, and identifying areas for further research and development. The book also offers exemplary works, highlighting novel approaches, innovative applications, and recent advances in machine learning in education research.

2 Foundations of Machine Learning

This section begins with a chapter on an introduction to machine learning for educational researchers. Professor DiStefano and her team from the University of South Carolina (Chap. 2) noted the increased access to large datasets and availability of analysis methods for researchers in the past 10 years, and machine learning is becoming a popular tool. The chapter defines the terms, explains the steps involved in conducting machine learning, types of machine learning, and choosing appropriate algorithms with applied examples. In Chap. 3, Fahd and Miah delve into machine learning applications in higher educational services. The chapter identifies the machine learning application trends with specific case studies. The chapter also highlights the machine learning algorithms used to address the research questions of the studies and identify the gaps for future work.

Pallipuram and his team from the University of the Pacific (Chap. 4) present the Camelot framework, a collection of supervised and unsupervised machine learning strategies to facilitate an automatic and effective analysis of both quantitative and qualitative teaching evaluations. Camelot frameworks combine the results from machine learning techniques to dispense effective improvement strategies. The authors assert that while the case study in the chapter deals with teaching evaluations, it can be adapted for other areas of education including student learning outcomes and accreditation studies. In Chap. 5, Yoo describes penalized regression in large-scale data analysis. The chapter explains the basics and extensions of penalized regression that help to reduce overfitting. Overfitting occurs when a model is too closely fit to the training data, and as a result, it does not generalize well to new data. The author notes that penalized regression as a linear method in large-scale data analysis is a promising area of continued growth that can yield interpretable prediction models.

3 Predicting Student Performance

Part 3 includes studies related to predicting student performance with machine learning. This part begins with the study by Hennebelle, Ismail, and Linden on school students' performance with machine learning (Chap. 6). The authors propose student performance prediction methods and evaluate the efficacy of such methods. The chapter provides a taxonomy of performance factors that help to gauge students' performance from different perspectives and gives insights into the categories and features that mostly affect students' performance. In Chap. 7, Denner and her team from the University of Notre Dame investigate how to predict response latencies on test questions based on the features of the questions. The study identifies the characteristics of a written test question including coverage of topics within the knowledge domain, use of content vocabulary, linguistic features, and question design features that were most likely to predict response times for each test question. The chapter also discusses the implication of their findings.

Bognár in Chap. 8 shares his work on predicting student attrition in university courses in Hungary. The author emphasizes that building a reliable machine learning model usually requires a large sample and as many predictors as possible. The chapter presents model building in detail that considers factors that influence dropout, models at different levels (e.g., degree level, year level, course level, etc.), and embedding in existing Learning Management System. The author concluded that addressing student dropout in universities requires a deep understanding of the underlying causes and the implementation of multifaceted solutions. In Chap. 9, Luo and Cui explore how to improve students' achievement prediction in the blended learning environment with integrated machine learning methods. The study measures students' learning styles, collects students' online learning behavior data and offline classroom voice data, and applies an integrated machine learning method to recognize students' data patterns. The study found that 79% of students in the class follow the recommended learning pattern. The authors also note that compared to collaborative learning integrated method is more suitable for discussion-based learning. The study by Al Braiki and the team (Chap. 10) involves using innovative ways to identify at-risk students in higher education by augmenting traditional machine learning classifiers with topological features and node embeddings derived from graph-based representations of student data. They find improvement in prediction when using a Multi-Layer Perceptron (MLP) classifier with the original dataset supplemented with both topological features and node embeddings with 93.5% accuracy. Their study lays a foundation for building a comprehensive and effective early-warning system and enhancing student support services.

4 Machine Learning in Assessment Processes

In Part IV, Chap. 11, Ober and her team provide an account of how topic modeling can be applied to understand the assessment practices of US college instructors in response to the COVID-19 pandemic. The study involves 145 faculty members teaching at over 80 different institutions in the US. The survey responses were analyzed with a topic modeling approach, using the Latent Dirichlet allocation (LDA) algorithm. The authors note that the topic models based on instructors' responses to open-ended questions provided insights into how instructors prioritized content, adjusted grading policies, prepared students for examinations, and overcome the challenges of administering online assessments while addressing the concerns of academic integrity. The findings from this study show an application of machine learning to understand assessment practices during the pandemic. Dai in Chap. 12 investigates how machine learning, such as topic modeling, Gaussian Mixture Model (GMM), and Artificial Neural Networks (ANN), can be applied to augment the design and assessment of immersive learning experiences. The author notes that when learners are engaged in such an environment, a vast amount of data was generated that can be analyzed by machine learning.

5 Machine Learning in Educational Research

The chapters in Part V present various findings in educational research using machine learning as an analytical tool. Alex Yu (Chap. 13) and his team demonstrate how machine learning is used to analyze the relationship between well-being and academic performance with large-scale assessment data. In their chapter, the authors discuss the measurement of well-being, and its relation to academic achievement, and the use of different machine learning models to predict test performance. In Chap. 14, Kieser and Wulff from Heidelberg University of Education report the use of large language models to examine cognitive constructs, augment data, and design instructional materials for teaching Physics. The authors describe the use of ChatGPT to generate novel learning materials dealing with physics problems. The authors note that large language models like ChatGPT can be used to supplement human researchers with machine learning capabilities.

Another application of machine learning in education research is presented by Musso and Cascallar in Chap. 15. In their chapter, the authors describe the prediction of academic performance at students and environmental levels. While the student level consists of cognitive, demonographies, and self-regulated learning factors, as well as the child's health, school trajectory, routines, and habits, the environmental level includes social-economic status and institutional characteristics. The authors suggest that machine learning methods, such as Artificial Neural Network (ANN) could help in designing interventions at individual and group levels. The generative IA models are a powerful tool that has the potential to disrupt a wide range of areas. The recent introduction of ChatGPT and Bard demonstrates how such models can be used to perform a variety of tasks, from creating new original content and generating new ideas.

Asthana and her team report the text classification for question generation using a machine learning method (Chap. 16). The questions are labeled according to Bloom's taxonomy, and classification models are developed using the Bidirectional Long Short-Term Memory Networks (BiLSTM) and Convolutional Neural Networks (CNN). The study used MATLAB to analyze the data and it was found that the CNN model provides more accurate results with the high validation accuracy. This work has provided an efficient method for designing computerized automated adaptive assessment where questions were classified based on Bloom's learning stages. The generative IA models are a powerful tool that has the potential to disrupt a wide range of areas. The recent introduction of ChatGPT and Bard demonstrates how such models can be used to perform a variety of tasks, from creating new original content and generating new ideas. Xin Miao, Brooker, and Monroe in Chap. 17 explore the possible applications in K12 education from industry perspectives. The authors also discuss ethical and practical challenges associated with its implementation and provide ideas for successful adoption.

6 Conclusion

This book is a collection of works by educators who believe in the potential of machine learning to improve educational research. The chapters in this book summarize recent advances in machine learning research in education, including theory building, systematic research methods, and multidisciplinary approaches. They each make a significant contribution to our understanding of how machine learning can be used to improve education processes. In addition to summarizing existing research, the book also presents new approaches, innovative applications, and advances in machine learning in education research.

It is an ardent desire that these works will spur other researchers to continue exploring the capabilities of machine learning in educational research. The studies in this book are essential reading for anyone who wants to stay up-to-date on the latest advances in machine learning in educational sciences.

References

1. Gemini (2024) Differences between artificial intelligence and machine learning. Google AI. Retrieved from https://gemini.google.com/
2. Samuel AL (1959) Some studies in machine learning using the game of checkers. IBM J Res Dev 3(3):210–229

Foundations of Machine Learning

An Introduction to Machine Learning for Educational Researchers

Christine DiStefano, Geophrey Odero, and Angela Starrett

Abstract The past 10 years have produced enormous changes by providing researchers with increased access to large datasets as well as the methods available to analyze these large datasets. Techniques such as data mining, data analytics, artificial intelligence, and related methods have become commonplace. Machine learning is one area under this broad umbrella of methods that is rapidly gaining popularity. Given that many educational researchers are interested in understanding the basics of machine learning and exploring how these methods may be useful in education, the goal of this chapter is to introduce educational researchers to this area. The relationship of machine learning to related fields and ways to conduct machine learning are discussed. Finally, the intersection of machine learning and educational research is explored with an applied example. Recommendations for future research are discussed.

Keywords Machine learning · Supervised learning · Decision trees · Naïve Bayes · Support vector machines

The last decade has led to enormous changes in researchers' ability to collect large datasets as well as an increased ease to obtain large amounts of collected data. Databases relevant to education and including thousands of cases are readily accessible from a variety of sources such as the National Center for Educational Statistics (NCES) and the Inter-university Consortium for Political and Social Research (ICPSR). In addition, the increasing use of technology and the Internet has allowed additional and diverse data sources to be collected (e.g., process data, eye-tracking data), resulting in new findings about learning and performance. Methods for analyzing large datasets, such as newer algorithms and analysis methods (e.g., data mining, data analytics), and increased computing power and capability have created new opportunities for educational researchers to obtain, process, and model data.

C. DiStefano (✉) · G. Odero · A. Starrett
University of South Carolina, Columbia, SC, USA
e-mail: DISTEFAN@mailbox.sc.edu

M. S. Khine (ed.), *Machine Learning in Educational Sciences*,
https://doi.org/10.1007/978-981-99-9379-6_2

Along with the increased prevalence of data sources and analysis, terms such as artificial intelligence, big data, and data mining are commonplace in research and appear in everyday conversations; not surprisingly, more educational researchers are adopting these methods in their investigations. Given the increasing popularity of such techniques and the applicability of the procedures to many areas of education, many educational researchers may be interested in learning more about machine learning, how these techniques function, and how the methods intersect with other big data analysis techniques. This chapter introduces selected machine learning methods for applied educational researchers. A discussion of machine learning methods used for data analysis is provided, along with an illustrative example compared to a commonly used analysis method. Avenues and extensions for future research in machine learning for education are provided.

1 Defining Terms

There are many terms that are often used when dealing with large datasets and machine learning. Terms such as data science, big data, artificial intelligence, deep learning, data mining, and machine learning are often discussed when considering this general area. To provide a perspective of how machine learning fits into the broader framework, these terms are broadly defined.

Data science is a multidisciplinary field that combines principles and practices from many areas (e.g., statistics, mathematics, artificial intelligence, computer science) to extract relationships between variables with estimates of likelihood (and error) from large amounts of data [9]. In other words, data science is the approach to knowledge creation [11], where the major goal is extracting information and knowledge from data [37]. Data science is often concerned with analyzing large datasets.

Big data is just what it sounds—enormous amounts and various types of data collected for a specific purpose. Typically, big data is characterized using the three Vs: volume (consisting of enormous quantities of data), velocity (created in real-time), and variety (characteristics of the data include may include structured, semi-structured, and unstructured cases) [28]. The National Institute of Standards and Technology defines big data as datasets "where the data volume, acquisition velocity, or data representation limits the ability to perform effective analysis using traditional relational approaches or requires the use of significant scaling (more nodes) for efficient processing" [13, p. 4–5]. In educational research, understanding the learning processes, student characteristics, and educational standards usually involve analysis and interpretation of big data. Examples of major decisions may include student recruitment to higher education sources, student admissions, budgeting, monitoring student progress, and evaluation of student performance.

Data mining methods are often used with big data. This is an umbrella term for techniques used to discover novel and potentially useful information from large amounts of data [5]. Educational data mining is a subdomain where data mining methods are used to draw conclusions from large educational datasets while incorporating predictive and psychometric modeling techniques [42]. Educational data mining generally uses modeling strategies to test learning theories and inform educational practice. Education data, such as grade profiles, administrative data from schools/universities, and test results may be analyzed for patterns and trends. Many of the educational data mining techniques, such as dimension reduction, regression analysis, cluster analysis, and data visualization techniques (e.g., [26]) are familiar to educational researchers. In other areas of educational data mining, data from computer-assisted collaborative learning environments, such as hybrid learning [30, 39], blended learning environments [7], or smart learning environments [48] may be studied to examine influences of setting to achieve more effective, better-quality learning.

Artificial intelligence (AI) refers to an umbrella of approaches that enable computers/machines to emulate human intelligence [12, 15, 23]. Some common examples of uses of AI are commonplace in daily life, seen in digital personal assistants like Apple's *Siri* and Amazon's *Alexa*. Other advanced applications are evident in robotics and natural language processing. Robotics is a branch of engineering that involves the conception, design, manufacture, and operation of robots [18]. Natural language processing, also known as computational linguistics, is a subfield of AI concerned with using computational techniques to learn, understand, and produce human language content [20]. Two major branches of AI include machine learning and deep learning. These areas differ from data mining in that the computer "learns" from the data to improve the estimation of an outcome, whereas data mining focuses more on uncovering patterns of relationships.

Machine learning (ML) is a subfield of AI that derives knowledge from examples, data, and experience to make predictions [10]. That is, ML processes enable computers to execute tasks by learning to predict an outcome accurately from available data rather than from using preprogrammed rules. Deep learning is a subset of ML that relies on data representations that emphasize learning through successive layers of increasingly meaningful data representations [12, 36]. These procedures combine advances in computing power and analysis methods to learn complicated patterns in large amounts of data. Figure 1 illustrates the relationship between artificial intelligence, machine learning, and deep learning [11, 23].

Given our focus on ML, a detailed discussion of the steps in the ML process, with reference to educational research situations, is provided. The topics included in our discussion are meant as an introduction and are in no way exhaustive. Many discussions, videos, and explanations of ML and related techniques are readily available on sites such as YouTube or Khan Academy and through massive open online course (MOOC) platforms.

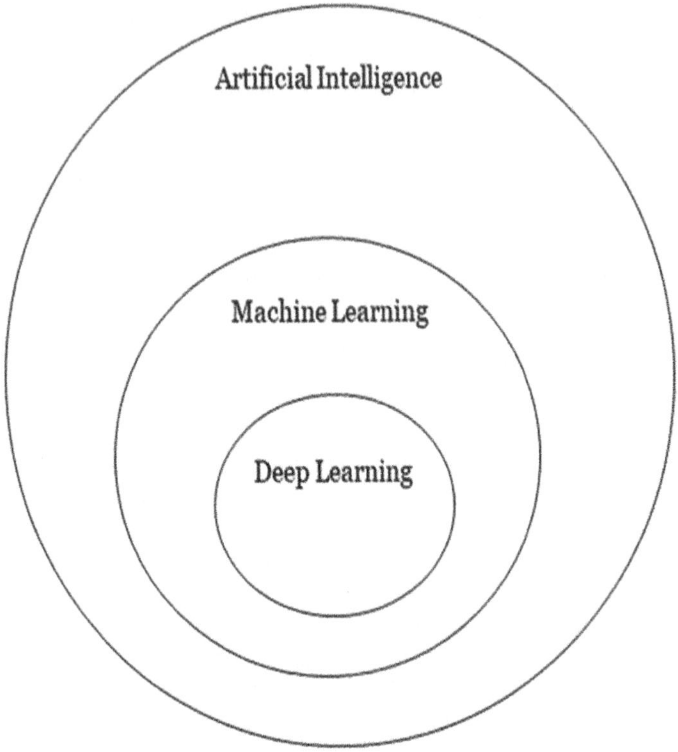

Fig. 1 Relationship of artificial intelligence, machine learning, and deep learning

2 Steps to Conduct ML

ML is based on the idea that people should not have to teach computers, but rather could let the computers learn on their own [45]. A core function of ML methodology is the development of algorithms that build useful models from data [51]. That is, ML aims to use information from an existing dataset to forecast or make future decisions, where the machine (e.g., computer) is given the ability to learn (i.e., build algorithms) without being explicitly programmed. [45]. Some general steps are recommended when applying ML methods.

Data Collection. As with any research study, researchers interested in developing an ML model should have a clear idea of the research question and the task(s) prior to data collection to focus on what data to collect before the start of the study. Data for ML studies may be structured, unstructured, or a mix of both. Structured data are typically more quantitative, such as numbers, dollar amounts, dates, and character strings (e.g., gender, race/ethnicity codes), which can be displayed in a database format (i.e., row and column description). Unstructured data include sources that cannot be displayed in a relational database and require much more storage space.

These include digital file types such as images, video, emails, audio files, etc., which may need to be screened, coded, and put in a uniform format prior to analysis. Putting data into a common format may be tedious and time-consuming; however, the process is a crucial step for developing and testing a ML model.

Data Preparation. As with other research projects, data preparation in ML refers to cleaning, manipulating, and organizing data before conducting analyses. Usually, at this stage, reformatting data, making corrections to data, and combining datasets are carried out. When using selected ML methods, the data are "labeled," meaning one (or more) informative or meaningful labels are attached to the raw data. This process is needed to provide context to the data for an ML model to "learn" from the data; typically, labels are provided with the outcome variable. Labeling can be challenging and time-consuming when there are inconsistencies in recording of the raw data, the concepts are defined differently by different people, or there are many unstructured data elements to examine [46]. Additionally, any data screening methods typically conducted with research projects, such as outlier detection, deciding how to treat missing data, and considering how to deal with skewed distributions, are completed at this stage.

As part of the data investigation process, researchers will want to select the input variables (called features) that are important to include in the model. Feature selection is the process of selecting the essential variables. If too many features or irrelevant features are included in a model, the ML algorithm mistakenly interprets and may identify unimportant patterns as meaningful. In essence, the model would be trained on "noisy" data. As part of the feature selection process, researchers may need to transform features (due to distributional anomalies) or reduce dimensionality using a method such as principal component analysis (due to large numbers of features and/or collinearity between input variables) as well as needing to examine additional (previously non-included) features as part of the process of building an optimal model [46].

Selecting a Model. To achieve the desired purpose, there are many ML methods from which a researcher may choose. The first step is needed to ensure the model selected meets the researchers' purpose (e.g., [14]). The primary functions of ML are for description (i.e., describing a set of data more thoroughly), classification (i.e., grouping new cases into homogenous clusters), prediction or regression (i.e., predicting/regressing an outcome on a set of variables), or discovery (i.e., identification of new groupings or theory building) (e.g., [14, 24, 31, 46]).

Supervised Learning. The most commonly used methods in ML fall under the heading of supervised learning [24]. In general, this method asks the machine to learn a system for mapping input(s) (i.e., data sources) to achieve a desired outcome based on example input–output pairs. There are several ways supervised learning ML can be utilized with educational data. For example, supervised learning can be used to predict student performance on exams or other assessments to identify students who may need additional support or to provide students with personalized learning plans. Supervised learning may also provide information to help students determine the courses that will help to achieve desired goals by recommending courses based on an individual's interests and academic needs. Lastly, supervised learning can be used to identify students who may be at risk of dropping out of school or failing classes and provide timely information to stakeholders (e.g., administrators, teachers, parents, and school counselors) to provide students with needed supports.

In supervised learning, a labeled training dataset that includes both the input and the output variable(s) to be predicted/classified is iteratively fit (i.e., trained) to a set of data [46]. Under the fitting process, different variables and/or model configurations are applied, and results are compared to determine which set of variables results in the least amount of predictive error. As ML algorithms learn from the dataset provided, a better training dataset will subsequently produce a better model. Moreover, good data training techniques can help address data bias which may affect the algorithm [2]. A general workflow of supervised machine learning algorithms is shown in Fig. 2.

Many ML models may use a full data set for training; however, if a large, labeled data set is readily available, a 'hold-out sample' may be kept separate for model evaluation. Recommendations suggest somewhere between 15 and 30% of cases to be set aside for model evaluation (e.g., [14, 46]). As training progresses, the model is refined or "tuned" (in some cases, on a separate subset of "tuning" data) with changes made to the model parameters.

Besides supervised learning, other methods are available to conduct ML. We provide a brief presentation of some of the more commonly used methods and then

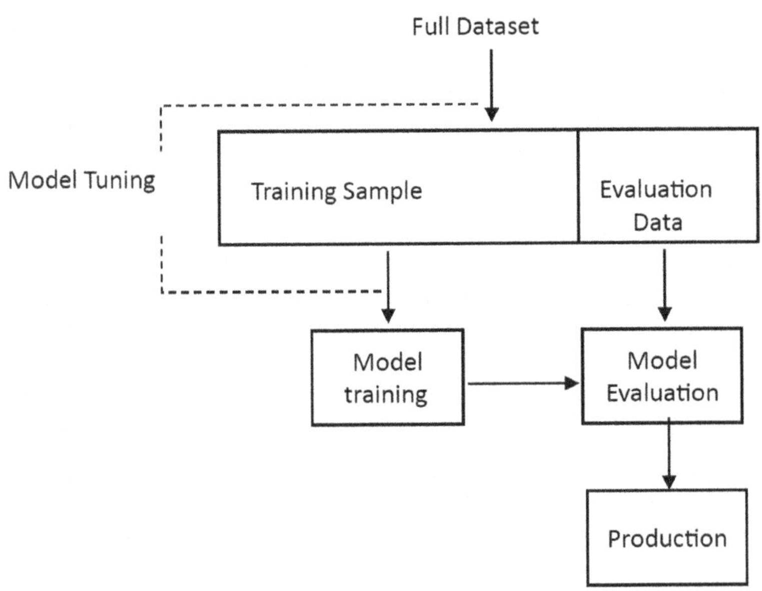

Fig. 2 Machine learning, supervised learning algorithm workflow

concentrate discussion on supervised learning and related algorithms. We also recognize that there may be an overlap between the methods and analysis strategies used across the areas described below.

Unsupervised Learning. Unsupervised learning aims to uncover meaningful patterns and conclusions underlying a dataset when a labeled outcome is not present at the start of the analysis [22]. The goal is to uncover underlying dimensions, components, clusters, or trajectories within a data structure to create subtypes or subgroups. With these methods, there is no training dataset, algorithms discover and present interesting features underlying the dataset [31]. Unsupervised learning differs from supervised learning ML techniques which use labeled data (i.e., data where the correct outcome is identified) to "learn" from a set of training datasets that fine-tune the performance of the algorithm (e.g., [4, 14, 31]). For example, unsupervised learning can be used to group students based on their academic performance, interests, and personality traits. Similarly, the methods can be utilized to recommend resources based on students' interests or academic needs or to identify patterns in student data to improve the educational experience for all students.

Unsupervised learning may be useful in ML description tasks, that is, those tasks that focus on providing a summary of certain variables [24]. New data are classified using previously learned features to recognize and match cases to previously uncovered groups or features. Unsupervised learning algorithms may be used to identify unobserved dimensions or trajectories, categorize dimensions into subtypes, or identify heterogeneity over time [24]. Analysis strategies that are often used with unsupervised learning, such as k-means cluster analysis, principal components analysis, factor analysis, and mixture modeling, are familiar to many psychometricians and educational researchers.

Reinforcement Learning. Reinforcement learning is a learning approach in which the model learns a desired behavior through trial-and-error interactions in a dynamic environment [25, 50]. The approach is conducted via an iterative process that trains on examining the identification of desired behaviors (i.e., correct outcomes) and penalizing the process when an incorrect response is made. In reinforcement learning, the machine receives input that is then evaluated to determine the state of the environment. Based on the evaluation of the stimuli, an action is determined, and the environment is reevaluated as needed [25]. Through these actions, the machine "learns" and is rewarded either positively or negatively based on the feedback. These methods may be useful in situations of prediction or regression. Two widely used reinforcement learning analysis strategies are the Markov Decision Process and Q learning [6, 49].

Reinforcement learning can be used with education data in several ways, including personalized learning, game-based learning, and adaptive tutoring. First, reinforcement learning models can be trained on data which represent a student's learning style, interests, and academic needs. The model can then use this information to recommend resources, activities, and assessments tailored to one's individual needs. Additionally, educational games can be made more engaging and challenging by using the reinforcement learning process to reward students for taking actions that lead to desired outcomes. As a final example, adaptive tutoring systems can be more effective by using reinforcement learning to track student progress and adjusting the level of difficulty of the instruction to meet student needs.

Ensemble Learning. Ensemble learning techniques relate to an umbrella term referring to techniques that examine a group of models that combines results across the set of models to produce the final predictions [16]. Generally, classification or prediction (or regression) is the desired outcome of an ensemble learning process. The method is sometimes categorized as a supervised ML task, as labeled data are used to learn from and improve the process (e.g., [24, 43]). Ensemble learning has been used for predicting student failure and facilitating more personalized educational paths [47]. The researchers used student performance, demographics, and other data stored in a learning management system with different ensemble learning algorithms, including Random Forest, AdaBoost, and Gradient Boosting, to train prediction models. The authors demonstrated that ensemble learning algorithms outperform individual models in predicting student failure, allowing for more accurate identification of students at risk.

Under these models, the ensemble learning takes a set of labeled examples as input and produces a model that generalizes these examples to create a model that may be used to classify new, unlabeled examples. The method has been described intuitively as modeling the human decision-making process where a person faced with a difficult decision may gather different opinions from many individuals, weigh the opinions, and combine the information to make one decision. The main idea is that weighing and aggregating several individual opinions will produce a better result than choosing the opinion of any one individual. In ensemble models, combining information from multiple models classifying cases under the same algorithm (e.g., decision trees), the errors of any one analysis are likely to be mitigated, and the overall prediction performance of the ensemble would benefit [43]. The two most commonly used methods for reducing error across ensemble techniques are bagging and boosting.

Bagging was proposed by [8], the name comes from the abbreviations for bootstrap and aggregating and is aptly termed, as bagging applies the bootstrap re-sampling method [17] to obtain the training data sets, the ouptuts are then aggregated across the set of learning datasets. Here, classification or regression models are fit to bootstrap samples from the data and combined by voting (classification) or averaging (regression). The main objective is to improve the accuracy of classification models by combining classifications of the same type from randomly generated training sets.

Boosting is a method used to reduce errors in predictive data analysis. Any ensemble model may make a prediction error, depending upon the accuracy of the training dataset used. The model is trained to convert the weak learners (i.e., training models with higher levels of prediction error) into strong learners. The models are trained sequentially, each time focusing on training examples that were found incorrect with previous models [51].

Both boosting and bagging are used to improve prediction accuracy; however, the methods differ in how they undergo training. With bagging, the accuracy is improved by conducting the training simultaneously on multiple datasets, whereas boosting is conducted using a sequential process. Bagging helps reduce decision variability while boosting appears to reduce bias and variance [27]. Most ensemble methods use decision trees, which model a series of decisions, related probabilities, and possible consequences in a flowchart pattern to determine the probability of arriving at a certain decision. Some commonly used techniques for ensemble ML techniques include random forest (bagging), and AdaBoost or gradient boosting (boosting).

Choosing a Model Algorithm. After choosing a model, an algorithm needs to be selected. Before diving more deeply into ML, we note that disciplines often "call" the same techniques/procedures differently, contributing to confusion when learning a new field. As our focus is on supervised learning techniques, we recognize that this branch of ML also uses unique terms that refer to procedures commonly encountered in educational research or statistics. Without a clear understanding of the relationship between terminology across fields, similarities in the practices and applications could be overlooked. Table 1 (adapted from [24] compares terms used in statistics and supervised machine learning. For example, statistics typically discusses the use of independent variables, where in supervised machine learning, these are termed features, outcome variables are termed targets; and a contingency table is termed a confusion matrix. Thus, concepts familiar to many educational researchers are identical in ML, only the terminology varies.

Focusing on supervised learning methods, three popular analysis strategies are decision trees, naïve Bayes, and support vector machine (SVM) algorithm. These three strategies are detailed below.

Decision Trees. The decision tree algorithm uses the notion of a graph to represent choices and the results by splitting the data into smaller and smaller subsets of similar

Table 1 Comparison of Common Terms Used in Statistics and Machine Learning Fields

Statistics	Machine learning
Prediction	Supervised learning
Independent variables/predictors	Features
Dependent variable/outcome	Output or target
Prediction of categorical outcome	Classification
Groups	Classes
Prediction of a continuous outcome	Regression
Number of predictors	Dimensionality
R-squared	Coefficient of determination
Residual	Error
Sensitivity	Recall
Positive predictive value	Precision
Contingency table	Confusion matrix

Note Adapted from Jiang et al. [24]

classes. A primary benefit of this method is the ability of decision trees to simplify complex decision-making processes for users to examine and interpret many possible solutions to a problem [32].

Each decision tree consists of nodes and branches. Classification begins at the root node, which is the overarching event or choice from which the tree is produced. Data are passed through the various decisions in the tree according to the values of its features. The branches are the values a node can take. Based on probabilities associated with the paths taken, the predicted class membership is assigned [1, 29, 40].

Figure 3 provides a simplified example of a decision tree; the included features are based on a study by [3], who used decision trees to assist with the examination of which aspects should be included to determine the university students most likely to succeed if awarded a scholarship. Here, a decision tree was used as the number of features available for evaluation for scholarship award decisions may create difficulties for candidate selection and result in a longer time for students to wait for a decision. From a larger list of inputs, the features of grade point average (GPA), participation in standardized testing, school rating, and the outcome of the scholarship award were considered essential features. In the graph, the "boxes" may be considered nodes and the arrows as branches. A subtree is shown in the shaded circle, and the ending decisions are the leaf nodes on the decision tree.

Decision trees are a popular ML algorithm for classification and regression problems and the method offers many advantages. Conceptually, the method is easy

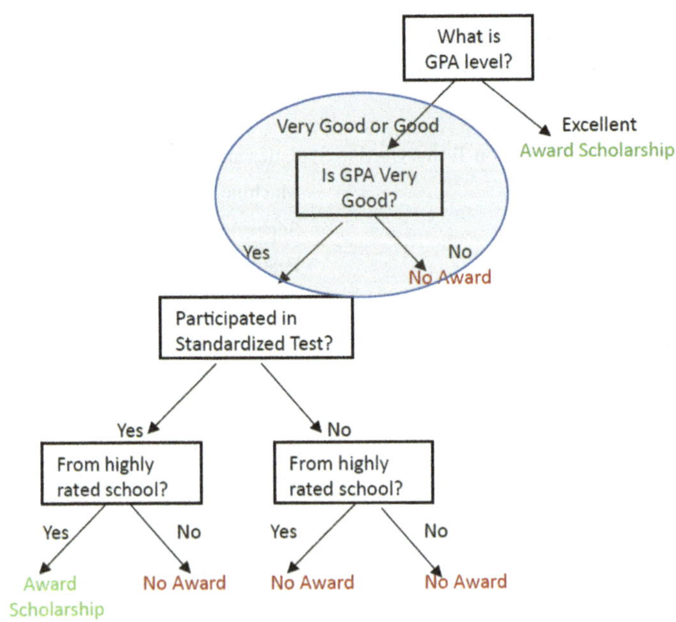

Fig. 3 Decision tree example, awarding a university scholarship

to visualize, and the picture may clarify the decision-making process [41]. Decision trees can be easily interpreted and can be used to model complex relationships between features and labels [22, 38]. Decision trees can be trained on both categorical and continuous data, and can be used to make predictions for new data points [22]. Further, [46] suggests that decision trees may work well with many problematic data situations, such as the presence of outliers, missing data, and nonlinear relationships. However, there are some drawbacks when using decision trees. Variability in the data may result in different solutions (trees) generated. Also, overly complex trees (i.e., using too many features) may create a tree that does not generalize well to new situations.

Naïve Bayes. The naïve Bayes technique is a simple but powerful method for addressing classification problems in ML. This approach is based on Bayes' theorem, a mathematical formula for calculating the probability of an event occurring given the probability of other events occurring. Naïve Bayes works by assuming independence, i.e., the presence of a particular feature in a class is unrelated to the presence of any other feature (e.g., [21]). For example, the conditional probability of membership in a class, c, gives a set of features $x_1, x_2, \ldots, x_n$ that can be expressed as $P(c_i \mid x_1, x_2, \ldots, x_n)$. The assumption of independence is called "naïve" because it is unlikely to be true in reality; however, it makes the algorithm much simpler to train, and the procedure is still effective in many cases.

The posterior probability is a key concept in naïve Bayesian classification. Bayes' theorem provides a method for computing the conditional probability of being in a class, c_i, given feature, x. The theorem is typically defined as:

$$P(c_i|x) = \frac{P(x|c_i) * P(c_i)}{P(x)}$$

where:

- $P(c_i|x)$ is the posterior probability of class c_i given feature x
- $P(c_i|x)$ is the likelihood of feature x given class c_i
- $P(C_i)$is the prior probability of class c_i
- $P(x)$ is the marginal probability of feature x

The prior probability is the probability of class c_i, constructed before seeing the data. This probability value can be estimated from the previous data or set to a default value (e.g., equal across all classes). The likelihood of feature x given class c_i is the probability of seeing feature x if class c_i is the optimal class. The marginal probability of feature x is the probability of seeing feature x regardless of class membership. The resulting posterior probability can be used to make predictions about the class of a new observation. As an oversimplified example, this type of ML could be used to classify students into different categories, such as at risk for dropout and not at risk for dropout, using results from a a standardized test. We would first need to train a Bayesian classifier on a dataset of students who drop out and graduate. Once the classifier is trained, we can use it to calculate the posterior probability of a new student being at risk of dropping out. The new student is predicted to be a member of the class with the highest posterior probability. While only one input variable (standardized test result) was used, a model can be extended to include more than one input variable.

To simplify the Bayes Theorem, the naïve Bayes Classifier considers each input variable as independent from each other (instead of dependent), reducing complexity in the calculations. The denominator is removed, and the conditional probabilities are constructed for each input variable within a class. Given that the features (x's) are independent, the conditional probabilities of the individual classifications can be multiplied (e.g., [21, 31]):

$$P(c|X) = P(x_1|c) * P(x_2|c) * \cdots * P(x_n|c) * P(c).$$

The calculation above may be performed for each class, and the class with the largest probability of membership is selected. This is called the maximum a posterior (MAP) decision rule.

The naïve Bayes classifier has some advantages as the procedure is easy to use and does not require much training data. Naïve Bayes can be used to solve both classification and regression problems, and, like decision trees, it can be used with both categorical and continuous data. However, there are some disadvantages to using naïve Bayes in educational research. While assuming conditional independence simplifies computations, this is a strict assumption. Poor performance of the naïve Bayes algorithm may be due to correlated input variables (e.g., [46]). It is also sensitive to outliers in the data [22], and it can be less accurate than other ML algorithms for some problems [35].

Support Vector Machines. Support vector machine (SVM) algorithms are used for classification. The procedure aims to identify the boundary that maximally separates the cases into distinct groups (e.g., [24]). This boundary is called a line (in two-dimensional space) or a hyperplane (in multidimensional space.) The objective is to find an optimal boundary (line/hyperplane) that minimizes misclassification [34].

To demonstrate the SVM algorithm, a simplistic two-group example (red cases and black cases) is provided in Fig. 4. The SVM takes the data points and finds the equation for the hyperplane that best separates the cases. This hyperplane (or, here, a line as there are only two features) is termed the decision boundary. Any case that falls "below" the line would be classified in Group 1 (red cases), and anything "above" the line would be classified as Group 2 (black cases). The hyperplane found is the one with the maximum margin (i.e., the space around a hyperplane equation). SVM tries to have the largest margin or distance with the cases closest to the boundary, termed support vectors.

While two groups are typically used with SVMs, the algorithms can accommodate any number of predefined classes labeled in the training data [34]. When there are two features, the hyperplane will be linear. When there are more dimensions and the model is not linear, SVMs separate the data into a higher dimensional space with a transformation of the data using what is called a kernel function, where the kernel

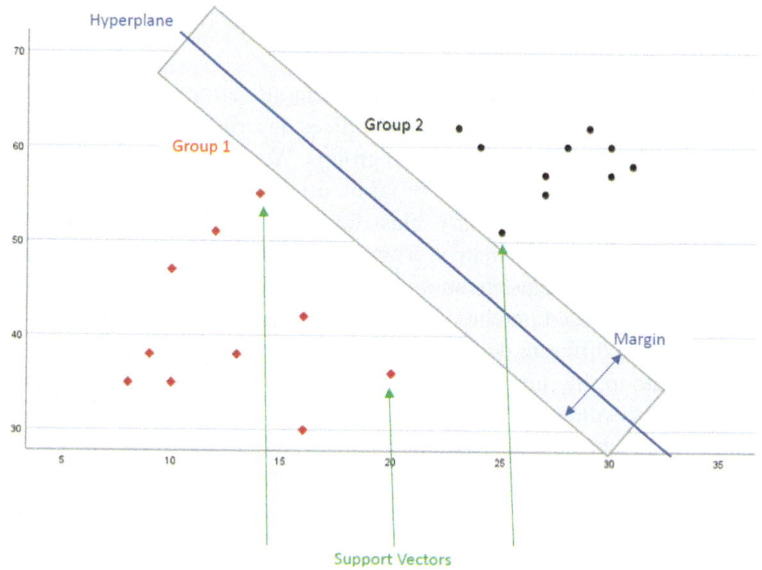

Fig. 4 Support vector machine, two group hypothetical model

can be linear if there are only two dimensions and is nonlinear (e.g., polynomial) if many features are included. Overall, the exact equation of any hyperplane can vary based on the number of features included and/or if the data follow a linear or non-linear pattern.

When constructing a classification algorithm, there are other choices a researcher may make. During tuning of an algorithm, the margin can be varied to impose a penalty (i.e., cost parameter) for misclassifying cases into the "wrong" class, where higher values represent a bigger penalty or "cost" of misclassification (e.g., [34]). Further, when using more than two features that are measured differently, data can be placed on the same scale so that variables with larger variability will not dominate the algorithm. Centering and scaling are commonly conducted before building the SVM classification function.

SVMs may be beneficial over other options for reducing model "overfitting," when the model gives accurate predictions for training data but not new or test data. In terms of limitations, SVMs may be seen as a "black box" approach in the sense that metrics are not provided for how predictors are combined to optimize the hyperplane. [24]. In addition, choosing which kernel function to use may be challenging for researchers [34].

3 Evaluating Model Fit

With ML, indices to measure model fit judge the classification accuracy of the model; with supervised learning techniques, these indices describe how accurately the final model works to classify cases into labeled groups. We note that the indices obtained may vary depending on the statistical program used.

If the response variable is binary, classification values can be computed using elements from the confusion matrix. This is constructed as a 2 × 2 table of the Actual Group (using the classification from the labeled dataset) by the Predicted Group (membership based on classification from the ML algorithm). The elements in Table 2 illustrate different decisions that may be made, with cells noted as a, b, c, or d, that relate to the intersection between the rows and columns. Elements on the confusion matrix diagonal (a, d) represent correct decisions, and elements on the off diagonal (b, c) represent incorrect decisions. When evaluating an ML algorithm, different choices for algorithm calculations are evaluated to determine the effect on correct decisions and errors in classification. These classification metrics are popular in health applications arising from looking for the presence (or absence) of a disease state [33].

When evaluating classification, performance is more often evaluated using classification characteristics such as sensitivity, specificity, positive and negative predictive value, and area under the receiver operating characteristic curve (AUC); however, as noted in Table 2, these terms are named differently within the ML literature. Sensitivity, or *Recall,* is the proportion of subjects correctly predicted to have a certain outcome occur (i.e., are positive) among all those who truly possess the outcome in question. Using the values from Table 2, recall is calculated as TP/(TP + FN). Specificity is the proportion of subjects correctly predicted not to have the outcome occur (i.e., are negative) among cases that do not possess the outcome in question. The specificity rate, calculated as TN/(TN + FP), provides information on the algorithm's ability to correctly identify subjects who are negative relative to the characteristic in question.

Other indices computed to judge the accuracy of the classifications include the positive predictive power or precision in ML literature. Calculated as TP/(TP + FP), positive predictive value (PPV) reports the proportion of subjects who are truly positive among all those who are predicted to be positive. The negative predictive value (NPV), TN/(FN + TN), is the proportion of subjects who are truly negative among

Table 2 ML confusion matrix, binary response variable

Observed outcome			
Predicted outcome	Present	Not present	Row totals
Positive	True Positive (TP)	False Positive (FP)	→PPV
Negative	False Negative (FN)	True Negative (TN)	→NPV
	↓ Sensitivity	↓ Specificity	

all those predicted to be negative. Accuracy refers to the cases that are correctly classified (along the diagonal) from the total (N). For classification, sensitivity indices between 70 and 80% and specificity close to 80% are recommended (Glascoe 2005). Area under the curve (AUC) may also be used. This method provides a value from 0 to 1 detailing how much of an improvement using the model is over a chance classification (i.e., 50%).

When the responses are continuous, the performance of a supervised learning algorithm is not assessed using the classification metrics. The overall amount of variance explained by the regression equation (i.e., R-squared) referred to in ML as the *coefficient of determination is examined.* In addition, residual or error values that examine how "far off" a point is from the regression line are summarized to report model performance. The Mean Squared Error (MSE), Root Mean Squared Error (RMSE), and Mean Absolute Error (MAE) are three indices that may be used to evaluate ML model quality, with smaller amounts of error (i.e., better prediction capability) desired.

4 Applied Example

To illustrate ML techniques, the three supervised learning methods detailed earlier (i.e., decision tree, naïve Bayes, and support vector machines) are compared with an example. The data source of this study is the High School Longitudinal Study of 2009 (HSLS:09), conducted by the National Center for Education Statistics. The HSLS:09 is a nationally representative, longitudinal dataset that follows a sample of students in ninth grade in the fall of 2009 (21,444 students from 944 schools) through high school and postsecondary education. Data used in this example from the publicly available HSLS:09 dataset. Our analytic sample is restricted to first-time 9th graders whose information was available in at least one of the two survey rounds ($n = 8{,}149$).

The objective of this example is to build a model to predict a student's aspiration to pursue a STEM (i.e., Science, Technology, Engineering, Mathematics) career in the future. The response variable, STEM career aspiration, includes two classes (Yes/No). The sample from the HSLS:09 dataset includes roughly 40% of students stating a desire to pursue a STEM career and 60% of the students not interested in a STEM career. Seven features were used in the ML process: student gender (male or female), family socio-economic status, mathematics self-efficacy, mathematics interest, science self-efficacy, science interest, and school engagement. With the exception of gender, composite values were created through principal component analyses and are on a continuous metric, with higher scores representing more of the variable in question (i.e., higher family SES, higher science interest, higher level of school engagement).

HSLS:09 data was used with the decision tree, naïve Bayes, and SVM algorithms to predict STEM aspiration. For a familiar comparison, logistic regression was included. The caret package in R was used for all ML models; logistic regression was run using the base R package (version 4.2.0). The total sample of 8,149 was randomly split into a 75% training sample (N = 6,122) for ML and applied to a test sample of 25% (N = 2,027 cases). To assess performance, classification indices of accuracy, recall, specificity, PPV, and NPV provided as part of the output were examined along with the confusion matrix for each method. While logistic regression does not readily compute classification values, a confusion matrix was these values, classification was made using the predicted model (with a threshold value of 0.42).

Table 3 provides the results for the HSLS:09 dataset across the four techniques. In general, all ML methods produced similar results, which may be partly due to use of the same R package to compute all ML algorithms. What is noteworthy is that all ML-based methods outperformed logistic regression, with higher levels of sensitivity (ability to correctly identify a student as STEM aspirant when the student actually has STEM aspirations).

Starting at the top of the table, the first row provides the confusion matrix. This shows the cross-classification of the actual student responses recorded in the HSLS:09 dataset (Yes = interested in pursuing a STEM-related career vs. No = not interested in pursuing a STEM career) and the outcome predicted by the different algorithms. While classifications are similar, SVM reported the fewest number of students to be interested in a STEM career. Accuracy reports the percentage of cases overall that were correctly classified to pursue a STEM career or to non-pursue a STEM career (main diagonal element sum) from the total number of cases. As shown, all methods reported similar accuracy rates (0.61–0.64), showing roughly 60% accuracy in the ability of the method in question to predict a "correct" response. The ML methods reported higher recall than logistic regression. In other words, ML-based

Table 3 Comparison of ML Methods and Logistic Regression: HSLS:09 Test Data (N = 2,037)

	Logistic Regression			Decision Tree			Naïve Bayes			Support Vector Machine		
Confusion Matrix	Actual			Actual			Actual			Actual		
		Yes	No		Yes	No		Yes	No		Yes	No
Predicted values (rows)	Yes	443	397	Yes	303	247	Yes	278	203	Yes	190	136
	No	432	845	No	537	1,030	No	528	1,029	No	616	1,095
Accuracy	0.61			0.63			0.64			0.63		
Recall (Sensitivity)	0.40			0.81			0.84			0.89		
Specificity	0.34			0.36			0.35			0.24		
PPV	0.53			0.66			0.66			0.64		
NPV	0.66			0.55			0.58			0.52		

Notes PPV = Positive Predictive Value; NPV = Negative Predictive Value. Formulas to compute classification accuracy can be examined with the caret program documentation as these may vary slightly from definitions provided in Table 2

algorithms had roughly were able to classify a roughly 80% of students interested in a STEM career correctly. Of the ML methods, SVM yielded the highest sensitivity rate. Specificity denotes the chance of correctly identifying a student who wants a non-STEM career. Among all methods, these values are much lower than the sensitivity values showing that it was harder for all algorithms to correctly choose students who do not want to pursue STEM in the future. Values were approximately at 30%, with SVM reporting the lowest specificity (24%).

PPV	0.53	0.66	0.66	0.64
NPV	0.66	0.55	0.58	0.52

The PPV reports the percentage of students predicted to be interested in STEM who are actually interested in pursuing a STEM career. The ML-based methods are able to do this for roughly 65% of the cases; logistic regression yielded a lower PPV of 0.53. The NPV reports the percentage of students who are not STEM aspirant from those predicted as non-interested in a future STEM career. Here, logistic regression performed higher, correctly identifying roughly 66% of the non-STEM aspirant cases however, as there were more students who were not interested in a STEM career, this could be partly due to the large number of "No" cases. Other ML methods were around 55%, with naïve Bayes reporting the highest result.

5 Future Research

We recognize that this chapter provides only an introduction to ML techniques. While numerous applications can be incorporated into the field of education to study a wide variety of problems, we offer a few areas that may be useful for future study. There are many additional methods under the ML umbrella that can be explored and applied to educational research. For example, there are many additional modeling techniques (e.g., neural network analysis, instance-based learning, semi-supervised learning), numerous choices within each modeling technique to assist with optimization (e.g., customizations to the tuning processes, various percentage splits for training/testing datasets, etc.) as well a wide variety of software programs (e.g., Python, R, SAS, SPSS) for interested researchers to investigate.

Given that many of the techniques used within ML are familiar to educational researchers, readers may wonder how these two areas are distinct. While the analytic methods do overlap the underlying philosophy between these two branches of social science differ. ML-based algorithms and similar methods (e.g., data mining, AI) are largely data-driven methods, meaning that the goal is to take existing data that are used to build the model. The model is trained and tested based on the information included in the dataset at hand. Educational statistics and psychometrics start with a strong theoretical perspective and select variables to include which are first and foremost important to the underlying theory. The difference may be considered as methods which, from a philosophical perspective, create a model through inductive reasoning vs. creating a model through deductive reasoning. What is clear is that future research may maximize both approaches together to build new theories, create new models, and uncover new findings, through capitalizing on the strengths of both theoretical perspectives more effectively and efficiently.

Besides additional involvement in ML methods, there are many applications in which educational researchers may be interested in applying these techniques. Process data may be a fruitful area for ML techniques. Briefly, process data refer to the collection of all information retained by the computer when an examinee is completing an assessment online (e.g., [19, 44]). Besides providing a scored response (i.e., correct or incorrect), process data can include additional information such as the time required to solve a problem, the number of times options are selected before submitting a response, and the order of responses selected. Previous studies have examined process data to understand patterns of students' test-taking behaviors, inform test administration and item creation, and provide information about students' use of universal design features (e.g., [19]). By examining process data, educational researchers can gain a better understanding of the processes that are used when taking an exam and how the information contributes to an examinee's score.

In summary, as artificial intelligence becomes more and more engrained, educational researchers will be able to access and analyze big data. ML methods provide an opportunity for researchers to streamline the prediction of outcomes important to educators, students, and families, and other interested stakeholders. Given the rapid advances and popularity of the techniques, ML has great potential for educational researchers to learn more about how students use, process, and learn information.

References

1. Ahsan MM, Luna SA, Siddique Z (2022) Machine-learning-based disease diagnosis: a comprehensive review. Healthcare 10(3):541
2. Akter S, McCarthy G, Sajib S, Michael K, Dwivedi YK, D'Ambra J, Shen KN (2021) Algorithmic bias in data-driven innovation in the age of AI. Int J Inf Manage 60:102387
3. Aldino AA, Sulistiani H (2020) Decision tree C4. 5 algorithm for tuition aid grant program classification (Case Study: Department Of Information System, Universitas Teknokrat Indonesia). Jurnal Ilmiah Edutic: Pendidikan dan Informatika (1):40–50
4. Baig MI, Shuib L, Yadegaridehkordi Y (2020) Big data in education: a state of the art, limitations, and future research directions. Int J Educ Technol Higher Educ 17(44). https://doi.org/10.1186/s41239-020-00223-0
5. Baker R et al (2010) Data mining for education. Int Encyclopedia Educ 7(3):112–118
6. Barto AG, Dietterich TG (2004) Reinforcement learning and its relationship to supervised learning. Chapter in Handbook of learning and approximate dynamic programming,9780470544785
7. Bonk CJ, Graham CR (2012) The handbook of blended learning: global perspectives, local designs. Wiley, Hoboken, NJ
8. Breiman L (1996) Bagging predictors. Mach Learn 24:123–140
9. Brodie ML et al. (2015) Understanding data science: an emerging discipline for data intensive discovery. DAMDID/RCDL, 238–245

10. Campesato O (2020) Artificial intelligence, machine learning, and deep learning. VA Mercury Learning Press, Dulles
11. Cao L (2017) Data science: a comprehensive overview. ACM Comput Surv (CSUR) 50(3):1–42
12. Ceron R (2019) AI, machine learning and deep learning: what's the difference. IBM
13. Chang WL, Grady N (2019) NIST big data interoperability framework: Volume 1, definitions. Gaithersburg, MD: NIST Publishers
14. Crisci C, Ghattas B, Perera G (2012) A review of supervised machine learning algorithms and their applications to ecological data. Ecol Model 240:113–122
15. Dhande M (2017) What is the difference between AI, machine learning and deep learning? Geospatial World (blog), May 6, 2017. https://www.geospatialworld.net/blogs/difference-between-ai%EF%BB%BF-machine-learning-and-deeplearning/
16. Dong X, Yu Z, Cao W, Shi Y, Ma Q (2020) A survey on ensemble learning. Front Comp Sci 14:241–258
17. Efron B, Tibshirani RJ (1994) An Introduction to the Bootstrap. Chapman & Hall/CRC Press, New York
18. Fuller JL (1998) Robotics: introduction, programming, and projects. Prentice Hall, Upper Saddle River, NJ
19. Hicks JC (2019) The use of process data to examine reading strategies (Doctoral dissertation, The University of North Carolina at Greensboro)
20. Hirschberg J, Manning CD (2015) Advances in natural language processing. Science 349(6245):261–266
21. Islam MJ, Wu QJ, Ahmadi M, Sid-Ahmed MA (2007) Investigating the performance of naive-bayes classifiers and k-nearest neighbor classifiers. In: 2007 international conference on convergence information technology (ICCIT 2007), pp 1541–1546. IEEE
22. James J, Sandhya L, Thomas C (2013) Detection of phishing URLs using machine learning techniques. In: 2013 International conference on control communication and computing (ICCC), pp 304–309. IEEE
23. Jeffcock P (2018) What's the difference between ai, machine learning, and deep learning. Eri¸sim Adresi: https://blogs.oracle. com/bigdata/difference-ai-machinelearningdeeplearning#:~: text= Machine% 20learning% 20is% 20a% 20subset, to% 20solve% 20more% 20complex% 20problems. Eri¸sim Tarihi, 13, 2021
24. Jiang T, Gradus JL, Rosellini AJ (2020) Supervised machine learning: a brief primer. Behav Ther 51(5):675–687
25. Kaelbling LP, Littman ML, Moore AW (1996) Reinforcement learning: a survey. J Artif Intell Res 4:237–285
26. Kanimozhi G, Kumaragurudasan P, Velmurugan T (2021) Educational data mining in blended learning: an updates survey on EDM techniques and learning tools used in BL. J Fundamental Comparative Res 7(8):85–91
27. Kotsiantis S, Pintelas P (2004) Combining bagging and boosting. Int J Comput Intell 1(4):324–333
28. Laney D (2001) 3D data management: controlling data volume, velocity and variety. META Group Research Note 6(70):1
29. Lantz B (2019) Machine learning with R: expert techniques for predictive modeling. Packt publishing LTD, Birmingham, UK
30. Linder KE (2017) Fundamentals of hybrid teaching and learning. New Dir Teach Learn 2017(149):11–18
31. Mahesh B (2020) Machine learning algorithms-a review. Int J Sci Res (IJSR). [Internet] 9:381–386
32. Mesarić J, Šebalj D (2016) Decision trees for predicting the academic success of students. Croatian Oper Res Rev 7(2):367–388
33. Metz CE (1978) Basic principles of ROC analysis. Chapter in Seminars in nuclear medicine. WB Saunders 8(4):283–298
34. Mountrakis G, Im J, Ogole C (2011) Support vector machines in remote sensing: a review. ISPRS J Photogramm Remote Sens 66(3):247–259

35. Murphy KP (2012) Machine learning: a probabilistic perspective. MIT Press, Cambridge, MA
36. Pedro F, Subosa M, Rivas A, Valverde P (2019) Artificial intelligence in education: challenges and opportunities for sustainable development
37. Provost F, Fawcett T (2013) Data science and its relationship to big data and data-driven decision making. Big Data 1(1):51–59
38. Quinlan JR (1986) Induction of decision trees. Mach Learn 1(1):81–106
39. Raes A (2022) Exploring student and teacher experiences in hybrid learning environments: does presence matter? Postdigital Sci Educ 4(1):138–159
40. Raju D, Schumacker R (2015) Exploring student characteristics of retention that lead to graduation in higher education using data mining models. J College Student Retention: Res Theory Practice 16(4):563–591
41. Rismayanti R (2018) Decision Tree Penentuan Masa Studi Mahasiswa Prodi Teknik Informatika (Studi Kasus: Fakultas Teknik dan Komputer Universitas Harapan Medan). Query J Inf Syst 2(1):16–24
42. Romero C, Ventura S (2013) Data mining in education. Wiley Interdisciplinary reviews: data mining and knowledge discovery 3(1):12–27
43. Sagi O, Rokach L (2018) Ensemble learning: a survey. Wiley interdisciplinary reviews: data mining and knowledge discovery 8(4):e1249
44. Sahin F, Hicks J, Ji SG (2020) Exploring examinees' perspective in a digitally based assessment using process data
45. Samuel AL (1959) Machine learning. Technol Rev 62(1):42–45
46. Scott IA (2021) Demystifying machine learning: a primer for physicians. Intern Med J 51:1388–1400
47. Smirani LK, Yamani HA, Menzli LJ, Boulahia JA (2022) Using ensemble learning algorithms to predict student failure and enabling customized educational paths. Sci Program 2022:1–15
48. Spector JM (2014) Conceptualizing the emerging field of smart learning environments. Smart Learning Environ 1(2):1–10, http://www.slejournal.com/content/1/1/2
49. Szepesvári C (2009) Reinforcement learning algorithms for MDPs. Unpublished manuscript. Retrieved from era.library.ualberta.ca
50. Uprety A, Rawat DB (2021) Mitigating poisoning attack in federated learning. In: 2021 IEEE symposium series on computational intelligence (SSCI), pp 1–7. IEEE
51. Zhou Z-H (2021) Machine learning. Springer Nature, Singapore

Machine Learning Applications in Higher Education Services: Perspectives of Student Academic Performance

Kiran Fahd and Shah J. Miah

Abstract Artificial Intelligence (AI) offers new technical reality in driving innovations across service sectors in areas of human societies. The application of AI has shown a profound impact for transforming key practices, including in the Higher Education (HE) sector. The presence of AI in the HE sector is already widely acknowledged for playing a pivotal role and potentiality in transforming process, policy and practices of the HE landscape. This chapter presents a contemporary overview of Artificial Intelligence applications to provide top trends in current AI in HE literature. The main objective of this chapter is to identify emerging directions of current AI applications in HE from the latest and existing studies in this domain. We studied and analyzed 10 existing case studies as representative samples of AI in HE. The analysis provided insights about the current trend and revealed the topic areas and themes that would be of paramount significance for HE stakeholders.

Keywords Machine learning · Higher educations · Learning management · Student performance

1 Introduction

Artificial Intelligence (AI) is considered the key driver of innovation across all areas of human society and has shown a profound impact, and the Higher Education (HE) sector is no different. The presence of AI in the HE sector is already widely acknowledged and plays a pivotal role in the transformation of the HE landscape. Application of AI in the HE sector has shown a potential to address educational issues and challenges and enhance and transform teaching and learning by leveraging emerging AI-enabled technologies [1]. The benefits of AI in HE have been highlighted in

K. Fahd · S. J. Miah (✉)
Newcastle Business School, College of Human and Social Futures, University of Newcastle, Callaghan, NSW, Australia
e-mail: Shah.Miah@newcastle.edu.au

M. S. Khine (ed.), *Machine Learning in Educational Sciences*,
https://doi.org/10.1007/978-981-99-9379-6_3

many studies, for instance, a few examples of the application of AI in the educational sector are focused on effective applications of data analytics, predicting student enrolments, a recommendation system for career pathway or resource management, adaptive tutoring, prediction of student readiness for employment, monitoring and predicting student academic performance or identifying struggling students in their progression.

The potential applications of AI to support educational process cover [2–4]:

- Content development and teaching method—AI supports the teaching and learning process by providing AI-infused technologies, e.g., personalized learning, improved instructional content development, and educational robot systems like a virtual teaching assistant.
- Communication—AI strengthens the communication and interactions within education, e.g., effective, timely, and targeted feedback or response to students, learning gap identification, chatbots, and intelligent tutoring systems.
- Task automation—AI supports educational stockholders in reducing repetitive and burdensome tasks, e.g., providing tools for data accumulation, automatic grading, and intelligent proctoring.
- Inclusive education—AI has made education more accessible for students living with disabilities, e.g., AI assistive technology assists students with speech impairments to get their message across using AI translation to support deaf students.

Few mentionable examples of AI-infused technologies in HE are [3]: Jill Watson, which is virtual teaching assistance and chatbot, Pearson AIDA provides personalized learning and feedback to calculus students; IBM Watson education classroom [5], which is a cloud-based AI-enabled solution to assist educators to deliver personalized learning for students; Microsoft AI-based plugin presentation translator; IRIS intelligent proctoring system to support remote invigilation and mitigate academic integrity misconduct.

2 Background of the Research

There are multiple definitions of AI in the literature but one of the key definitions of AI is [6]:

> Artificial intelligence is that activity devoted to making machines intelligent, and intelligence is that quality that enables an entity to function appropriately and with foresight in its environment.

In layman's terms, Artificial Intelligence is giving the ability to a machine for performing a task that reduces human effort. AI-powered innovations are becoming a significant part of our daily lives, from AI-enabled drone deliveries to AI-embedded virtual assistants (Fig. 1).

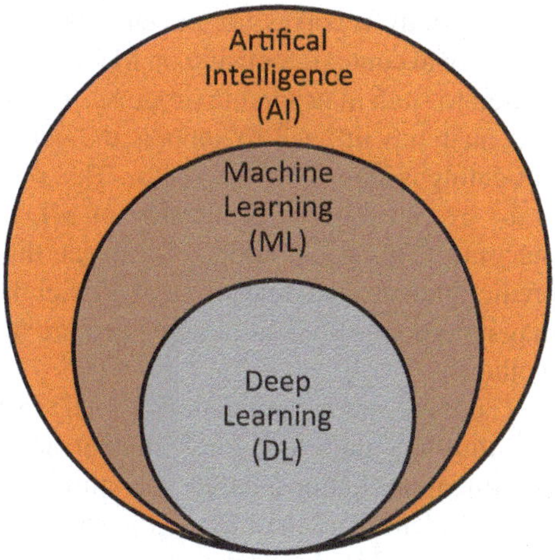

Fig. 1 **a** Relationship between AI, ML, and DL

ML, DL, and NLP are a subset of AI as shown in Fig. 2a. ML learns from data and uses data as examples to solve the given problem by applying rules and parameters defined in the examples. In this modern era of data analytics, ML is the key to application intelligence. The three subfields of ML are unsupervised learning; supervised learning; semi-supervised learning; and reinforcement learning based on different mathematical modeling as shown in Fig. 2b.

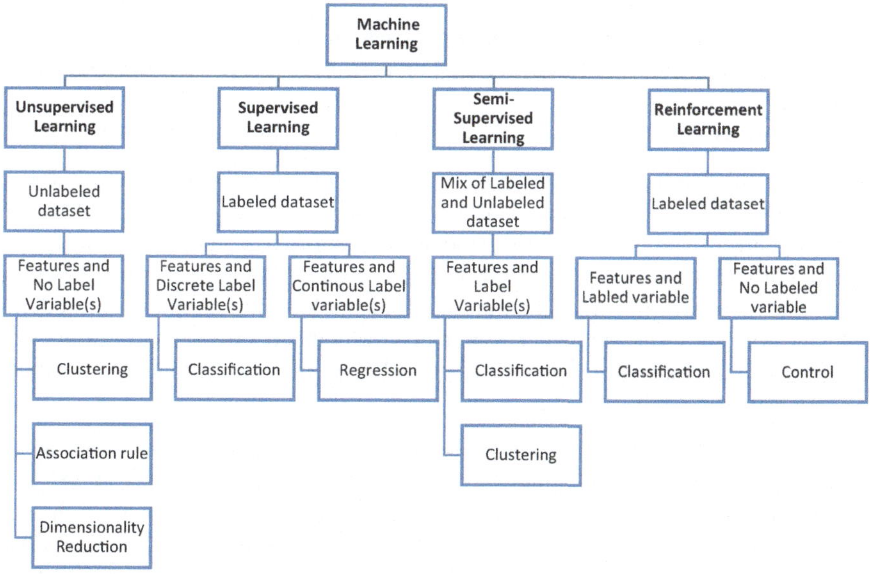

Fig. 2 Different types of ML learning types and models

This study revolves around the application of AI in tertiary institutions in the educational context, focusing on Machine Learning (ML) applications, most notably related to student academic progress. This comes under the umbrella of educational data analytics utilized to support the general objective of supporting learning and mitigating issues in the HE sector. This objective forms the research question, what is the overall state of application of ML in HE with a focus on student academic outcome? This paper adopts the stance that ML algorithms have more potential to predict student academic progress efficiently than manual methods. These algorithms may incorporate student features that are not even possible or practical with manual methods.

There are many applications of ML in HE, including, but not limited to, adaptive tutoring, data analytics, student academic performance monitoring, student enrolment predictions, career recommendations, resource management recommendations, identification of struggling students, and prediction of student readiness for employment. ML-based data analytics assists in better and more informed decision-making from the data generated within the HE environment. ML-based data analytics solutions support:

- Improved learning path with adaptive learning.
- Better course delivery strategy with student academic performance prediction.
- Mitigate breaching academic integrity by plagiarism detection.
- Reduce attrition rate by predicting student academic progress and predicting dropouts.
- Actively engaged students by predicting preferences with course or major selection or recommendation systems.
- Enhance learning experience with appropriate and timely intervention by identifying students at risk of failing based on their behavior pattern or engagement pattern.
- Better strategic planning to achieve HE goals by predicting resource management.

These data analytics solutions are trained on the data generated within an educational setting. A huge amount of educational data is produced throughout different phases of educational processes, e.g., student personal data or student academic data. Figure 3 provides the overview of ML in HE and shows a real potential to leverage ML to transform teaching and learning for all stakeholders and support both students and educators and make it a better experience for both students and educators.

The next paragraphs discuss the different ML algorithms utilized to develop these ML applications in HE.

Decision Tree (DT): A DT is a tree-like structure of decisions, their reasoning, and their corresponding outcomes based on if–then conditional rules. A DT is a commonly and widely used supervised data analysis technique that covers both classification and regression. A DT is designed from top to bottom and left to right. Each internal node represents a decision node for a condition based on a feature of the dataset called a rule. Each internal node splits the tree into further branches based on these rules, also known as edges. The first node at the top is known as the root node

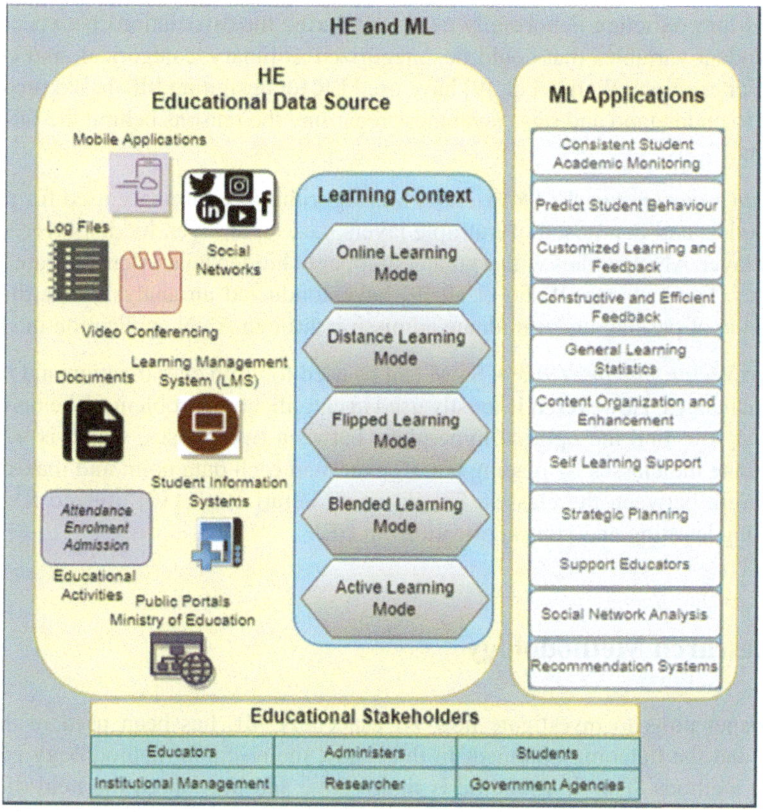

Fig. 3 Overview of ML in HE

that marks the start of the tree. The leaf node is the end node of a branch that does not split into further branches and represents each labeled class. Each path from the root to the leaf represents a specific classification rule. For instance, [7] have explored the application of DT and advanced statistics to predict student academic performance in HE.

Random Forest (RF): RF is an ensemble technique that combines multiple models. Random Forest is an algorithm that generates multiple DTs from a different set of samples from the original dataset. In other words, it combines multiple trees constructed from subsets of the main dataset. Once built, each tree produces a classification outcome and the aggregate number of votes given by the set of trees. It covers both classification and regression task [8] for example, predicting student dropout in HE courses.

Logistic regression (LR): Regression methods for predicting student performance use a finite set of relationships among the dependent and independent variables, generating a predictive function that models these associations. The logistic regression method for predicting is normally used to describe the associations between a few independent variables that could be categorized as binary, categorical, and continuous. For example, Walid et al. [9] have used LR for designing ML-based predictive model to understand and raise awareness regarding the reasons behind the failure of students.

Artificial neural network (ANN): ANN ML algorithm is commonly used for prediction models and composed of multiple layers, i.e., input layer, hidden layer(s), and output layer. ANN models with a greater number of hidden layers demonstrate higher accuracy. For instance, Walid et al. [9] have conducted an analysis regarding the prediction of passing undergraduate admission through ANN application in HE.

Support vector machine (SVM): SVM can be used for both classification and regression challenges. However, it is mostly used in classification problems. The main idea of SVM is to find the optimal hyperplane between two classes, which is working to separate the classes by placing a margin around each data point and maximizing the margin between the classes. For instance, Veluri et al. [10] utilized SVM for designing learning analytics applications in HE.

3 Research Methodology

This paper aims to investigate how AI especially ML has been used in the HE sector and the outcome obtained. In this study, the research methodology consists of two methods. The first method is interpretive and introduces the field of AI in Education. The second method is to review and identify top trends of AI in education by conducting reviewing and synthesizing the relevant scholarly publications. The rest of the paper is organized as: the next section presents the background of this study in terms of AI and ML. In the third section, the paper discusses the latest trends of application of AI in HE with a focus of student academic progress. The fourth section discusses the findings of the trends reviewed in the previous section. Finally, the last section concludes the study.

4 ML Application Trend in HE

This study explores the application of ML techniques in HE with a focus on predicting student academic performance, which is a focal point for HE educators and researchers for the previous decade. Student academic progress is considered a vital element in the field of HE management, e.g., for addressing student attrition or managing institution resources, or for strategic planning. Predicting student academic progress supports educators to implement effective strategies to increase the student success rate. The following articles demonstrate the various trends in the application of different ML algorithms to enhance student academic performance. A summary of the articles is given in Table 1.

Representative case 1: *Learning analytics using deep learning techniques for efficiently managing educational institutes* [10]: The first article investigated ML-based framework to predict student performance, behavior, attitude, and success based on their pre-enrolment dataset. The secondary dataset used consists of student pre-enrolment data like gender, SAT scores, financial status, and academic emphasis. The study applied and compared three ML algorithms to train and test the ML-based framework and achieved 94% accuracy to forecast student academic performance to obtain a high education standard.

Representative case 2: *Analysis of machine learning strategies for prediction of passing undergraduate admission test* [9]: The study trained an ML-based predictive model to understand and raise awareness about the reasons behind the failure of students in the undergraduate admission test, minimize the failure rate, and improve student performance. The ML predictive model is trained on a dataset consists 10 features including the result of previous exam(s), family economical and educational background, average time spend on social media or gaming, and average duration of the study. They compared five ML algorithms with other ML techniques to train and test the ML predictive model and achieved 98% classification accuracy.

Representative case 3: *Predicting student dropout in university classes using two-layer ensemble machine learning approach: A novel stacked generalization* [8]: The study model has developed a ML model based on stacking ensemble technique to predict student dropout. The study has used the dataset to train and test the model consisting of academic data like test grades, project grades, exams, and graduation year. The study has applied four ML algorithms in two layers to compare the prediction results, i.e., Random Forest (RF), Extreme Gradient Boosting (XGBoost), and Gradient Boosting (GB) in the first layer and Feed-forward Neural Networks (FNN) and achieve the accuracy of nearly 93% of predicting dropout.

Representative case 4: *Machine learning strategy for enhancing academic achievement in a private university* [11]: The study developed an ML model to predict student academic progress in a private HE institution to address the student dropout problem and improve decision-making. ML predictive model is trained on a dataset comprised of student academic characteristics including assignment grades, exam scores, test grades, and final exam scores. Two ML algorithms are compared to develop ML predictive model and achieved 84% accuracy to accurately identify a student at risk of failure and assist HE institution in estimating the attrition and retention rate for better management.

Representative case 5: *Predicting student performance in a blended learning environment using learning management system interaction data* [12, 13]: This study proposed an ML-based predictive model to identify potential students at risk of failing to study student attrition. The study trained and tested the ML-based predictive model to predict student academic progress from LMS interaction data like count

Table 1 Summary of the articles reviewed

Article no	Target	Dataset type and source	Dataset features	ML algorithms	Other ML techniques	Metric	Prediction accuracy (%)
Article #1	Efficient HE management and high education standard	Secondary dataset (UCI Machinery repository)	Enrolment student dataset (Features:17 Instances:285)	Artificial Neural Network (ANN), Nave Bayes, and Support Vector Machine (SVM)		Accuracy	94
Article #2	Efficient HE processes and improve student performance	Primary dataset (Interviewing Method, Email Questionnaire Method)	Personal and enrolment academic data (Features:10 Instances:343)	Logistic Regression (LR), Support Vector Machine (SVM), Random Forest (RF), Artificial Neural Network (ANN), K-Nearest Neighbor (KNN)	Stratified K-fold cross-validation, SMOTE, AdaBoost algorithm	Accuracy, Precision, Recall, F-Measure	98
Article #3	Reduced student dropout and enhanced student academic performance	Secondary dataset (HERI repository)	Student academic data (Features:12 Instances: 261)	Random Forest (RF), Extreme Gradient Boosting (XGBoost), Gradient Boosting (GB), and Feed-forward Neural Networks (FNN)	1o-fold cross validation	Accuracy, Precision, Recall, F1-Score, and AUC	93
Article #4	Enhance student academic progress and reduce student dropout	Primary dataset	Student academic data (Features:12 Instances: 4500)	Naive Bayes, Decision Tree		Accuracy, F1, Precision, Recall, FPR	84
Article #5	Reduce student attrition and enhance student academic performance	Secondary Dataset (UCI Machinery)	LMS interaction dataset (Features:8 Instances: 230,318)	Random Forest (RF), J48, NBTree, OneR, DecisionStump	k-Fold cross-validation	Accuracy, Precision, Recall, F-Measure	86

(continued)

Table 1 (continued)

Article no	Target	Dataset type and source	Dataset features	ML algorithms	Other ML techniques	Metric	Prediction accuracy (%)
Article #6	Enhance student academic performance and high education standard	Primary dataset (Moodle)	LMS (Moodle) interaction dataset, student personal and past results (Features:16 Instances: 802)	Regression and Cluster Analysis		RMSE (Root Mean of Square Error)	-
Article #7	Efficient HE management and enhanced student academic performance	Primary dataset (student database)	Academic dataset (Features:13 Instances: 1100)	K-Means algorithm, Random Forest (RF),	tenfold cross-validation	Accuracy, Precision, Recall, F-Measure	80
Article #8	Better education standards and enhance student academic performance	Primary dataset	Enrolment and semester academic dataset (Features:1000 Instances: 150)	Random Forest (RF)	Wrapper technique	Accuracy, Precision	85

(continued)

Table 1 (continued)

Article no	Target	Dataset type and source	Dataset features	ML algorithms	Other ML techniques	Metric	Prediction accuracy (%)
Article #9	Efficient HE resources management and enhanced student academic performance	Primary dataset	Student demographic characteristics and enrollment academic dataset (Features:13 Instances: 150)	Logistic regression (LR)		Accuracy, AUC	85
Article #10	Efficient HE resources management and enhanced student academic performance	Primary dataset	Student demographic characteristics and enrolment academic dataset (Features:13 Instances:1491)	Artificial Neural Network (ANN), K-Nearest Neighbors (KNN), K-Means Clustering, Naïve Bayes (NB), Support Vector Machine (SVM), Logistic Regression (LR), Decision Tree (DT)	SMOTE, 10 folds cross-validation, Ensemble Technique	Accuracy	83

of right click, count of left clicks, or idle time and identify a student at risk to reduce student attrition. They applied and compared five ML algorithms and gained 85.7% prediction accuracy in identifying students at risk of failing.

Representative case 6: *Early prediction of undergraduate student's academic performance in completely online learning: A five-year study* [14]: The study explores regression and cluster analysis to predict student academic progress early from LMS interaction data and student academic data. The dataset consists of 16 features including The GPA and enrolment test data. The data collected over 5 years consist of LMS log files data like the total number of assessments in a subject, the number of times student access the teaching material, forum, or glossaries, and demographic characteristics like age, and past academic achievements. Categorized into three factors: Access Factor (AF), Questionnaire Factor (QF), and Task Factor (TF). The study has designed and analyzed the regression and clustering to improve the student academic performance earlier in their studies.

***Representative case* 7:** *A two-phase machine learning approach for predicting student outcomes* [15]: The study developed and compared ML models to predict student progress to make better decisions concerning their academic pathways, e.g., better schedule or better resource allocation. The dataset consists of data collected from student admission to graduation like date of admission, Time to Degree, GPA, and entrance exam scores. The study has applied the ML clustering technique to obtain student clusters and utilized these clusters to train ML-based predictive model.

***Representative case* 8:** *Multi-class assessment based on random forests* [16]: The study developed a ML approach to predict student academic performance to set up specific support measures to improve student success. The dataset consists of attributes related to the student's pre-enrolment academic characteristics and post-enrolment academic characteristics including high school information, average marks in scientific and non-scientific units, average in all units, and range of highest and lowest marks in each unit. The study applied the wrapper technique for feature selection,

***Representative case* 9:** *Applying a machine learning approach to predict students' performance in higher educational institutions* [17]: The study developed a supervised ML approach to predict student academic performance to improve the success rate and better resource management in HE institutions. The ML model is developed on a dataset containing student demographic, pre-enrolment academic, and course academic datasets including gender, address, state of origin, birthdate, exam score, and current course cumulative GPA. The ML predictive model is trained and tested by using the logistic regression algorithm on this dataset to accurately 84.7% of student academic success.

***Representative case* 10:** *Enhancing prediction of student success: automated machine learning approach* [7]: The study explored the application of ML and advanced statistics to predict student academic performance to efficiently manage HE institution resources, i.e., efficiently supporting student success while employing least resources. The study utilized 13 student demographic attributes and pre-enrolment dataset features like HE enrolment program, gender, ethnicity, scholarship, and Mathematics and English level. The study achieved 83% accuracy by comparing eight different ML models to predict the failing students and offer them immediate attention and support to improve their academic performance.

5 Findings

Most studies have presented their unique ML-based solutions in the HE field to address the challenges faced by the HE sector. These studies developed ML predictive models with a vast difference of input datasets to mitigate the problem(s) these studies addressed by predicting student academic performance. All studies embrace that the outcome of using ML techniques to predict student academic performance is either more accurate than the traditional practice or is not possible with traditional predictions. This section presents an overview of the findings of the review of AI applications in HE with ML focus as a promising technology to support the HE environment.

process

- Most of the studies reviewed are quantitative studies. These studies used standard methods and ML models in various phases of the study. Except for one study, all studies used experimentation research methodology to train and test the ML models to obtain the objectives of the study. All the studies used an enhanced/customized approach of the generalized computational model to execute the experiment for data analysis based on ML and consist of basic four phases: Data collection, Data pre-processing, ML Model Training, and testing and statistical evaluation as presented in the following Fig. 4.

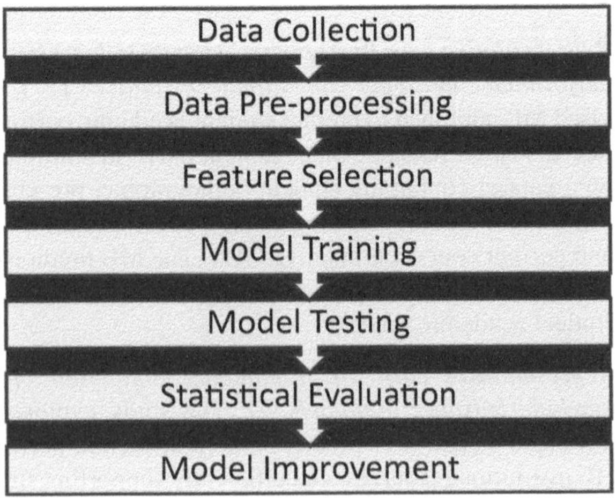

Fig. 4 Generalized computational model of ML

- Various ML algorithms have been used to train and test the ML predictive model to predict student academic performance to enhance resource management and efficient processes of the HE sector. Mostly supervised ML algorithms are utilized, namely, Random Forest (RF), Decision Tree (DT), Logistic Regression (LR), Support Vector Machines (SVM), Naïve Bayes (NB), Artificial Neural Network (ANN), and K-Nearest Neighbor (KNN).
- The studies investigate the ML prediction models to predict student academic progress with a focus on identifying and selecting the characteristics that influence the student academic performance and achieving the highest prediction accuracy. Student personal and socio-demographic characteristics, student enrolment (academic or non-academic), and student academic attributes are the most common features used in training and testing ML predictive models.
- Several performance metrics are used to assess the trained ML model performance namely Precision, Accuracy, and Recall. However, most of the studies focus on prediction accuracy to define the performance of the ML model.
- Most studies provided information about the dataset and its features. The features of the datasets are based on student demographic information, student socio-economic background, pre-enrolment academic information, and university academic records, Learning Management System (LMS) interaction data.
- In the reviewed articles, both the primary dataset and secondary dataset were used to train and test the ML models. The secondary data are collected from online repositories, e.g., UC Irvine Machine Learning Repository (https://archive-beta.ics.uci.edu/). This finding highlights the hindrance to ML application is HE, i.e., availability or accessibility to the educational dataset. Most HE institutes are often reluctant to provide educational data to confidentiality and privacy concerns.
- Most of the reviewed studies did not identify post-implementation changes attributed to the ML models themselves. A challenge in measuring the effectiveness of ML model implementation lies in the difficulty of identifying the extent to which any change after the ML application implementation is attributed to the ML model improvement itself.
- Most studies discussed and explained the main ML algorithm used to train the ML model in detail and did not discuss in-depth the other ML techniques like feature selection techniques or resampling methods used to develop the ML model and achieve high accuracy.
- Most of the studies have considered dataset as a possible limitation focusing on two fronts. The limited or small size of the dataset (instances) and the limited number of features. Most of the studies explore the demographic characteristics, academic attributes, and LMS logs to collect student interaction with the LMS to train the ML models. However, student attributes like personality traits, psychographic characteristics, attitude, motivation, stress, and anxiety are not captured.
- The majority of the studies, as future work, have recommended exploring other ML algorithms and a combination of ML algorithms and techniques (other than the ones utilized in the study) to improve the results obtained.

6 Conclusion

This book chapter has highlighted the current state of the application of AI in HE revealing the main ML algorithms utilized and characteristics of the dataset utilized to identify the trends and direction. The findings have addressed the research question of the study. The study not only informs educators regards the latest ML algorithms and technologies but also highlights the research gaps for future work recommendations. For example, our aim is to promote practical problem-solving studies through the applications of ML and other emerging technologies [18, 19] in different problem domains. These types of studies provide huge opportunities for the creation of new domain knowledge, although it is recommended that adequate research methodology such as design science [20–22] can be of paramount practice that offers guidance and supportive framework for research operations. Such design research can be facitalated for academic record management [23] or expert system application [24] for assisting HE providers for effective practice improvements.

References

1. Masters K (2019) Artificial intelligence in medical education. Med Teacher 41(9):976–980. https://doi.org/10.1080/0142159X.2019.1595557
2. Bulathwela S, Pérez-Ortiz M, Holloway C, Shawe-Taylor J (2021) Could AI democratise education? socio-technical imaginaries of an EdTech revolution
3. Chassignol M, Khoroshavin A, Klimova A, Bilyatdinova A (2018) Artificial Intelligence trends in education: a narrative overview. Procedia Comput Sci 136:16–24. https://doi.org/10.1016/j.procs.2018.08.233
4. Pokrivcakova S (2019) Preparing teachers for the application of AI-powered technologies in foreign language education. J Lang Cult Educ 7:135–153. https://doi.org/10.2478/jolace-2019-0025
5. Rajeshwari M, Krishna Prasad K (2020). IBM watson industry cognitive education methods. Int J Case Studies Bus IT Educ (IJCSBE) 4(1):38–50. https://doi.org/10.5281/zenodo.3766900
6. Ensmenger N (2011) Nils J Nilsson: the quest for artificial intelligence: a history of ideas and achievements 102(3):588–588
7. Zeineddine H, Braendle U, Farah A (2021) Enhancing prediction of student success: Automated machine learning approach. Comput Electr Eng 89:106903. https://doi.org/10.1016/j.compeleceng.2020.106903
8. Niyogisubizo J, Liao L, Nziyumva E, Murwanashyaka E, Nshimyumukiza PC (2022) Predicting student's dropout in university classes using two-layer ensemble machine learning approach: a novel stacked generalization. Comput Educ Artif Intell 3:100066. https://doi.org/10.1016/j.caeai.2022.100066
9. Walid MAA, Ahmed SMM, Zeyad M, Galib SMS, Nesa M (2022) Analysis of machine learning strategies for prediction of passing undergraduate admission test. Int J Inf Manag Data Insights 2(2):100111. https://doi.org/10.1016/j.jjimei.2022.100111
10. Veluri RK, Patra I, Naved M, Prasad VV, Arcinas MM, Beram SM, Raghuvanshi A (2022) Learning analytics using deep learning techniques for efficiently managing educational institutes. Mater Today Proc 51:2317–2320. https://doi.org/10.1016/j.matpr.2021.11.416
11. Cam HNT, Sarlan A, Arshad NI, Thanh VVT (2022) Machine learning strategy for enhancing academic achievement in Private University. 2022 International conference on digital transformation and intelligence (ICDI)

12. Fahd K, Miah SJ, Ahmed K (2021) Predicting student performance in a blended learning environment using learning management system interaction data. Appl Comput Inf, ahead-of-print(ahead-of-print). https://doi.org/10.1108/ACI-06-2021-0150
13. Fahd K, Venkatraman S, Miah SJ, Ahmed K (2022) Application of machine learning in higher education to assess student academic performance, At-risk, and attrition: a meta-analysis of literature. Educ Inf Technol 27:3743–3775
14. Bravo-Agapito J, Romero SJ, Pamplona S (2021) Early prediction of undergraduate Student's academic performance in completely online learning: a five-year study. Comput Human Behav 115:106595. https://doi.org/10.1016/j.chb.2020.106595
15. Iatrellis O, Savvas IK, Fitsilis P, Gerogiannis VC (2021) A two-phase machine learning approach for predicting student outcomes. Educ Inf Technol 26(1):69–88. https://doi.org/10.1007/s10639-020-10260-x
16. Berriri M, Djema S, Rey G, Dartigues-Pallez C (2021) Multi-class assessment based on random forests. Educ Sci 11(3)
17. Yakubu N, Abubakar A (2021) Applying machine learning approach to predict students' performance in higher educational institutions. Kybernetes, ahead-of-print. https://doi.org/10.1108/K-12-2020-0865
18. Sabharwal R, Miah SJ (2022) An intelligent literature review: adopting inductive approach to define machine learning applications in the clinical domain. J Big Data 9:53. https://doi.org/10.1186/s40537-022-00605-3
19. Shee H, Miah SJ, de Vass T (2021) Impact of smart logistics on smart city sustainable performance: an empirical investigation. Int J Logist Manag 32(3):821–845
20. Miah SJ, Ahamed R (2011) A cloud-based DSS model for driver safety and monitoring on Australian roads. Int J Emerg Sci 1(4):634–648
21. Miah SJ (2009) End user as application developer for decision support. In: The Proceedings of the fifteenth Americas conference on information systems (AMCIS 2009), San Francisco, California August 6th-9th 2009, 142
22. Miah SJ, McGrath GM, Kerr D (2016) Design science research for decision support systems development: recent publication trends in the premier IS journals. Australas J Inf Syst 20:1–14
23. Miah SJ, Samsudin AZH (2017) EDRMS for academic records management: a design study in a Malaysian University. Educ Inf Technol 22:1895–1910
24. Genemo H, Miah SJ, McAndrew A (2016) A design science research methodology for developing a computer-aided assessment approach using method marking concept. Educ Inf Technol 21:1769–1784

Camelot: A Council of Machine Learning Strategies to Enhance Teaching

Vivek Pallipuram, Gafur Mammadov, Brittany Ho, Manohar Dhulipala, and Sebastian Dziallas

Abstract Several educational institutions including schools, colleges, and training institutes use teaching evaluations to improve instruction. In due course, the massive data overwhelms educators, leading to sub-optimal analysis and marginal instructional improvements. This book chapter presents Camelot, a collection of supervised and unsupervised machine learning strategies to facilitate an automatic and effective analysis of both quantitative and qualitative teaching evaluations. Camelot inputs the quantitative evaluations (numerical ratings for various instructor attributes) and provides insights with varying granularity. The coarse-grained insights comprise categorical ratings for instance, 'exceeds expectations', 'at par', 'below par', etc. The fine-grained insights include the identification of crucial instructor attributes for improvement and pertinent strategies. Camelot uses a deep semantics model to analyze the qualitative teaching evaluations (free-form textual responses) and identify keywords that characterize an educator's teaching methodology. Camelot combines the results from machine learning techniques to dispense effective improvement strategies. Although we describe Camelot using teaching evaluations as a case study, we assert that it can be easily adapted for other areas of education including student learning outcomes and accreditation studies. We envision Camelot to be a useful automation for seasoned educators, new teachers, school leaders, and policymakers.

Keywords Supervised machine learning · Natural language processing · Ensemble learning · Model selection · Instructor evaluation

1 Introduction

Teaching introspection and continued instructional improvement is the hallmark of effective education. From elementary schools to universities, many institutions employ teaching evaluations as a means to assess instructional effectiveness and

V. Pallipuram (✉) · G. Mammadov · B. Ho · M. Dhulipala · S. Dziallas
School of Engineering and Computer Science, University of the Pacific, 3601 Pacific Avenue, Stockton, CA, United States 95211
e-mail: vpallipuramkrishnamani@PACIFIC.EDU

M. S. Khine (ed.), *Machine Learning in Educational Sciences*,
https://doi.org/10.1007/978-981-99-9379-6_4

identify areas for improvement. Typically, these evaluations include several quantitative questions with Likert scaling (1 to 5) to assess instructional effectiveness. These questions may inquire about a teacher's ability to engage students with the subject matter, ability to effectively convey ideas, and foster student collaboration, among others. The evaluations may also include qualitative textual response questions. For instance, 'What worked and what did not work for you (the student) in this course?' Providing both quantitative and qualitative feedback to a teacher's instruction can be a powerful tool for their ongoing professional development. However, educators and institutions may find it challenging to analyze the vast amount of feedback information available to them. Although tools like Microsoft Excel can help with analyzing small data in spreadsheets, dealing with hundreds of thousands of spreadsheets can make these tools impractical.

With the advancements in the fields of machine learning and computer hardware, several powerful tools are now available for analysts to quickly study their large data. Additionally, the ever-growing user-friendliness of programming languages like Python has brought machine learning to general desktops. Albeit, machine learning can be challenging due to its various intricacies, such as data preprocessing, model fitting, and inferring results using complex metrics.

This chapter presents Camelot, an easy-to-use machine learning (ML) framework to analyze large sets of quantitative and qualitative teaching evaluations. The framework studies the objective teacher evaluations to offer course-grained insights. Specifically, the framework classifies the quantitative evaluations into multiple categories including 'excellent', 'above average', 'average', 'below average', and 'unsatisfactory'. The framework also includes a semantic analyzer that studies qualitative textual feedback and identifies the sentiment levels (positive, negative, neutral, etc.). This level of analysis provides a well-rounded insight into a teacher's instructional ability, thereby identifying avenues for continued improvement.

The framework comprises three phases including data preprocessing, model training, and model testing and inference. The data preprocessing phase provides users with tools and techniques to curate the data and make them ready for ML analysis. The model training phase includes several classification-based ML models including the Na¨ıve Bayes, K Nearest Neighbors, Support Vector Machine, logistic regression, decision trees, and neural network. In the model testing and inference phase, Camelot outlines an effective method to test the ML models. This phase also includes a combiner that uses ML evaluation metrics to select an optimal model for prediction.

The main objective of this chapter is to explore the use of machine learning in teaching evaluation. We provide readers with a range of ML techniques and code snippets that can be customized for their own projects. We envision that educators and school/university leaders will use Camelot for educational research purposes.

The rest of the chapter is organized as follows. Section 2 provides a brief background on ML models for classification. Section 3 describes the Camelot framework in detail with ready-to-use Python code snippets wherever appropriate. We assume that readers have some programming background, although it is not entirely necessary. The chapter concludes in Sect. 4 with conclusions and a discussion on educational services that may benefit from the Camelot framework.

2 Machine Learning Models: Theory

This section provides a brief theoretical background on commonly used supervised machine learning (ML) models. We encourage readers to refer to machine learning textbooks for a detailed explanation [1]. Sections 2.1–2.6 discuss the supervised models including the Na¨ıve Bayes classifier, K Nearest Neighbors (KNN), Support Vector Machines (SVM), logistic regression, decision trees, and neural networks. We use these six models as a part of Camelot's prediction framework.

2.1 Supervised Model: Na¨ıve Bayes Classifier

The Na¨ıve Bayes classifier is a supervised machine learning model that employs Bayes theorem for classification. Consider a feature point with *n* features given by $x = [x1, x2,.., xn]$. We aim to classify this feature point into one of the *k* classes given by $Ck = [1, 2, 3,.., k]$. The classifier's goal is to compute the probability of a class, *Ck*, given the feature point, *x*. To find this probability, the classifier applies Bayes theorem as shown in Equation 1a.

$$P(C_K|X) = \frac{P(X|C_K) - P(C_K)}{P(X)} \tag{1a}$$

$$P(X|C_X) = \frac{1}{2\pi\sigma_K^2} e^{-\frac{(X-\mu_K)^2}{2\sigma_K^2}} \tag{1b}$$

$$P(X) = P(X_1) \times P(X_2) \times \cdots P(X_n) \tag{1c}$$

In Eq. 1a, P(*Ck* |*x*) is the posterior probability, which is to be found. The terms *P*(*x*|*Ck*), *P*(*Ck*), and *P*(*x*) are the likelihood, prior, and the evidence, which are obtained using a labeled training data set.

The classifier is na¨ıve because it assumes that features, × 1, × 2, …*xn* are independent. Further, the classifier assumes their normal distribution. With these assumptions, the classifier obtains the likelihood using Eq. 1b by pooling together the feature attribute values for a given class, *Ck*. Equation 1c employs the independence assumption to obtain the probability (*P*(*x*)) of a feature point, *x*.

2.2 Supervised Model: K Nearest Neighbors

K Nearest Neighbors (KNN) classifier employs distance measures to classify a feature point. The classifier first inputs a labeled training data set with classes (Ck) assigned to each feature point (x). For classification, the classifier inputs a feature point and considers K feature points in the training data set that are close to the input feature point. The classifier pools the class labels of the nearest K neighbors and determines the most frequent class. This class denotes the classification of the input feature point.

2.3 Supervised Model: Support Vector Machines

The goal of the Support Vector Machine (SVM) is to find the best separable hyperplane in the feature space that effectively separates feature points into multiple classes. SVM constructs a set of hyperplanes in a high-dimensional space for classification. The SVM for a multi-class problem implements a one-versus-one approach. In this approach, SVM generates a hyperplane to separate every two classes while ignoring the feature points of other classes. A detailed mathematical discussion can be found in [2].

2.4 Supervised Model: Logistic Regression

The multinomial logistic regression (LR) model employs the *softmax* function [2] (characterized by the weight vector, β) to determine the probability that the class (Cx) of the feature point $x = [\times 1, \times 2,.., xn]$ is class, k. Using a labeled training data set, LR estimates β using the Newton-cg algorithm. After training, the classifier uses the *softmax* function given by Eq. 2 to identify the class of the feature point with the highest probability.

$$P(y_i = k_1 X) = \frac{e^{\beta_K X_i}}{k = 1\, e^{\beta_K X_i}} \tag{2}$$

i

2.5 Supervised Model: Decision Trees

The decision tree classifier creates a tree-like structure for classification. In this tree, the internal nodes denote the feature attributes ($\times 1, \times 2,..$), and the leaf nodes denote the classes (Ck where k is the number of classes). Using the concept of information gain [1], the classifier recursively splits the data at the internal nodes to create new nodes. This splitting process stops when no more information gain can be achieved.

To classify a feature point, the classifier traverses the tree with the guidance of internal nodes. The terminal leaf node is the classification of this feature point. We recommend [1] for more insights into decision trees.

2.6 Supervised Model: Neural Networks

The multi-layer perceptron (MLP) neural network classifier comprises neurons arranged in multiple successive layers. The first layer (called the input layer) contains as many neurons as the number of features. The last layer is the output layer, which contains as many neurons as the number of classes. In between these two layers are several hidden layers; their number and the number of neurons in them are determined using empirical methods. Typically, all the neurons in the ith layer are connected with the neurons in the $(i + 1)$th layer. The hidden layer neurons work with an activation function that determines their activation. After training, the neural network inputs a feature point. An activated neuron in the output layer determines the class for this input feature point.

3 The Camelot Framework

Figure 1 shows the Camelot framework comprising three phases namely data preprocessing, model training, model testing, and inference. Phase-1, data preprocess- ing, contains techniques that clean or transform the data to make it amenable to machine learning. Phase-2, model training, consists of supervised machine learning (ML) and sentiment analysis models. The supervised models analyze the instructor's objective teaching evaluations to categorize them into one of the multiple classes, which denote an evaluation's quality. The sentiment analysis model inputs the qualitative data (free-form textual response) to characterize the students' sentiments into emotional attributes such as positive, negative, neutral, etc. Phase-3, model testing and inference, evaluates the quality of ML models using metrics including the confusion matrix, accuracy, precision, recall, and F1-score. Sections 3.1–3.3 describe the Camelot framework's three phases. These successive phases are commonly used in machine learning and therefore, we advise readers to adopt them for any ML-based data analysis project. In what follows, we present Camelot's implementation of these phases in a generic manner so that readers may employ them in their own projects.

Table 1 explains the instructor's teaching attributes (features) using which, we classify the instructor's evaluations into one of the five categories: 'excellent' (E), 'above average' (AA), 'average' (A), 'below average' (BA), and 'unsatisfactory' (U). These features take on continuous values between 1 (lowest rating) to 5 (highest rating). Camelot's model training and testing are based on these features. Throughout

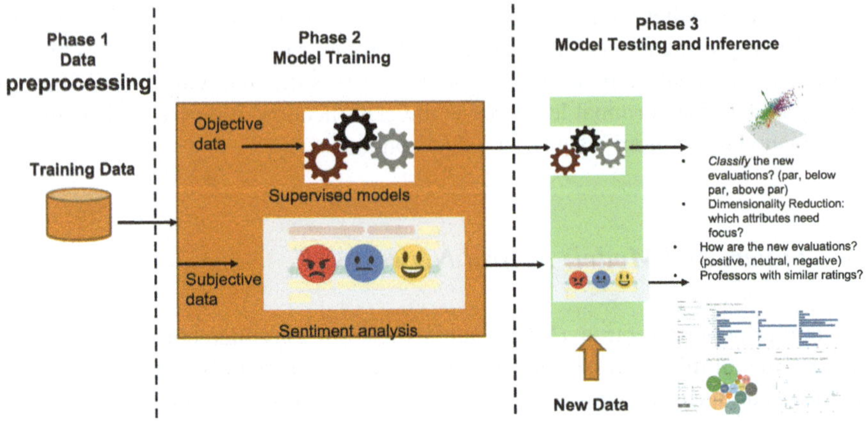

Fig. 1 A top-level view of the Camelot framework comprising three phases: data preprocessing, model training, model testing, and inference

Table 1 The nine instructor attributes and their meaning

Teaching attribute	Meaning
Knowledge	Instructor's knowledge of the given subject matter
Enthusiasm	Instructor's enthusiasm to teach in the class
Organization	Instructor's class organization
Preparedness	Instructor's lecture preparedness
Explanation	Instructor's ability to explain difficult concepts
Speaking ability	Instructor's ability to speak clearly during the lectures
Approachability	Denotes how approachable an instructor is for doubt clarification, general questions and answers, etc
Discussion	Instructor's ability to foster students' discussion
Effectiveness of Examples	Instructor's ability to provide effective examples pertaining to the subject

the rest of this chapter, we use a subset of these features to explain Camelot's characteristics.

3.1 Phase-1: Data Preprocessing

The first step prior to machine learning is to *preprocess* the data. *Preprocessing* is an umbrella term for various data processing procedures. The Camelot framework addresses four common scenarios including (1) reading the data; (2) accounting for the missing values; (3) standardization; and (4) handling the categorical variables.

In what follows, we describe the techniques to resolve them along with the pertinent Python code snippets. To demonstrate Camelot's selected functionality, we use the data excerpt provided in Table 2. This table includes sample values of six features (independent variables) including instuctor's knowledge, enthusiasm, class organization, lecture preparedness, explanation ability, and ability to foster student discussion. The seventh column, Rating, is determined by the user. For instance, consider a case where a department chair has certain expectations for quality teaching. According to their department's vision, the chair may rate (label) these features as shown in Table 2. The ML models *learn* from this labeled training data. This situation is akin to showing a bunch of fruits to a five-year-old child so they may recognize fruits in the future. Notice that the table includes a few missing values, which is a common artifact in data collection.

Table 2 An excerpt of the data file (data.csv) to demonstrate the selected functionality of the Camelot framework. The first six columns (Knowledge through Discussion) are the independent variables. A single row of these six attributes constitutes an evaluation. The rating column provides the user-defined ratings for the evaluations. This table is also used as the labeled training data set for ML training

Knowledge	Enthusiasm	Organization	Preparedness	Explanation	Discussion	Rating
3.82	3.34	3.72	3.48	3.71	3.28	Unsatisfactory
3.30	3.83	3.58	3.84	3.55	3.63	Unsatisfactory
3.48	3.36	3.58	3.19	3.51	3.81	Unsatisfactory
3.41	3.72	3.51	3.80		3.63	Below Average
3.46	3.98	3.59	4.1	3.76	3.74	Below Average
3.63	3.82	3.65	3.67	3.6	3.46	Below Average
4.14	4.35	4.24	3.93	3.57	4.27	Average
4.10	4.16	3.85	3.88	4.28	3.78	Average
3.85	4.08	4.13		3.63	3.70	Average
4.91	4.52	4.65	4.15	4.63	4.26	Excellent
4.66	4.38	4.56	4.53	4.43	4.3	Excellent
4.46		4.10	4.39	4.46	4.65	Above Average
4.40	4.45	4.21	3.93	4.42	4.35	Above Average
4.55	4.70	4.59	4.56	4.94	4.47	Excellent
4.12	4.51	4.35	4.31	4.45	4.53	Above Average

3.1.1 Reading the Data

The first step is to read the data and extract the dependent and independent variables. Camelot's preferred data file format is the comma-separated value (CSV) format. Camelot is written in the Python programming language that has rich support to handle CSV files. Listing 1 provides the pertinent code snippet to read the data and extract the dependent and independent variables. In this code, Line 1 im- ports the pandas library to process the data file. Line 3 invokes the read csv function to read the CSV data file. Lines 5–7 identify the column names to be used as independent variables (features) and the dependent variable (dep var). In this case, a professor's teaching attributes including *knowledge* of the subject matter, teaching *enthusiasm*, lecture *organization*, classroom *preparedness*, *explanation* ability, and the ability to facilitate *discussion* constitute the features. The instructor's overall teaching is rated as 'unsatisfactory', 'below average', 'average', 'above average', and 'excellent'. Lines 9–11 extract the values of features and the dependent variables into data frames, X and y, respectively. The lines starting with a pound (#) are comments that make the code self-explanatory.

Listing 1. Camelot code snippet demonstrating data file input and the extraction of dependent and independent variables

```
1  import pandas as pd
2  # read the data into a data frame
3  df = pd. read_csv ('data. csv')
4  # define features and dependent variable
5  features = ['Knowledge', 'Enthusiasm', 'Organization', \
6  'Preparedness', 'Explanation', 'Discussion']
7  dep_var = ['Rating']
8  # create a data frame of independent variable values
9  X = df [features]
10 # dependent variable
11 y = df [dep_var]
```

3.1.2 Accounting for the Missing Values

Data collection can be an arduous procedure and it is understandable that certain features may have missing values. Note that the data provided in Table 2 includes some missing values that need to be resolved prior to the analysis. There are several methods to handle the missing values; the choice of a method depends on the data set and the analyst. Camelot employs two methods to handle the missing values: (1) removing the data point altogether, and (2) replacing the missing values with their median values.

- Elimination of incomplete feature points—Camelot reads missing values as not-a-number (NaN). Because ML models for classification do not work well with

NaNs, Camelot provides an option to delete the feature points using dropna() function as shown in Listing 2.

Listing 2. Use of dropna() to remove data points with missing values

```
df = pd. read_csv ('data. csv')
newdf = d f. dropna ()
```

• Replacing missing values with median—The deletion of feature points in a small data set may negatively impact the ML performance. To alleviate this issue,

Camelot enables users to replace the missing feature values with their median, which is the middle value of a given feature. The choice of using the median over other measures including average and mode is prudent because the median is an unbiased measure that is resilient to outliers. Listing 3 provides Camelot's code snippet that replaces the missing values with the median.

Listing 3. Replacing the missing values with the median

```
1 import numpy as np
2 from sklearn. impute import SimpleImputer
3 # Extract independent and dependent variables
4 # as before into X and y
5 impute = SimpleImputer (missing values = np. nan, \
6                 strategy = "median")
7 X = impute. fit_transform (X)
```

In Listing 3, Line 1 imports the numpy library to manipulate NaNs. Line 2 imports the SimpleImputer function to replace the missing values with the median. This task is accomplished in Lines 5-7 where X is the final transformed data set with missing values replaced with their respective medians.

3.1.3 Standardization

The Camelot framework applies standardization (also called feature scaling) when the magnitude of the feature values differ significantly. The ML models that work with the scale of data (KNN and SVM, among others) require standardization. The standardization process serves to bring the features to the same scale. This feature scaling prevents large feature values from skewing the ML prediction. For instance, consider an educator attempting to model the mathematics and reading aptitude of five students. Consider two independent variables (features): mathematics score ranging from 100 to 300 and reading aptitude score ranging from 0–10. Table 3 (on the left) provides the feature values for five students prior to standardization. As seen in this table, the mathematics score is an order of magnitude larger than the reading score—an issue that may lead to biased ML models.

Listing 4 provides the applicable code for feature standardization. In this code, Line 2 imports the StandardScaler function to enable feature standardization. Lines 7–11 apply the standardization procedure on X, the original feature data. Xscale is

Table 3 Mathematics and reading scores of five students prior to standardization

Math score	Reading score	Scaled math score	Scaled reading score
157	6	-1.42	-0.21
233	8	0.02	0.86
300	9	1.29	1.4
277	4	0.85	-1.29
192	5	-0.75	-0.75

the standardized feature data. We recommend readers refer to this listing, should they determine that standardization is necessary.

Listing 4. Standardizing feature values so that their magnitudes are similar

```
import pandas as pd
from sklearn. preprocessing import StandardScaler
df = pd. read_csv ('scaling. csv')
features = ['math score', 'reading_score']
X = df [features]
print (X)
# apply standardization
 scale r = StandardScaler()
 scaler. fit (X)
Xscale = scaler. transform (X)
print (Xscale)
```

Table 3 (on the right) shows the standardized mathematics and reading scores. Notice that the feature values are comparable, which removes any bias during ML training.

3.1.4 Handling Categorical Variables

The Camelot framework's primary goal is to classify teaching evaluations into one of the five categories: 'unsatisfactory' (*U*), 'below average' (*BA*), 'average' (*A*), 'above average' (*AA*), and 'excellent' (*E*). Consequently, the dependent variable (to be predicted) in this framework is the *instructor rating*, which is a categorical variable. A categorical variable takes on discrete values that may denote a class (in our case), an order (e.g. income groups), or an interval (e.g. height in inches). More information on categorical variables can be obtained here [2].

The machine learning models only understand numbers and therefore, the categorical variables must be transformed into numbers. This process is accomplished by simply assigning numbers to the classes that the categorical variable may assume. In Camelot's case, instructor rating is a categorical variable that takes on five categories: *U*, *BA*, *A*, *AA*, and *E*. Camelot employs the code in Listing 5 to convert the instructor rating classes to numbers. Line 1 creates a user-defined dictionary that assigns numbers to the classes. Line 3 implements the map function that replaces the classes with their assigned numbers by simply looking up the dictionary. Line 3 finalizes the transformation of the classes to numbers.

Listing 5. Assigning numbers to five categories for the instructor rating

```
1 d = {'Unsatisfactory': 0, Below Average': 1, \
2 'Average': 2, 'Above Average': 3, 'Excellent': 4}
3 df ['Rating'] = df ['Rating']. map (d)
```

3.2 Phase-2: Model Training

Figure 1 shows Camelot's Phase-2, model training, which includes supervised machine learning models and a sentiment analyzer. In what follows, we describe model training by providing pertinent Python code snippets. Section 3.2.1 describes the training of supervised models. Section 3.2.2 explains the sentiment analysis of textual data in Python.

3.2.1 Supervised Model Training

As shown in Fig. 2, the Camelot framework employs six machine learning (ML) models to classify instructor ratings into one of the five categories: U, BA, A, AA, and E where U and E are the lowest and highest ratings, respectively. The framework also employs a combiner that combines the results from the six ML models. To perform model training, the framework inputs training data that first undergoes labeling—a process requiring human intervention. Specifically, the user must categorize (label) each feature point in the training data by applying domain knowledge. For instance, consider a case where a feature point comprises three instructor teaching features including knowledge of the subject matter (K), enthusiasm for the subject (En), and ability to explain the concepts (Ex). Further, assume that these feature values are between 1–5. Using domain knowledge, a user may label a feature point with feature values ($K = 3.8$, $En = 4$, $Ex = 3.5$) as average (A). In contrast, a feature point with values ($K = 4.4$, $En = 4.3$, $Ex = 4.6$) may be labeled as excellent (E). Table 2 provides an excerpt of a labeled training data where feature points comprising six features are labeled as excellent, above average, average, below average, or unsatisfactory (see Rating column). In what follows, we explain how the six MLs learn in Python using the labeled training data.

Na¨ıve Bayes Classifier—Listing 6 provides the pertinent Python code to train the Na¨ıve Bayes classifier. The snippet assumes that the readers have already performed Phase-1, data cleaning, as described in Sect. 3.1. For this classifier, the necessary preprocessing steps are reading the labeled training data (Sect. 3.1.1), accounting for the missing values (Sect. 3.1.2), and handling the categorical variables (Sect. 3.1.4).

Listing 6. The code snippet to train the Na¨ıve Bayes classifier

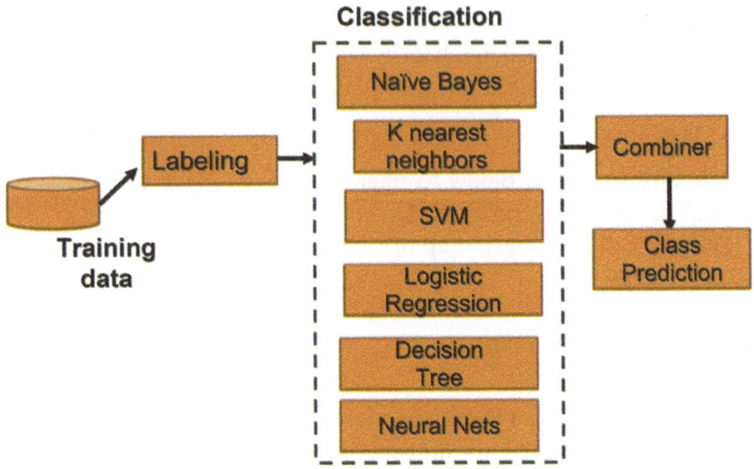

Fig. 2 The six ML models along with a combiner to classify the instructor rating into one of the five categories

```
1 from sklearn. naive bayes import GaussianNB
2 gnb = GaussianNB (). f i t (X, y)
3 gnb_predict = gnb. predict (X test)
4 classprob = (gnb. predict_proba (X test))
```

In Listing 6, Line 1 imports the GaussianNB function from scikit-learn library[1] to create the model. In Line 2, the GaussianNB model employs the independent variable data frame (X) and the dependent variable frame (y) to create a fitted model, gnb. Line 3 employs this model to generate a prediction for the input,

X text. Line 4 employs the fitted model to generate the probability of each class. Readers may utilize this code listing to fit a Na¨ıve Bayes model to their own data.

K Nearest Neighbors (KNN)—KNN also requires labeled training data and the data cleaning steps elucidated in Sect. 3.1. Readers should note that KNN requires a special process called hyperparameter tuning. This process selects effective classifier parameters for a quality classification. The Camelot framework applies the grid search method for KNN's hyperparameter tuning. This method inputs multiple classifier parameter values (selected empirically by the analyst) and tests various combinations. The method selects a parameter combination that performs with the highest accuracy.

Listing 7. The code snippet to tune and train the K nearest neighbors classifier

```
from sklearn. neighbors import KNeighborsClassifier
from slearn. model_selection import GridSearchCV
from sklearn. model_selection import \
RepeatedStratified K Fold
```

[1] https://scikit-learn.org/stable/.

```
5  # Hyperparameter Tuning
6  n_neighbors = range (1, 2 0)
7  weights = ['uniform', 'distance']
8  metric = ['euclidean', 'manhattan', 'minkowski]
9  grid = dict (n _neighbors = n _neighbors, \
10 weights = weights, metric = metric)
11 cv = RepeatedStratifiedKFold (n splits = 1 0, \
12 n_repeats = 3, random _state = 1)
13 grid _search = GridSearchCV (KNeighborsClassifier (), \
14 param _grid = grid, n _jobs = −1, cv = cv, \
15 scoring = 'accuracy', error _score = 0)
16 grid _result = grid _search. fit (X, y)
17 # Fitting the model using the best results
18 knn = KNeighborsClassifier (\
19 weights = grid_result. Best _params _['weights'], \
20 n _neighbors = grid _result. best _params ['n _neighbors'], \
21 metric = grid _result. best_params ['metric']). fit (X, y)
22 knn _predictions = knn. predict (X t e s t)
```

Listing 7 provides the code snippet to conduct KNN's hyperparameter tuning and training. Lines 1–4 import the pertinent functions for tuning and model training. Lines 5–10 create a grid of KNN parameters including the number of neighbors (n neighbors), their weights (weights), and the distance metric (metric). Lines 11–12 invoke the StratifiedKFold function that creates indices to split the data into training/testing sets. Lines 13–15 invoke the GridSearchCV function that tests various combinations of the grid parameters. This function uses accuracy as the scoring method. A grid parameter combination with the highest accuracy is selected for the model. Lines 18–21 fit the KNN classifier model using the identified parameters. Line 22 employs the model to perform a prediction on the test data (X test).

Support Vector Machine (SVM)—Listing 8 provides the Python code snippet to tune and train the support vector machine (SVM) classifier. Lines 1–2 import the support vector machine model and grid search function to search for effective model parameters. Similar to KNN, SVM undergoes hyperparameter tuning in Lines 3–6 where we create multiple SVM model parameters. Line 7 fits the SVM model with the best parameters identified in Line 5. Line 8 performs prediction on a test feature point, X test.

Listing 8. The code snippet to tune and train the SVM classifier

```
1 from sklearn. svm import SVC
2 from sklearn. model selection import GridSearch C V
3 param grid = {'C': [0. 1, 1, 1 0, 1 0 0, 1 0 0 0], \
4 'gamma' : [1, 0. 1, 0. 0 0 1, 0. 0 0 0 1], 'k e r n e l' : ['r b f'] }
5 SVM Model = GridSearchCV (SVC (), param_grid, \
6 refit = True, verbose = 3)
7 gridresult = SVM_Model. fit (X, y)
```

```
y_pred = SVM Model. predict (X t e s t)
```

Logistic Regression (LR)—Listing 9 provides the code to tune and train the logistic regression classifier. Lines 1–3 import the pertinent functions for hyperparameter tuning and training. Lines 4–13 perform LR's hyperparameter tuning. Line 14 trains the LR model with optimal parameters and Line 15 performs prediction for a test feature point, xtest.

At this point, readers should observe that the model tuning and training methodologies are uniform across the ML models. Readers must refer to skit-learn's website to identify individual ML model parameters for tuning. Listings 10 and 11 pro- vide the hyperparameter tuning and training for the decision tree and neural network classifiers, respectively. We expect readers to read these code snippets and explain them on their own. In what follows, we conclude our model training discussion by instructing readers when to use these six ML models.

Listing 9. The code snippet to tune and train the logistic regression classifier

```
1 from sklearn. linear model import LogisticRegression
2 from sklearn. model selection import GridSearchCV
3 import numpy a s np
4 param_grid = {'penalty': [' 1 1', '1 2', \
5 'elasticnet', 'none'],
6 'C': np. logspace (− 4, 4, 2 0), \
7 'solver' : ['lbfgs', 'newton − cg', 'liblinear', \ 8 's a g', 's a g a'], \
9 'maxiter' : [1 0 0, 1 0 0 0, 2 5 0 0, 5 0 0 0] }
10 LogModel = LogisticRegression ()
11 classifier = GridSearchCV (LogModel, \
12 param_grid = param_grid, cv = 3, \
13 verbose = True, n j o b s = − 1)
14 classifier. fit (x train, y)
15 y pred = classifier. predict (x t e s t)
```

Listing 10. The code snippet to tune and train the decision trees classifier

```
from sklearn. tree import Decision TreeClassifier
from sklearn. model_selection import GridSearchCV
clf = DecisionTreeClassifier ()
params = {'min_samplesleaf': [1, 2, 3], \
'max depth' : [1, 2, 3, 4, 5] }
dtree = GridSearchCV (estimator = clf, param_grid = params, \
cv = 10, n_jobs = 1, verbose = 2)
dtree. fit (X, y)
ypred = dtree. predict (X t e s t)
```

Listing 11. The code snippet to tune and train the neural network classifier

```
from sklearn. neural_network import MLP Classifier
from sklearn. model_selection import GridSearchC V
```

```
paramgrid = {'hidden_layer_sizes': [(5, 5, 5), (7, 7, 7)], \
'activation': ['identity', 'relu'], \
'solver' : ['lbfgs', 'sgd'], 'alpha' : [0. 0 1, 0. 1, 1, 1 0] }
clf = GridSearchCV (MLPClassifier (), param grid, \
cv = 5, n _jobs = −1, refit = True, verbose = 3)
clf. fit (Xscale, y)
clfpredict = clf. predict (X testscale)
```

When to use Na¨ıve Bayes Classifier—This classifier works well even for small training data and trains very quickly (runs fast on the computer). However, the classifier assumes that the features are independent, which is a simplistic assumption. The readers are encouraged to apply a correlation analysis to remove any correlated features. The readers should note that this classifier does not allow for any parameter tuning.

When to use KNN—The KNN classifier works best when the number of features is small. As the number of features increases, it takes longer time for KNN to train. Additionally, the prediction accuracy may also drop with an increasing number of features. We recommend using KNN when the data set is balanced, meaning that there are sufficient feature points for each class in the training data set. An imbalanced training data (one or more classes are either over or underrepresented) may throw off KNN's predictions. To overcome this artifact, the analyst has to either collect more data or apply other principles such as under/oversampling (currently out-of-scope).

When to use SVM—Unlike KNN which suffers in performance when the number of features increases, SVM is found to do well with more features. SVM also performs effectively when there is a clear separation between multiple classes. We advise readers to pay careful attention to training data labeling. We also recommend users to employ unsupervised labeling techniques including K means clustering [1] to create a clear separation between multiple classes.

When to use logistic regression—Logistic regression may work well with an imbalanced training data set where one or more classes are either under or overrepresented. The model works effectively when the data set is simple and there is no collinearity between features. The L1 and L2 regularization techniques [2] may help avoid over or underfitting. Readers are advised to avoid logistic regression when the data set is small, or if there exists feature collinearity.

When to use decision trees—The decision tree usually does not require significant data preprocessing such as scaling or removal of missing feature values. It is also intuitive to explain. However, the decision trees may change significantly if the training data is changed. A decision tree, although a visual technique, may be difficult to read if the tree is deep and/or if there are too many classes. Figure 7 in the Appendix shows an example of a decision tree generated by Camelot. The figure caption explains the decision tree.

When to use neural network classifiers—The neural network classifiers work well for cases where the relationship between the independent and dependent variables is complex [1]. This classifier also works effectively for large data sets. How- ever, given that this classifier allows for multiple hidden layers with multiple neurons, its hyperparameter tuning and training may take time on a desktop.

3.2.2 Sentiment Analysis

Camelot's sentiment analyzer studies an instructor's textual feedback response to underscore the emotions of their students. By using students' emotions, an educator can employ intelligent instructional decisions for a better learning outcome. Listing 12 provides an example code snippet that deduces the level of emotions in a given textual feedback. Line 1 imports the Natural Language Toolkit (nltk) library[2]. Line 2 imports SentimentAnalyzer, a pre-trained model capable of recognizing emotional levels. Camelot requires the textual feedback responses stored in a text file with one line per response. Line 3 reads each line from the text file. Line 5 invokes the sentiment analyzer model and Line 7 classifies each line in the text file as positive, negative, or neutral (stored in scores).

Listing 12. Assigning numbers to five categories for the instructor rating

```
1 import nltk
2 from n ltk. sentiment import SentimentIntensityAnalyzer
3  lines = open ("subjective_eval. txt"). read (). splitlines ()
4 # Initialize the SentimentIntensityAnalyzer
5  s i a = SentimentIntensityAnalyzer ()
6 # Get the sentiment scores for each rating
7  scores = [sia. polarityscores (line) for line in lines]
```

Because the sentiment analyzer is a pre-trained model, we do not present its testing. We advise readers to use visualization techniques such as pie charts to view the level of user sentiments in the textual feedback response. Figure 3 shows an ex-ample. An analyst may use this feedback to support (or refute) the classifications given by the ML models. An instructor (or the department chair) may study from this chart that $\approx 50\%$ of the students have neutral or negative emotions. This observation may prompt some interventions, for instance, involving the student life department and psychological services to improve student satisfaction, identifying student pain points and working on them, among others. Users may also generate a list of student comments that received various sentimental levels to determine what worked and what did not work. Table 4 shows some example comments rated by the sentiment analyzer. From this table, an educator may derive fine-grained insights such as working on active learning and being more available for students.

[2] https://www.nltk.org/.

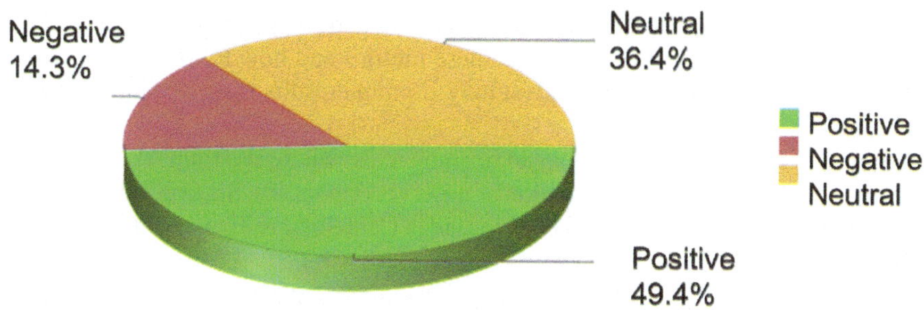

Fig. 3 The pie chart showing sentiment levels: positive, negative, and neutral

Table 4 Some example comments rated by the sentiment analyzer. Educators may study comments rated as negative or neutral to devise strategies for improvement

Comment	NLP rating
'Professor Smith is extremely knowledgeable in the subject matter and presents the material in a clear and organized manner. They are approachable and always willing to help students outside of class. The assignments and exams are fair and reflective of the material covered. Highly recommended!'	Positive
'Professor Johnson has a unique teaching style that may not be suitable for everyone. Their lectures can be a bit disorganized at times, making it challenging to follow along. However, they are passionate about the subject and are available during office hours to clarify any confusion. The grading criteria for assignments could be more transparent.'	Positive
'Professor Thompson lacks enthusiasm in the classroom, which can make the lectures dull and monotonous. The explanations are often unclear, and it is difficult to get clarification during office hours. Grading seems inconsistent, and the feedback provided on assignments is minimal. It would be helpful if they were more accessible and provided additional resources.'	Negative
'Professor Davis is an exceptional educator. They have a deep understanding of the subject and go above and beyond to ensure that students grasp the material. The lectures are engaging, and they incorporate real-world examples to make concepts more relatable. Professor Davis provides timely feedback on assignments and genuinely cares about their students' success.'	Positive
'Professor Wilson's teaching style is confusing and lacks organization. Their lectures often stray off-topic, making it difficult to understand the core concepts. They are rarely available outside of class and provide limited support. Assignments and exams are poorly structured, and the grading criteria are unclear. Overall, it is challenging to succeed in this class due to the lack of guidance.'	Negative

3.3 Phase-3: Model Testing and Inference

This section describes Camelot's testing methodology. Because Camelot's approach aligns with the state of practice, we explain it in a way that readers are able to apply them in their own ML projects. First, we discuss ML evaluation terms including the confusion matrix, true/false positives, true/false negatives, precision, recall, and F1-score. We instruct readers when to use these metrics and how to infer results using them. We also describe how an analyst may combine results from ML models using these metrics.

To describe the metrics, we use an example case where we train ML models to classify a professor's course evaluation as good (1) or poor (0). Section 3.3.1 explains the evaluation metrics and Sect. 3.3.2 provides Camelot's testing results with these metrics.

3.3.1 Evaluation Terminology

Confusion Matrix—The confusion matrix is a commonly used technique to visualize the prediction performance of ML models. Specifically, it shows how many correct and incorrect predictions are for a given class. Figure 4 shows the structure of the confusion matrix. An effective ML performance would entail most of the predictions along the diagonal (green boxes). Using this confusion matrix, the ana lyst evaluates other metrics including positives, negatives, accuracy, precision, and recall. These terms are the most commonly used and unfortunately, most misunderstood terms in the area of machine learning. In what follows, we explain them using intuitive examples.

True Positives (TP) denote how many positive predictions are actually positive. For instance, in our case of professor's evaluation, this term highlights how many courses were predicted to have a good evaluation and in fact, they were good. *False Positives* (*FP*) denotes how many positive predictions are actually incorrect. For instance, in our case of professor's evaluation, this term highlights how many courses obtained a good evaluation, however, they turned out to be poor.

True Negatives (TN) depicts how many predictions were given 0 and in fact, they were 0. In our professor evaluation's case, TN is the number of courses that were deemed poor and in fact, they were poor. Similarly, *False Negatives* (*FN*) denote how many times the ML labeled evaluations as 0, however, their true labels were 1.

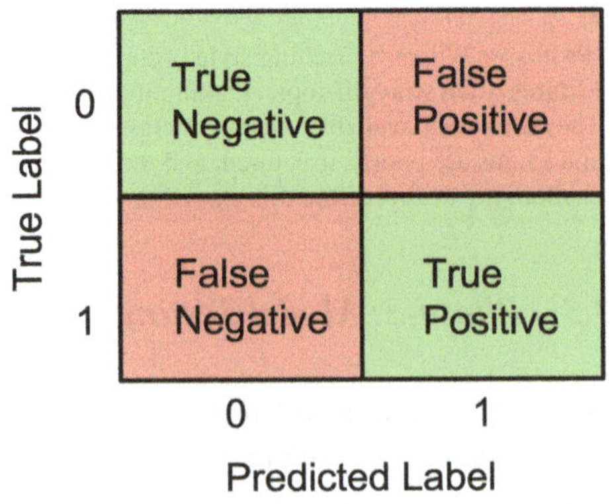

Fig. 4 The structure of a confusion matrix denoting true/false positives and true/false negatives

In our analogy, *FN* represents the number of course evaluations predicted as poor, however, they were actually good.

Accuracy—The accuracy metric denotes how many predictions were correct out of all the predictions made by the ML model; the metric is evaluated using Eq. 3a. Accuracy is usually the most favored metric in order to select an ML model for prediction.

Precision—This metric is used to evaluate an ML's ability to detect positive cases. Specifically, this term answers the question: out of the cases labeled as positives by the model, how many are truly relevant? We employ Eq. 3b to evaluate this quantity. Readers should pay special attention to precision in case the false positives are costly for the analysis. As an additional example, consider an ML model assisting an engineering school dean with labeling candidates for tenure-track positions into two categories: hire (1) and pass (0). If the model recommends an under-qualified candidate to be hired, then this prediction is a false positive which may be costly for the school. In this case, the ML model should have high precision for reliable predictions.

$$\text{Accuracy} = \frac{TP + TN}{FP + FN + TP + TN} \tag{3a}$$

$$\text{Precistion} = \frac{TP}{TP + FP} \tag{3b}$$

$$\text{Recall} = \frac{TP}{TP + FN} \tag{3c}$$

$$F1\text{score} = \frac{2 \times \text{precision} \times \text{recall}}{\text{precision} + \text{recall}} \tag{3d}$$

Recall—This metric evaluates the model's capability to identify truly relevant cases and it is given by Eq. 3c. Specifically, it penalizes an ML model for incorrectly dropping the positive cases. Let us use the previous example of our engineering school dean attempting to hire tenure-track professors using machine learning. A model with poor recall may not directly incur losses, but it will prevent any gains by ignoring good candidates (false negatives). Therefore, the dean should opt for a model with a high recall to avoid any loss of gains.

In summary, readers should view precision as a metric to help avoid direct losses and recall as a metric to help avoid losses due to missed gains.

F1-score—An analyst may determine that both precision and recall are important for their analysis. F1-score, which is the harmonic mean of these two terms given in Eq. 3d, accounts for both. Our engineering school dean's goal is to avoid under-qualified candidates and at the same time, not to miss out on the good ones. Therefore, the dean should focus on F1-score of the ML models to make reliable predictions. In what follows, we describe how to interpret these tools and metrics to evaluate an ML model's predictions regarding instructor evaluations.

3.3.2 Camelot's Testing Results

Camelot's goal is to classify instructor's teaching evaluations comprising multiple attributes (features) into one of the five categories: 'excellent' (*E*), 'above average' (*AA*), 'average' (*A*), 'below average' (*BA*), and 'unsatisfactory' (*U*). To explain the testing results, we consider the nine teaching attributes provided in Table 1. To test the supervised machine learning models, we employ labeled testing data disjoint from the labeled training data. For each instructor evaluation (a feature point comprising nine attributes), we compare the classification (predicted label) generated by the ML models with the previously assigned label (true label) in this testing set.

Figures 5a–f provide the confusion matrices for the six classification ML models. A superior ML model has a diagonal confusion matrix, meaning that all of the prediction values are along the diagonal. Out of the six ML models, KNN, LR, and SVM are close to this ideal scenario, denoting superior prediction capability. In contrast, the Na¨ıve Bayes, decision tree, and the neural network classifiers have many off-diagonal values, implying some errors in prediction. Specifically, notice that Na¨ıve Bayes has five (out of 40) false negative predictions (more predictions are above the diagonal). In contrast, the neural network classifier has eight false positive predictions (more predictions below the diagonal vs. above it). The decision tree classifier has more or less equal numbers of false positives and negatives.

Using the confusion matrices and by applying Eqs. 3a, 3d, we obtain the accuracy, precision, recall, and the F1-score for the ML models; the values appear in Table 5. As seen in this table, KNN, logistic regression, and the support vector machine have among the highest prediction accuracies in contrast with the remaining models. However, notice that even a highly accurate model may have sub-optimal precision and recall, which may adversely affect the model's performance. Notice that SVM's precision and recall are less than 90%. In such a case, a combiner can pool the individual model performances and select the most effective model for predictions. The discussion follows next.

Combiner—An occasional tie may arise among the ML models with respect to their prediction accuracy. In such a case, the analyst must resort to precision, recall, and F1-score to determine an appropriate tie-breaking metric that works best for their analysis. Let us consider our teaching evaluation classification analysis. Although Table 5 identifies SVM as one of the accurate models, notice its low precision and recall scores. In our study of instructors' teaching evaluation classification, it would be less service if our ML models were to overestimate (false positives) the evaluations. The false negatives (good evaluations classified as poor) are not particularly detrimental in this case. Therefore, we may argue that precision (a metric that penalizes false positives) is of more importance versus recall (and by extension, F1-score). Camelot's combiner works as follows. Camelot shortlists KNN, LR, and SVM among the highly accurate models with accuracy over 90%. In our case, since precision is more important than recall, Camelot selects the KNN model for future predictions.

A generic combiner algorithm appears in Fig. 6. The analyst must determine if the precision, recall, or the F1-score is the correct metric for their analysis by studying the impacts of false positives and negatives. Note that the combiner's results will

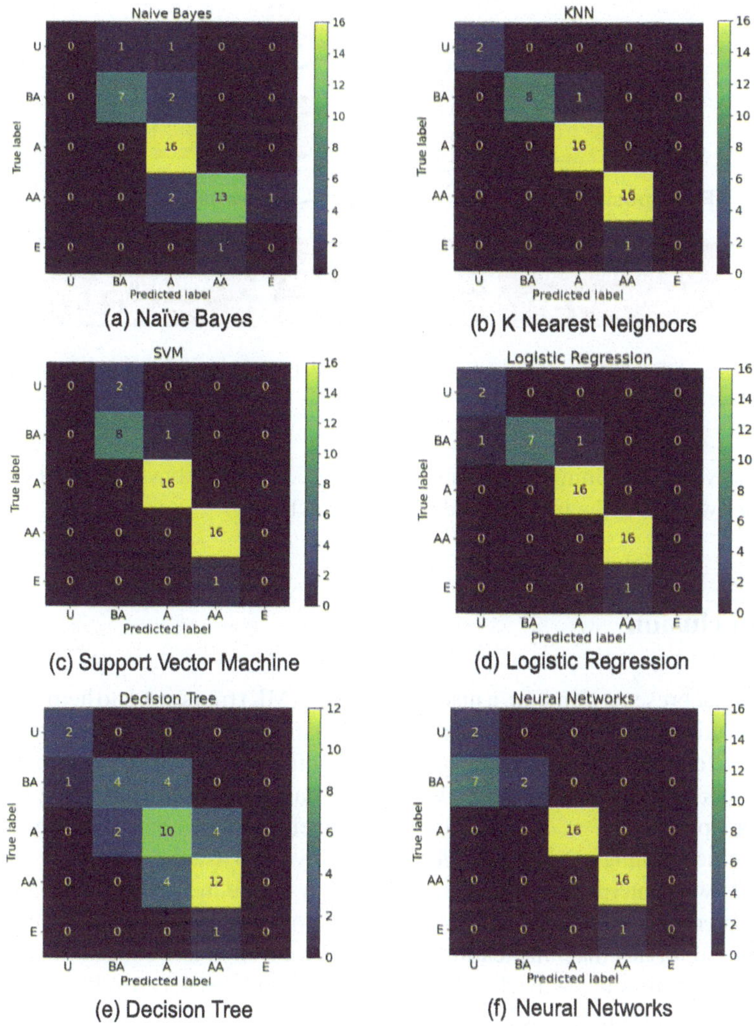

Fig. 5 The confusion matrices for the six classification ML models. The more diagonal values, the better the predictions

Table 5 Accuracy, precision, recall, and F1-scores of the six classification ML models

ML model	Accuracy	Precision	Recall	F1-score
Na¨ıve Bayes	0.818	0.794	0.818	0.943
KNN	0.955	0.934	0.955	0.943
SVM	0.909	0.848	0.577	0.57
LR	0.932	0.919	0.932	0.921
Decision tree	0.636	0.625	0.636	0.624
NN	0.818	0.921	0.818	0.807

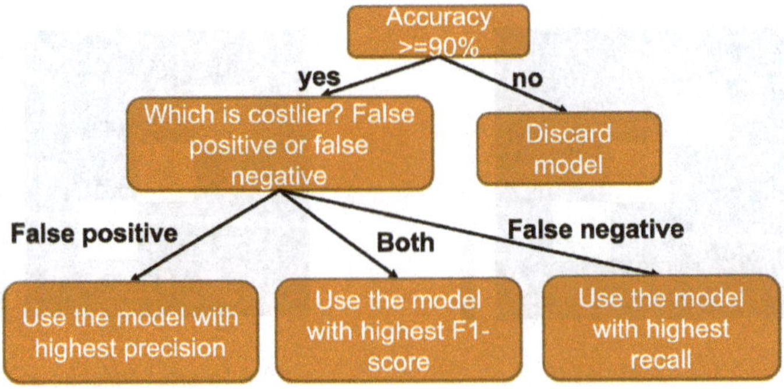

Fig. 6 A generic combiner to select an effective ML model based on its accuracy, precision, recall, and F1-score

vary with varying training and testing data. Therefore, we advise readers to use the combiner whenever the training and testing data change.

4 Conclusion

This chapter presents Camelot, a machine learning (ML) framework to help educators and institutions study their large teaching evaluations data. The framework's three phases including data preprocessing, model training, model testing, and in- ference guide users to effectively navigate through any data analysis project using machine learning. Specifically, we describe how to effectively curate the data for analysis, train the machine learning models, and when to employ them. Camelot's testing educates readers how to interpret the ML metrics and how to use them to select an effective ML for prediction. Throughout this chapter, we provide Python code excerpts that may be used for any machine learning project. We envision that educators, education leaders, and institutions will adapt Camelot's framework to analyze other educational data and make productive decisions.

We foresee two additional scenarios where Camelot, with some modifications, can be used by educators. Camelot may offer specific suggestions for improvement, including strategies for instructors seeking to adopt active learning approaches. For instance in computing education, approaches like peer instruction and pair programming have consistently been shown to be effective [3] and could be recommended where appropriate. In this context, the study of qualitative responses would be particularly powerful. Recent work in the UK has suggested that machine learning approaches can help to identify actionable insights from large-scale teaching evaluation surveys [4].

Second, while Camelot in its current form is focused on classification, it may in the future also be able to offer additional context. Research has shown that while active learning is more effective than traditional passive approaches, students perceive to be learning less [5]. Similarly, challenges remain with student evaluations of teaching more broadly. Researchers have argued that these evaluations do not necessarily accurately reflect teaching and are prone to gender bias [6]. Machine learning tools such as Camelot can be easily augmented to highlight such artifacts and provide deeper context to students' feedback.

Finally, we encourage readers to try Camelot by downloading it from GitHub.[3]

Acknowledgements The work presented in this chapter is supported in part by the Summer Undergraduate Research Fellowship (SURF) 2023 grant awarded by the undergraduate research office at University of the Pacific.

Appendix A

See Fig. 7.

References

1. Bishop CM, Pattern recognition and machine learning. Springer Science+Business Media, LLC
2. Ai X (2021) A tiered recommender system for cost-effective cloud instance selection. University of the Pacific, Thesis
3. Kleinbaum DG, Kupper LL, Muller KE, Nizam A, Applied regression analysis and other multivariate methods, 3rd Edition. Duxbury Press, Pacific Grove, CA
4. Brown NCC, Wilson G (2018) Ten quick tips for teaching programming. PLoS Comput Bio 14(4)
5. Nawaz R, Sun Q, Shardlow M, Kontonatsios G, Aljohani NR, Visvizi A, Hassan S-U (2022) Leveraging ai and machine learning for national student survey: actionable insights from textual feedback to enhance quality of teaching and learning in uk's higher education. Appl Sci 12(1)
6. Deslauriers L, McCarty LS, Miller K, Kestin G (2019) Measuring actual learning versus feeling of learning in response to being actively engaged in the classroom. The proceedings of the national academy of sciences (PNAS) 116(39)
7. Kogan LR, Schoenfeld-Tacher R, Hellyer PW (2010) Student evaluations of teaching: perceptions of faculty based on gender, position, and rank. Teaching Higher Educ 15(6)

[3] https://github.com/vkpallipuram/ML Education Services.

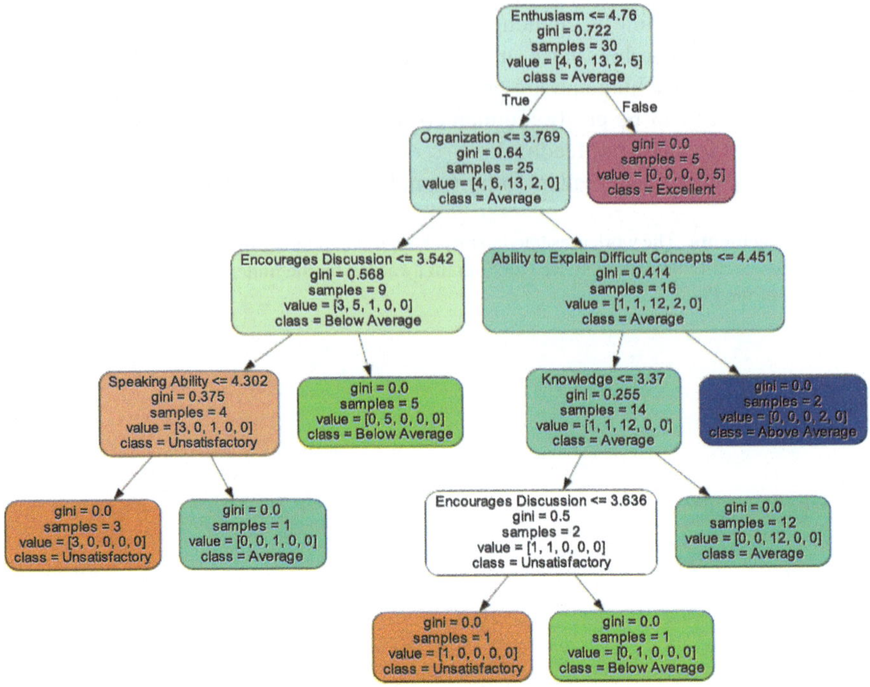

Fig. 7 An example decision tree generated by the decision tree classifier. In this tree, the gini score at each node depicts the impurity in classification; the closer it is to 0, the better the classification performance. To classify a new feature point, we simply traverse this decision tree and identify the terminal node (leaf). The class at the leaf node is the classification of the new feature point. For instance, consider a feature point with the following feature values: Enthusiasm equal to 4.3, Organization equal to 4.4, Ability to explain difficult concepts equal to 4.1, Organization equal to 3.8, Speaking ability equal to 4.2, Encourages discussion equal to 4.4, and Knowledge equal to 4.5. For this feature point, the final class is 'Average' (7th node from the left).

Penalized Regression in Large-Scale Data Analysis

Jin Eun Yoo

Abstract Among machine learning methods, penalized regression provides interpretable predictive models, which increases its importance and usability in educational research in which explanation has been valued over prediction. Particularly coupled with large-scale data, the sparsity assumption of penalized regression is most likely to be met, and penalized regression contributes to the exploration and identification of yet uninvestigated variables or relationships. Penalized regression methods such as LASSO, elastic net, and group Mnet have been widely applied to educational large-scale data. Recently, penalized regression has been extended to various statistical models, including significance testing and multilevel models. In this chapter, we will overview predictive modeling, and explain the basics of penalized regression and model assessment, followed by extensions of penalized regression. Finally, future research topics are addressed, and R coding examples are provided for reference in the context of educational large-scale data analysis.

Keywords Penalized regression · Machine learning · Large-scale data · Model assessment · R programming

1 Introduction

Depending on the purpose, statistical modeling can be categorized as explanatory and predictive [1]. Education has a long tradition of explanatory modeling that is to conduct significance tests on a set of a few pre-selected variables based on theory and/or previous research [2]. In the field of machine learning (ML), on the other hand, 'prediction' is the operative word in which a statistical model is developed to predict a given outcome, thus being called a predictive model. In particular, many predictive models are nonlinear and complex, and such modeling has been effective in applications including face recognition and automatic driving which typically analyze big data of low signal-to-noise ratio (SNR). In such ML applications with

J. E. Yoo (✉)
Korea National University of Education, Cheongju, Republic of Korea
e-mail: jeyoo@knue.ac.kr

M. S. Khine (ed.), *Machine Learning in Educational Sciences*,
https://doi.org/10.1007/978-981-99-9379-6_5

nonlinear predictive modeling, interpretability is sacrificed over predictability. The purpose of face recognition research for instance is not identifying what pixels are crucial to prediction, nor is the relationship among pixels likely to be explained. In the same vein, nonlinear predictive models are difficult to interpret.

While prediction is the foremost goal in ML, educational research typically demands a greater level of explanation than face recognition does [3]. For example, researchers as well as practitioners in education seek to understand what self-regulated learning variables relate to students' achievement. A typical process of explanatory modeling might be as follows: Researchers review existing theoretical models on self-regulation, set research hypotheses from the literature review, either collect data or search databases (e.g., large-scale data such as TIMSS or PISA), and test hypotheses using statistical methods. Such explanatory models are interpretable. From the analysis results we know which self-regulation variables significantly explain students' achievement, and the knowledge can be used in the later steps such as designing intervention plans. Compared to predictive models, however, explanatory models are less likely to predict new data, and may also suffer from overfitting [1, 4]. In other words, explanatory models are commonly difficult to generalize. For instance, it is not surprising to come across structural equation models (explanatory models) that fail to fit other samples. The significant self-regulation variables in the aforementioned example of explanatory modeling might not be that much effective to other groups of students.

Given the issues of difficult-to-interpret nonlinear predictive modeling and difficult-to-predict explanatory modeling, this chapter attends to *linear predictive modeling* and its application in educational research. Linear predictive models are interpretable, and are preferred in ML applications in which they show similar prediction performance to the nonlinear counterparts. Among linear predictive modeling, we focus on penalized regression, or regularization. Some recent studies in education reported that penalized regression outperformed nonlinear methods such as random forest (e.g., [3, 4] and deep learning [42]. Beyond providing interpretable predictive models, penalized regression can handle high-dimensional data without convergence problems. Even within the realm of large-scale educational data analysis, what was initially tall data (i.e., more observations than predictors) frequently transforms into a high-dimensional dataset due to the necessity of data split for model validation and comparison in ML.

Furthermore, when combined with large-scale data, penalized regression has the following advantages [2–6]. First, the sparsity assumption in penalized regression [7] is most likely to hold true when working with data from large-scale studies than smaller-scale studies. Second, by encompassing a multitude of variables connected to large-scale educational assessments within a single statistical model, it becomes possible to discover and identify new variables or relationships [1]. This is a capability that is often challenging to achieve using traditional methods. Large-scale data are commonly a product of carefully planned multi-year studies, and provide hundreds or thousands of variables ready for analysis. Nevertheless, it is the convention of educational research that only a small set of variables is investigated for its statistical significance in light of theory or previous research. This convention partly relates to the methodological limitations of the conventional techniques (e.g., nonconvergence). By employing penalized regression, the multitude of possible variables of large-scale data can be explored and important predictors can be identified, minimizing convergence and overfitting issues.

Penalized regression approaches such as Least Absolute Selection and Shrinkage Operator (LASSO; [8], elastic net [9], adaptive LASSO [10] and Mnet [11], have evolved as valuable ML tools. They are adept at managing a substantial number of predictors to facilitate variable selection for model development. Accordingly, studies analyzing large-scale educational data have started to employ penalized regression. For instance, LASSO [5], elastic net [2, 5], adaptive LASSO [12], and group Mnet [6, 13] have been employed to predict students' anxiety, mathematics achievement, career decision; school belonging; and teacher cooperation as well as teacher job satisfaction, analyzing PISA (Programme for International Student Assessment), Trends in International Mathematics and Science Study (TIMSS), KCYPS (Korea Children and Youth Panel Survey), and Teaching and Learning International Survey (TALIS) data. Details of the penalized regression methods are explained in the subsequent sections.

To summarize, penalized regression techniques have been widely applied to social science (educational) large-scale data. Recently, penalized regression has been extended to various statistical models, including glmmLasso [14] and statistical inference after selection [15]. In particular, glmmLasso, a combination of LASSO and Generalized Linear Mixed Models (GLMM), handles multilevel data which are common in educational large-scale data. Statistical inference after selection or post-selection inference (hereafter referred to as PSI) conducts significance testing on the predictors selected by LASSO. While large-scale applications of the newer penalized regression techniques appear limited in literature, large-scale educational data applications of glmmLasso and PSI are emerging, including [16] and [17] for glmmLasso and [18] and [19] and [20] for PSI. In this chapter, we will overview predictive modeling, and explain the basics of penalized regression and model assessment, followed by extensions of penalized regression. Finally, this chapter provides concluding remarks, along with coding examples in the R programming language for reference within the context of social science large-scale data analysis.

2 Predictive Modeling

The fundamental purpose of ML is to construct predictive models. While traditional statistical techniques primarily aim to build models that provide a good explanation for existing data, ML places a strong emphasis on creating models that are effective in making accurate predictions on new, unseen data. For this purpose, data are split into training data and test data in ML. With training data, models are built, and with test data, models are assessed in terms of prediction errors. As a result, models of low prediction errors are selected, and such models are called as predictive models.

If the model obtained from the training data fits the new, unseen data (test data) well, then the model is considered proficient at prediction.

To understand the importance of predictive models, let's consider an example from [21]. Imagine a student who takes a practice test, covering the same material as the upcoming final exam. If the student focuses on studying the questions that were answered incorrectly on the practice test, the performance on the same practice exam is likely to improve significantly. This is because the student has invested time and effort into the areas where she initially struggled. However, the crucial question remains: can this student achieve a high score on the final exam, which is entirely new and unfamiliar at the time of the preparation? In the final exam, new question types that the student did not encounter in the practice test may appear, making it uncertain whether the prior preparation will be sufficient. This scenario illustrates the importance of predictive models, which can help predict a student's performance on the final exam based on the past performance and other relevant factors.

In the example, we can draw parallels between mock exams and actual tests with training and test data in ML, respectively. The student practices with mock tests (akin to training) to achieve a high score on the actual test, but even if she consistently scores close to perfection on the mock test, there is no guarantee of achieving a high score on the actual test. This uncertainty is reminiscent of the concept of overfitting in ML. Overfitting occurs when a model becomes too tailored to the specific data it was trained on, explaining that data very well but struggling to predict new, unseen data. This situation leads to generalization problems. Conversely, if the student's knowledge and skills are effectively generalized, akin to a well-established predictive model in ML, she will excel at solving new types of questions on the actual test. This emphasizes the importance of creating models that generalize well beyond the training data.

To summarize, in order to assess the predictive power of ML models, new data which were not used in training serve as test data. Models built from the training data are fitted to the test data, and then are evaluated to select the best model of the least prediction error. In the section on Model Assessment, we will discuss techniques for model assessment and validation, including cross-validation and information criteria, as well as prediction errors and selection counts. The following section will cover the fundamental principles of penalized regression.

3 Basics of Penalized Regression

3.1 OLS Regression

Before we directly turn to penalized regression, let us briefly review the commonly encountered multiple linear regression model. Suppose a response variable y is estimated with an $\boldsymbol{X}$ matrix of n observations and p predictors, represented as $\boldsymbol{X}_j$. $\boldsymbol{X}^T = \left(\boldsymbol{X}_1, \boldsymbol{X}_2, \ldots, \boldsymbol{X}_p\right)$. Then, the i-th person's response variable is y_i, modeled by $\boldsymbol{x}_i = \left(x_{i1}, x_{i2}, \ldots, x_{ip}\right)^T$ and an error term ϵ. For simplicity, let's assume that there is no intercept in the linear regression model. The linear model is expressed as in Eq. 1.

$$y = f(\boldsymbol{X}) + \epsilon = \sum_{j=1}^{p} \boldsymbol{X}_j \beta_j + \epsilon. \tag{1}$$

The training data $\left(\boldsymbol{x}_1^T y_1\right) \ldots \left(\boldsymbol{x}_n^T, y_n\right)$ are used to estimate the regression coefficients β. When the error term follows a normal distribution, the optimization algorithm, Ordinary Least Squares (OLS), is typically utilized to estimate the β. Specifically, OLS estimates the β, which minimizes the sum of the squared values of the difference between estimated and observed values. Eq. 2 explains the OLS algorithm. When the predictors are represented using matrix notation as $\boldsymbol{X}$ (an n by p matrix), the OLS equation can be expressed as Eq. 3, and this is equivalent to Eq. 2. From the Eq. 3, it is obvious that the inverse matrix of $\boldsymbol{X}^T\boldsymbol{X}$ should be calculated to estimate the β [21].

$$\widehat{\beta^{OLS}} = argmin\left\{\frac{1}{2}\sum_{i=1}^{n}\left(y_i - \sum_{j=1}^{p} x_{ij}\beta_j\right)^2\right\}. \tag{2}$$

$$\widehat{\beta^{OLS}} = argmin\frac{1}{2}(y - \boldsymbol{X}\beta)^T(y - \boldsymbol{X}\beta) = \left(\boldsymbol{X}^T\boldsymbol{X}\right)^{-1}\boldsymbol{X}^T y. \tag{3}$$

In particular, high correlations among predictors can result in a singular $\boldsymbol{X}^T\boldsymbol{X}$ matrix, and the inverse matrix is difficult to be calculated. This issue is known as multicollinearity, a problem in OLS regression. Multicollinearity arises when two or more predictors are highly correlated, and can lead to inaccurate or unreliable statistical inferences. The early penalized regression, ridge, was developed in response to the multicollinearity problem [22], which will be elaborated in the following subsection.

3.2 Penalized Regression Using Convex Penalty Functions

This subsection consists of penalized regression methods using convex penalty functions. Among such methods, we will focus on ridge, LASSO, and elastic net. Penalized regression with convex penalties imposes the same amount of penalty on the coefficients. Ridge was originally invented by [22] for multicollinearity problems in regressions using ordinary least squares, but now is famous as an early penalized regression method. When the $\boldsymbol{X}^T\boldsymbol{X}$ matrix in Eq. 3 is singular, ridge adds a penalty term, λ, to the diagonals of the $\boldsymbol{X}^T\boldsymbol{X}$ matrix, resulting in an invertible matrix (Eq. 4).

The objective function in Eq. 5 is equivalent to Eq. 4, and shows that ridge consists of least squares and an L_2 penalty term (the sum of squared coefficients).

Due to the additional penalty term, ridge shrinks the regression coefficients. Specifically, the penalty (or regularization) term, λ, controls the amount of shrinkage. When λ is 0, Eqs. 4 or 5 reverts back to the OLS equation (Eq. 2 or 3). When λ is larger than 0, ridge imposes the value of λ to all the regression coefficients, resulting in smaller coefficients. However, ridge does not perform variable selection; all predictors are retained in the model, even though their coefficients may be significantly reduced.

$$\widehat{\beta^r} = \left(\boldsymbol{X}^T\boldsymbol{X} + \lambda\boldsymbol{I}\right)^{-1}\boldsymbol{X}^T y. \tag{4}$$

$$\widehat{\beta^r} = argmin\left\{\frac{1}{2}\sum_{i=1}^{n}\left(y_i - \sum_{j=1}^{p} x_{ij}\beta_j\right)^2 + \lambda\sum_{j=1}^{p}\beta_j^2\right\}. \tag{5}$$

On the other hand, LASSO, developed by [8], is the first and most popular penalized regression method for variable selection. Compared to ridge, LASSO employs a distinct penalty function, namely an L_1 penalty, represented as the sum of the absolute values of coefficients (as shown in Eq. 6). When visualized in a two-dimensional coefficients space, the penalty constraint of ridge appears as a circular disk, whereas LASSO's constraint resembles a diamond [7]. An essential feature of the L_1 LASSO penalty is the presence of corners in its error contours. If the elliptical contours intersect with a corner, the corresponding coefficient becomes zero, effectively enabling variable selection [7]. Consequently, LASSO offers an advantage of variable selection compared to the ridge regression [2].

Of note, LASSO does not handle multicollinearity, and thus [9] propose elastic net. LASSO is known for its capability in variable selection, while ridge regression is renowned for its effectiveness in addressing multicollinearity. Acting as a bridge between these two techniques, elastic net (hereafter referred to as Enet) combines the strengths of both LASSO and ridge. It achieves this by incorporating the L_1 and L_2 penalties of LASSO and ridge, respectively, within a single equation. In essence, Enet not only facilitates variable selection, courtesy of the LASSO component, but also excels in managing multicollinearity, thanks to the ridge component.

The objective functions of LASSO and Enet for a Gaussian family are illustrated in Eqs. 6 and 7, respectively. Notably, both equations share the same first term on the right-hand side, which represents the loss function of least squares. The second term in both equations corresponds to the penalty functions specific to LASSO and Enet. Both LASSO and Enet incorporate the L_1 norm (the sum of the absolute coefficients), thereby facilitating variable selection. Similar to ridge regression, the regularization or penalty parameter of λ plays a pivotal role in controlling the extent of penalty applied; larger values of λ result in more substantial coefficient shrinkage. In Eq. 7, the L_2 norm (the sum of squared coefficients) is added to incorporate the ridge penalty and address multicollinearity issues. Another tuning parameter, α, , bridges between ridge and LASSO. When α is set to 1, the equation becomes equivalent to LASSO (as seen in Eq. 6), and for α equal to 0, it reverts to the ridge equation (as depicted in Eq. 5). Therefore, the value of α, lying between 0 and 1, determines whether the model is closer to ridge or LASSO, offering the advantages of both techniques. The value of α is typically set to 0.5 [23].

$$\widehat{\beta^L} = argmin\left\{\frac{1}{2}\sum_{i=1}^{n}\left(y_i - \sum_{j=1}^{p} x_{ij}\beta_j\right)^2 + \lambda\sum_{j=1}^{p}|\beta_j|\right\}. \quad (6)$$

$$\widehat{\beta^E} = argmin\left\{\frac{1}{2}\sum_{i=1}^{n}\left(y_i - \sum_{j=1}^{p} x_{ij}\beta_j\right)^2 + \lambda\sum_{j=1}^{p}\left(\alpha|\beta_j| + (1-\alpha)\beta_j^2\right)\right\}. \quad (7)$$

Ridge, LASSO, and Enet are implemented in multiple R packages such as glmnet [23], grpreg [24], and penalized [25]. Among them, we recommend glmnet [23], of which the developers of LASSO and Enet are the maintainers. However, glmnet does not handle categorical predictors of more than two levels. The section on Extensions of Penalized Regression discusses group penalized regression methods for models handling both categorical and continuous predictors.

4 Model Assessment

We can draw a parallel between traditional statistical techniques, which typically do not incorporate test data and prediction errors, and the construction of models optimized for performance in practice exams. In contrast, ML models are designed as predictive models with the primary aim of performing effectively in real, unseen data situations, like actual exams. In this section, we will review model assessment techniques in the context of ML. Specifically, an overview of cross-validation and information criteria techniques will be provided along with a discussion on prediction errors and selection counts.

4.1 Cross-Validation

In penalized regression, the penalty (or regularization) parameter, λ, determines model complexity. To find the optimal λ that minimizes prediction error, K-fold cross-validation (CV) is a commonly employed approach. K-fold CV partitions training data into K sets of equal size, with K typically chosen to be 5 or 10. For each fold ($k = 1, 2, \ldots K$), the model is fitted with the training set that excludes the k-th fold, and the resulting model is used to generate predictions for the k-th fold. This process is repeated for every k-th fold, and the CV error for each fold is computed. The overall CV error is determined by averaging the errors from all K folds (Eq. 8), and its standard error is also calculated.

$$CV\left(\widehat{f}, \lambda\right) = \frac{1}{n}\sum_{i=1}^{n} L\left(y_i, \widehat{f}^{-k(i)}(x_i, \lambda)\right) \tag{8}$$

The model assessment process involves several steps [3, 4, 6]. First, the entire dataset is randomly divided into a training set and a test set, typically with the ratio of 7:3 or 8:2. Second, a 10-fold CV procedure is performed on the training data. For a given value of the penalty parameter, λ, the training data are further divided into a 9:1 ratio. Specifically, 9/10 of the training data are used for model training, and the remaining 1/10 serves for model evaluation. This process is repeated for every fold in the 10-fold CV, and the prediction error associated with the specific λ is calculated; this is termed as the CV error of λ (Eq. 8). Third, step 2 is repeated for every λ within a specified range. The λ value that corresponds to the lowest CV error is selected as the optimal penalty value. Fourth, the chosen λ value is then applied to the test data, as outlined in step 1, which enables the calculation of prediction errors for the test dataset.

While the best model with the least prediction error is typically accomplished through CV, the "one-standard-error rule" is often applied to select the most parsimonious model [7]. This rule entails choosing the least complex model from among the models falling within a one-standard-error range of the best model. The R package, glmnet, offers this option as well as the minimum CV errors, while grpreg does not.

4.2 Akaike Information Criterion and Bayesian Information Criterion

Other model assessment criteria include information criteria such as the Akaike Information Criterion (AIC) [26] and Bayesian Information Criterion (BIC) [27]. Particularly in glmmLasso, AIC or BIC is typically employed to tune the penalty parameter λ. Eqs. 9 and 10 explain AIC and BIC, respectively. In these equations, L, W, and n represent the maximized value of the likelihood function of the model, the number of parameters estimated by the model, and the number of observations, respectively.

$$AIC = -2log(L) + 2W. \tag{9}$$

$$BIC = -2log(L) + Wlog(n). \tag{10}$$

As smaller AIC and BIC are deemed better, the number of parameters is a penalty for both AIC and BIC, whereas the number of observations acts as a penalty only for

BIC in model selection. Particularly when n is sufficiently large, BIC not only raises the penalty for the increased complexity of a model [28] but also is known to result in consistent model selection [29].

The model selection and assessment steps using information criteria are as follows (Koo and Yoo under review). First, the entire dataset is randomly split with a ratio of 7–3 or 8–2, comprising training and test data, respectively. The training and test ratio can depend on the sample size, and in large-scale data studies, the ratio of 7:3 is typically utilized (e.g., [3–6, 13]). The training set is employed to fit a model, and the test set is used to estimate prediction error. Second, for λ in range, the information criterion is measured using the training data from Step 1. As a result, the λ of the lowest information criterion is identified as well as the W, the number of parameters. The steps are repeated independently for each of the penalized regression methods of comparison.

4.3 Prediction Errors

This subsection is largely from [21]. In the field of ML, prediction errors are crucial in deciding which ML techniques or models outperform others. Models of smaller prediction errors are typically preferred. Depending on the type of the response variables, different prediction errors are calculated and compared. For continuous response variables, the Root Mean Square Error (RMSE) of test data typically serves as a prediction error. In the following equation, y_i and $\widehat{y_i}$ denote the true and predicted responses for a participant i, respectively

$$RMSE = \sqrt{\frac{1}{n}\sum_{i=1}^{n}\left(y_i - \widehat{y_i}\right)^2}.$$

For categorical response variables, the prediction models are evaluated using measures such as prediction accuracy, specificity, sensitivity, precision, AUC (Area Under the ROC Curve), kappa, and FDR. For illustration purposes, a 2×2 confusion matrix is constructed with two axes, true and predicted, which consists of 4 cells; TP (True Positive), FP (False Positive), FN (False Negative), and TN (True Negative) (Table 1). Among them, TP and TN indicate cases of correct prediction. TP is actually $\oplus$ and the prediction is also $\oplus$, and TN is actually $\ominus$ and the prediction is $\ominus$. Conversely, FP is actually $\ominus$ but is predicted to be $\oplus$, and FN is actually $\oplus$ but predicted to be $\ominus$. Type I and Type II errors in hypothesis testing are FP and FN, respectively. $P, N, \widehat{P}, \widehat{N}$ are marginal probabilities. P is the probability of adding both TP and FN, which are predicted probabilities when it is actually $\oplus$. N is the probability of adding FP and TN, which are actually predicted probabilities when it is $\ominus$. Similarly, the probability of adding TP and FP, which is the probability when

Table 1 Confusion matrix

Pred	True		
	Positive(⊕)	Negative(⊖)	
Positive(⊕)	TP	FP (Type I error)	$\widehat{P}$
Negative(⊖)	FN (Type II error)	TN	$\widehat{N}$
	P	N	

the prediction is ⊕ , is expressed as $\widehat{P}$, and the probability of adding FN and TN, which is the probability when the prediction is ⊖ , is expressed as $\widehat{N}$.

We will explain prediction accuracy, specificity, and sensitivity using the confusion matrix in Table 1. Accuracy is the proportion of correct predictions out of the total number of cases. The denominator is the sum of all cases, TP, TN, FP, and FN, and the numerator is the sum of TP and TN, which are correctly predicted cases. At a superficial level, one might assume that presenting accuracy is sufficient for assessing a model's performance. However, this is not always the case, especially true when dealing with highly imbalanced datasets. To illustrate this point, let's consider the example of a COVID-19 test. As of March 2021, the number of confirmed COVID-19 cases in Korea was approximately 93,000. If you calculate the rate of confirmed cases by dividing this number by the total number of people tested, which is around 6.8 million, you will find the rate of confirmed cases is only 1.4%. From a hypothesis test standpoint, when dealing with highly imbalanced datasets, it may be difficult to reject the null hypothesis, while the alternative hypothesis may be rejected. This means that the model's performance may not effectively distinguish the positive cases (alternative hypothesis) from the negative cases (null hypothesis), leading to limitations in its predictive capability, especially for the minority class. In such cases, alternative evaluation metrics should be considered to gain a more accurate understanding of model performance.

Also, in situations with highly imbalanced category ratios such as 98.6% non-infected versus 1.4% infected cases, it might initially seem unnecessary to develop a predictive model. After all, if one were to predict that everyone is non-infected without conducting any tests, the accuracy would be exceptionally high at 98.6%. In other words, from the perspective of accuracy alone, data scientists might appear to have no need for intervention. However, it is crucial to recognize that in a scenario like this, where only a tiny proportion of cases are confirmed (1.4%), relying solely on accuracy can be perilous. Imagine the rapid spread of COVID-19 in such a situation. To detect the confirmed cases effectively, which make up just 1.4% of the total, it's imperative to invest substantial efforts, including providing accurate test kits and conducting extensive epidemiological investigations. During these circumstances, the FN, where a confirmed person is incorrectly predicted as non-infected, should be minimized. Naturally, efforts should also be made to reduce FP, where a non-confirmed person is mistakenly predicted as confirmed. However, the social cost of a confirmed individual moving freely in the community and potentially spreading the disease while thinking they are non-infected is often considered greater than the inconvenience of unjustly isolating a non-confirmed person. Therefore, it is not just accuracy that should be considered. Evaluation metrics that distinguish between FN and FP, such as sensitivity and specificity, are essential for providing a more nuanced and appropriate assessment of model performance in these critical situations.

Sensitivity is a metric used to quantify the degree of error in predicting that a person is not a confirmed case when, in fact, they are confirmed. It represents the proportion of predicted cases out of the actual cases. In other words, a high sensitivity indicates that the model is effective at predicting confirmed cases. Sensitivity is closely related to Type II error in hypothesis testing, and is also referred to as recall. On the other hand, specificity is a metric that measures the error in predicting a confirmed case when a person is not confirmed. It is the proportion of predicted non-confirmed individuals out of the actual non-confirmed individuals.

$$accuracy = \frac{TP + TN}{TP + TN + FP + FN}.$$

$$sensitivity = P\left(\widehat{Y} = \oplus | Y = \oplus\right) = \frac{TP}{P} = \frac{TP}{TP + FN}.$$

$$specificity = P\left(\widehat{Y} = \ominus | Y = \ominus\right) = \frac{TN}{N} = \frac{TN}{FP + TN}.$$

4.4 Selection Counts

Aforementioned, a well-constructed model should excel not only in fitting the available data but also in making accurate predictions on new, unseen data. To achieve this, the dataset is divided into two subsets: training data and test data. A sound model is developed using the training data, and its performance is assessed by evaluating it on the test data, which may be an independent dataset not used during the model training process. Of note, due to the data split, results may change, particularly when data are not huge. To control the bias resulting from data splitting, researchers suggest multiple iterations of data splitting and model fitting (e.g., [30]). When the goal of penalized regression is to obtain the best predictive models, different sets of selected predictors may not be an issue. The prediction errors of multiple iterations can be easily summarized (e.g., descriptive statistics of RMSEs).

However, in educational research pinpointing predictors contributing to prediction is of research interest. For this purpose, [4, 6] proposed selection counts of predictors. The idea is straightforward, which is to count the number of selections of each predictor in the model. Specifically, they presented predictors selected 1, 25, 50, 75, and 100 times after 100 iterations, and interpreted that predictors that were selected more often are more important to prediction. [4] after conducting simulation studies emulating social science large-scale data suggest that predictors selected 75% or more of the iterations are worth further investigation for convex penalty functions and 50% or more for concave penalty functions.

5 Extensions of Penalized Regression

5.1 *Group Penalized Regression*

Categorical predictors, measured in nominal or ordinal scales, are quite common in educational large-scale data (e.g., race, region of residence, occupation, and school type). In regression modeling, a categorical variable of more than two levels should be treated as a group after dummy coding. In penalized regression, it is also essential to consider a group of dummy-coded variables originating from a categorical variable as a collective unit. These types of models are often referred to as group penalized regression models.

In order to explain group penalized regression, the linear regression equations require different notations [6]. Consider a linear regression model with p predictors. These predictors are divided into K distinct, non-overlapping groups, and the model can be represented as described in Eq. 11. Y represents an n-dimensional vector that serves as a response variable. X_k represents the $n \times p_k$ design matrix associated with the p_k predictors in the k-th group. Furthermore, $\beta_k = \left(\beta_{k,1}, \ldots, \beta_{k,p_k}\right)^T$ is a p_k dimensional vector of regression coefficients of the k-th group. Finally, ϵ is an n dimensional vector of mean zero, representing the error term in the model. The following methods including concave penalty functions and glmmLasso are based on the regression model in Eq. 11.

$$Y = \sum_{k=1}^{K} X_k \beta_k + \epsilon. \tag{11}$$

5.2 *Penalized Regression Using Concave Penalty Functions*

In the previous section, we reviewed penalized regression using convex penalty functions such as ridge, LASSO, and Enet. Such methods impose the same amount of penalty on the coefficients regardless of the coefficient size, which results in inconsistent estimates in terms of variable selection [31–34]. Conversely, concave penalties gradually decrease in influence as the absolute values of coefficients become larger, ultimately leading to nearly consistent coefficient estimates [35]. Among such methods of concave penalty functions, we will focus on MCP (Minimax Concave Penalty, [35] and Mnet [11]. In particular, Mnet is a combination of MCP and ridge.

The relationship between MCP and Mnet mirrors that of LASSO and Enet. With the addition of the ridge component to LASSO, Enet handles possible multicollinearity. Likewise, Mnet yields nearly consistent estimates (a feature of MCP), effectively handling multicollinearity problems (a feature of ridge).

Group MCP and group Mnet are hereafter shortened as MCP and Mnet, respectively. The objective functions of MCP and Mnet are shown within the context of a Gaussian family in Eqs. 12 and 13 [6]. The equations are the same except that the ridge penalty, λ_2, is added to the Mnet equation (Eq. 13). In Eqs. 12 and 13, the first term on the right-hand side represents the least squares loss function. The subsequent term on the right-hand side pertains to the MCP penalty. The λ_1 parameter within the equations regulates the amount of penalty. The γ parameter, the concavity penalty, controls the rate at which penalization occurs. When the coefficients surpass the product of the two penalties, the MCP penalty's rate substantially diminishes, leading to a reduced degree of coefficient shrinkage and consequently producing less-biased estimates compared to LASSO. MCP and Mnet in the equations can be used to analyze mixed-format data; the L_2 norm ($\|\beta_k\|_2$) accounts for group membership of the variables.

$$\hat{\beta}^{MCP} = argmin_{\beta}\left[\frac{1}{2n}\|Y - \sum_{k=1}^{K} X_k\beta_k\|_2^2 + \sum_{k=1}^{K} J(\|\beta_k\|_2 \mid \lambda_1, \gamma)\right]$$

where

$$J(x|\lambda_1, \gamma) = \begin{cases} -\frac{1}{2\gamma}x^2 + \lambda_1|x|, & |x| \le \gamma\lambda_1 \\ \frac{1}{2}\gamma\lambda_1^2, & |x| > \gamma\lambda_1. \end{cases} \tag{12}$$

$$\hat{\beta}^{Mnet} = argmin_{\beta}\left[\frac{1}{2n}\|Y - \sum_{k=1}^{K} X_k\beta_k\|_2^2 + \sum_{k=1}^{K} J(\|\beta_k\|_2|\lambda_1, \gamma) + \lambda_2\sum_{k=1}^{K}\|\beta_k\|_2^2\right]. \tag{13}$$

The R package, glmnet, is for penalized regression with convex penalties, and only handles continuous or binary predictors. A few packages for group penalized regression are developed in R, including grpreg [24] and grplasso [43]. Among them, we recommend grpreg, which offers options for both convex and concave penalties, and also the developers of Mnet are the maintainers.

5.3 *GlmmLasso*

While LASSO and Enet neglect data structure, the glmmLasso algorithm proposed by [14] reflects the nested data structure, typically resulting from cluster sampling, imposing the LASSO penalty on GLMM (Generalized Linear Mixed Models). The following explanation on glmmLasso is from [16, 17], and Koo and Yoo (under review). Using the linear model in Eq. 11, a GLMM function of a response variable (y) is shown in Eq. 14. $\boldsymbol{g}(\cdot)$ is a link function, and the first and second terms in the right-hand side of the equation represent fixed and random effects, respectively. The fixed effects covariates matrix is denoted as $\boldsymbol{X}$ and the random effects covariates vector is denoted as $\boldsymbol{Z}$. $\boldsymbol{\beta}$ is the fixed effects coefficient vector for $\boldsymbol{X}$ and $\mathbf{b}$ is the subject-specific random effects for $\boldsymbol{Z}$. For the random effects $\mathbf{b}$ the mean is zero, and the variance–covariance matrix follows a normal distribution with a block-diagonal covariance matrix $\boldsymbol{\Sigma}$.

$$\boldsymbol{g}(\boldsymbol{\mu}) = \boldsymbol{X\beta} + \mathbf{Zb}, \mathbf{b} \sim N(0, \boldsymbol{\Sigma}). \tag{14}$$

As a combination of generalized linear models (GLM) and mixed models, GLMM can handle both multilevel data (e.g., teachers nested in schools) and categorical response variables (e.g., binary dependent variable such as success or failure). A specific multilevel data structure is reflected on the covariance matrix $\boldsymbol{\Sigma}$ in Eq. 14, which is connected to the estimation of the fixed and random effects of the model. The link function covers various types of categorical variables.

Equation 15 explains the glmmLasso algorithm. As the name denotes, the algorithm uses the L_1 norm or LASSO penalty. To maximize the complex log-likelihood of GLMM, penalized quasi-likelihood (PQL) is utilized; PQL approximates the log-likelihood using the Laplace method. The first term of the right-hand side of the equation denotes the approximated log-likelihood of GLMM, which contains two parameter vectors. Vector $\boldsymbol{\delta}$ comprises $\boldsymbol{\beta}$ and $\mathbf{b}$, the fixed and random effects, respectively. Another parameter, vector $\boldsymbol{\gamma}$, contains the dispersion parameter ϕ and an unknown parameter vector $\boldsymbol{\sigma}$. The parameter scalar ϕ is estimated when overdispersion can be an issue (e.g., binomial or Poisson cases; [36]. The covariance matrix $(\boldsymbol{\Sigma})$ of random effects $\mathbf{b}$ depends on the parameter vector $\boldsymbol{\sigma}$.

$$\hat{\delta}^{glmmLasso} = argmax_{\delta}\left[l^{app}(\delta, \hat{\gamma}) - \lambda \sum_{k=1}^{K} \|\boldsymbol{\beta}_k\|_1 \right]. \tag{15}$$

The main goal of the glmmLasso algorithm is to estimate the optimum $\boldsymbol{\delta}$, given $\boldsymbol{\gamma}$. To estimate the optimum $\boldsymbol{\delta}$, glmmLasso utilizes a gradient ascent algorithm; PQL computes starting values of parameters ($\widehat{\boldsymbol{\beta}}$, $\widehat{b}$, and $\hat{\boldsymbol{\gamma}}$) with an unconditional model (global intercept model), and the parameters are updated and re-estimated via the gradient ascent optimization with the Fisher scoring algorithm and Taylor series expansion [14]. The R package, glmmLasso, executes glmmLasso, as the package name denotes. Unlike glmnet, glmmLasso handles categorical predictors as well as continuous predictors.

5.4 Post-selection Inference (PSI)

Despite the advantages of penalized regression such as LASSO, statistical significance testing was a challenging task, which was a practical hindrance in the wider application of penalized regression, particularly in educational research in which explanation has been valued over prediction. Penalized regression aims at reducing variance by introducing a little bias, leading to smaller mean square errors (MSE) and prediction errors than those of unpenalized regression such as OLS regression. Specifically, the utilization of a penalty parameter in the objective function causes variables with nearly zero coefficient estimates to be excluded from the model, while retaining those with non-trivial effects, which naturally leads to variable selection. Therefore, the selected predictors are biased estimates. As significance testing is performed on unbiased estimates, traditional hypothesis testing cannot be conducted and the selected predictors by penalized regression should not be interpreted as statistically significant [3].

LASSO is the first penalized regression to select variables, and has been extensively utilized in various disciplines since its invention by [8]. PSI conducts significance testing on LASSO-selected predictors. PSI appears superior to other methods to serve similar purposes such as debiased lasso by Javanmard and Monatanari [37]. As penalized regression, both methods assume sparsity. One of the biggest differences between PSI and debiased LASSO hinges on the other assumptions. Debiased lasso assumes the correctness of the linear model, and uses a full model to obtain regularized estimates. By contrast, PSI assumes that LASSO has found all true variables of nonzero coefficients, and continues inference conditional on the LASSO-selected model [15]. Simulation studies indicate that LASSO or convex penalty regularization methods can select *all* true variables as well as false variables, when coupled with cross-validation [38, 39]. Relatedly, a single step of Newton-Raphson is sufficient, as PSI considers only a set of LASSO-selected variables in the consequent inference [40]. On the other hand, debiased lasso uses a full model to obtain regularized estimates, resulting in a more complex model and thus multiple iterations for convergence. Multiple comparison errors are another issue of debiased lasso.

The dominant practice of traditional significance testing has been frequently problematic in practice. For instance, the set of variables included in modeling often determines statistical significance; the same variable from the same data can be either statistically significant or non-significant, depending on the inclusion of other variables in the model; and sometimes even the signs of the coefficients can change. PSI can be a viable solution to this by serving the purpose of both prediction and significance testing. More specifically, PSI performs significance testing on the LASSO-selected predictors after exploring possible hundreds or thousands of predictor candidates of large-scale data. For a simplified explanation on PSI in the context of large-scale data analysis, refer to Yoo and Rho [19, 20].

6 Concluding Remarks

Penalized regression in large-scale data analysis is a promising area of continued growth. Compared to nonlinear ML methods such as deep learning, which produces prediction models notoriously difficult to interpret, penalized regression as a linear method yields interpretable prediction models; this brings great merit in educational research which particularly has valued explanation. Applying penalized regression to large-scale data, we can also find new important predictors which have been neglected in the literature. This can help researchers break free from the confines of traditional approaches and discover novel insights [2]. In the similar context, it is notable that penalized regression rests on the sparsity assumption, and may be less suitable for small-scale studies. Starting with LASSO for variable selection, variations of penalized regression have been developed including Enet for multicollinearity issues and Mnet for consistent coefficient estimates. Furthermore, the scope of penalized regression has been expanded to mixed models and statistical inferences, making it a versatile and powerful tool for a wide range of data analysis tasks. These developments have enriched the landscape of penalized regression and its capability in different research domains.

Future research topics are as follows. First, due to the data split requirement for model validation and assessment, the results may vary from iteration to iteration. Based on the practical needs of social science research to pinpoint important predictors, selection counts are proposed as a heuristic approach [4, 6]. Research on this topic is limited and should be furthered. Second, algorithmic advancement on the extensions of penalized regression is required as well as theoretical arguments. There are various R packages for penalized regression, but analysis packages of the newer techniques including mixed models and statistical inference have been developed only in the context of LASSO (i.e., glmmLasso and PSI, respectively). At this point, we understand that LASSO does not inherently address multicollinearity issues. However, it is important to note that educational large-scale datasets often grapple with multicollinearity problems. Additionally, many educational datasets are characterized by a multilevel structure, yet the application of penalized regression techniques with concave penalties has not been widely explored. Furthermore, the issue of significance testing in the context of penalized regression is a topic that warrants further research. This is especially important considering that explanatory modeling, rooted in significance testing, has been a long-standing tradition in the field of education. In light of these challenges and needs specific to the education domain, we strongly recommend the continued development of penalized regression algorithms and the subsequent theoretical discussions to address these prominent requirements. Such advancements have the potential to significantly benefit the field of education and enhance its data analysis capabilities.

7 Coding Examples in R

7.1 LASSO and Enet Using glmnet()

```
# cv.glmnet(): Cross-validation for glmnet
# alpha: Enet mixing parameter, with 0 ≤ α ≤ 1.
# alpha=1 (LASSO); alpha=0 (ridge)
# nfolds = 10 (default);
# family = "gaussian", "binomial" etc.;
# type.measure= "deviance" (default), "class" (for binomial) ;
# glmnet(): fit a GLM with lasso or Enet regularization
# coef(): Extract coefficients from a glmnet object
# predict(): Get predictions from a glmnetfit fit object
# lambda: penalty parameter

## LASSO ##

# run cv.glment to obtain lambda
 cv.fit <- cv.glmnet(x=train.x, y=train.y, family="gaussian")
 plot(cv.fit)

# extract coefficients #
coef(cv.fit, lambda=cv.fit$lambda.1se)

# prediction error with test data #
pred <- predict(cv.fit, newx=test.x, lambda=cv.fit$lambda.1se)
sqrt(mean((test.y-pred)^2))

## Enet ##

# run cv.glment to obtain lambda
 cv.fit <- cv.glmnet(x=train.x, y=train.y, family="binomial", alpha=0.5,
type.measure="class")
 plot(cv.fit)

# extract coefficients
coef(cv.fit)
```

```
# prediction error with test data #
pred <- predict(cv.fit, newx=test.x, type="class", lambda=cv.fit$lambda.1se)
xtab <- table(pred, test.y)
caret::confusionMatrix(xtab) # accuracy, sensitivity, specificity, Kappa etc.
roc_pred <- ROCR::prediction(as.numeric(pred), test.y)
auc_ROCR <- ROCR::performance(roc_pred, measure = "auc")
auc_ROCR@y.values[[1]]
```

7.2 MCP and Mnet Using grpreg()

```
# grpreg: Regularization Paths for Regression Models with Grouped Covariates
# grpreg can incorporate MCP as well as LASSO penalties, and handle categorical
predictors.
# Unlike glmnet, grpreg does not offer 1-se lambda.
# cv.grpreg(): Cross-validation for grpreg
# grpreg(): Fit a group penalized regression path
# penalty: "grMCP", "grLasso", "grSCAD"
# family: "gaussian", "binomial"
# alpha: default is 1(LASSO penalty)
# group: A vector describing the grouping of the coefficients.
# gamma: default is 3 for MCP
# lambda.min: The value of lambda with the minimum cross-validation error

## MCP
cv.rst <- cv.grpreg(X=train.x, y=train.y, group, penalty="grMCP", family="gaussian")
plot(cv.fit)

# extract coefficients
coef(cv.fit) # base CV error min

# prediction error with test data #
pred <- predict(cv.fit, X=test.x, lambda= cv.fit$lambda.min)
sqrt(mean(test.y-pred)^2)

## Mnet
cv.fit <- cv.grpreg(X=train.x, y=train.y, group=g.idx , penalty="grMCP",
                    family="binomial", alpha=0.5)
# extract coefficients
coef(cv.fit) ## Beta at minimum CV error
```

```
# prediction error with test data
pred <- predict(cv.fit, newx=test.x, lambda=cv.fit$lambda.min, type="class")
xtab <- table(pred, test.y)
caret::confusionMatrix(xtab) # accuracy, sensitivity, specificity, Kappa etc.
roc_pred <- ROCR::prediction(pred, test.y)
auc_ROCR <- ROCR::performance(roc_pred, measure = "auc")
auc_ROCR@y.values[[1]]
```

References

1. Shmueli G (2010) To explain or to predict? Stat Sci 25(3):289–310
2. Yoo JE (2018) TIMSS 2011 Student and teacher predictors for mathematics achievement explored and identified via elastic net. Front Psychol 9:317. https://doi.org/10.3389/fpsyg.2018.00317
3. Yoo JE, Rho M, Lee Y (2022) Online students' learning behaviors and academic success: an analysis of LMS log data from flipped classrooms via regularization. IEEE Access 10:10740–10753. https://doi.org/10.1109/ACCESS.2022.3144625
4. Yoo JE, Rho M (2022) Large-scale survey data analysis with penalized regression: a Monte Carlo simulation on missing categorical predictors. Multivar Behav Res. https://doi.org/10.1080/00273171.2021.1891856
5. Immekus JC, Jeong T, Yoo JE (2022) Machine learning procedures for predictor variable selection for schoolwork-related anxiety: evidence from PISA 2015 mathematics, reading, and science assessments. Large-scale assessments in education 10(30). https://doi.org/10.1186/s40536-022-00150-8
6. Yoo JE, Rho M (2020) Exploration of predictors for Korean teacher job satisfaction via a machine learning technique, group Mnet. Front Psychol 11:441. https://doi.org/10.3389/fpsyg.2020.00441
7. Hastie T, Tibshirani R, Friedman JH (2009) The elements of statistical learning: data mining, inference, and prediction, 2nd edn. Springer. https://doi.org/10.1007/978-0-387-84858-7
8. Tibshirani R (1996) Regression shrinkage and selection via the Lasso. J Roy Stat Soc Ser B (Methodol) 58(1):267–288. https://doi.org/10.1111/j.2517-6161.1996.tb02080.x
9. Zou H, Hastie T (2005) Regularization and variable selection via the elastic net. J R Stat Soc Ser B (Stat Methodol) 67(2):301–320. https://doi.org/10.1111/j.1467-9868.2005.00503.x
10. Zou H (2006) The adaptive lasso and its oracle properties. J Am Stat Assoc 101(476):1418–1429. https://doi.org/10.1198/016214506000000735
11. Huang J, Breheny P, Lee S, Ma S, Zhang CH (2016) The Mnet method for variable selection. Stat Sin 26(3):903–923. https://doi.org/10.5705/ss.202014.0011
12. Rho M, Yoo JE (2019) Exploration of variables relating to career decisions via adaptive LASSO. J Yeolin Educ 27(4):133–155
13. Lim HJ, Yoo JE, Rho M, Ryu JJ (2022) Exploration of variables predicting sense of school belonging using the machine learning method—group Mnet. Psychol Rep. https://doi.org/10.1177/00332941221133005
14. Groll A, Tutz G (2014) Variable selection for generalized linear mixed models by L1-penalized estimation. Stat Comput 24(2):137–154. https://doi.org/10.1007/s11222-012-9359-z
15. Lee JD, Sun DL, Sun Y, Taylor JE (2016) Exact post-selection inference, with application to the lasso. Ann Stat 44(3):907–927. https://doi.org/10.1214/15-AOS1371
16. Kim HG, Yoo JE (2020) ICILS 2018 variable exploration to predict computer and information literacy: variable selection in multilevel modeling via glmmLasso. J Educ Sci 22(4):1–21. https://doi.org/10.15564/jeju.2020.11.22.4.1

17. Koo M, Yoo JE (2021) Intraclass correlation and the performance of penalized regression: exploration of predictors for TALIS 2018 teacher cooperation. Asian J Educ 22(1):31–59
18. Rho M, Yoo JE (2021) Statistical inference after variable selection in penalized regression: focusing on variables relating to adolescents' smartphone reliance. Stud Korean Youth 32(1):147–174
19. Yoo JE, Rho M (2021) Statistical inference after variable selection via penalized regression: Focusing on variables predicting belonging to school. Paper presented at 2021 American Educational Research Association (AERA) Annual Meeting. Orlando, FL
20. Yoo JE, Rho M (2023) Penalized regression versus forward stepwise regression in variable selection and significance testing: Evidence from PISA 2015. Paper presented at 2023 American Educational Research Association (AERA) Annual Meeting. Online.
21. Yoo JE (2021) AI, big data, and machine learning. Hakjisa
22. Hoerl AE, Kennard RW (1970) Ridge regression: biased estimation for nonorthogonal problems. Technometrics 12(1):55–67. https://doi.org/10.1080/00401706.1970.10488634
23. Hastie T, Qian J, Tay K (2023) An introduction to glmnet. https://cloud.r-project.org/web/packages/glmnet/vignettes/glmnet.pdf
24. Breheny P, Zeng Y, Kurth R (2022) Regularization paths for regression models with grouped covariates. https://cloud.rproject.org/web/packages/grpreg/grpreg.pdf
25. Goeman J, Meijer R, Chaturvedi N (2022) L1 and L2 penalized regression models. https://cran.r-project.org/web/packages/penalized/vignettes/penalized.pdf
26. Akaike H (1973) Information theory and an extension of the maximum likelihood principle. In: Petrov BN, Csaki F (eds) Proceedings of the 2nd international symposium on information theory. Akademiai Kiado, pp. 267–281
27. Schwarz G (1978) Estimating the dimension of a model. Ann Stat 6:461–464. https://doi.org/10.1214/aos/1176344136
28. Kuha J (2004) AIC and BIC: comparisons of assumptions and performance. Sociol Methods Res 33(2):188–229
29. Shao J (1997) An asymptotic theory for linear model selection. Stat Sin 7(2):221–242
30. Meinshausen N, Bühlmann P (2010) Stability selection. J R Stat Soc Ser B (Stat Methodol) 72(4):417–473
31. Fan J, Li R (2001) Variable selection via nonconcave penalized likelihood and its oracle properties. J Am Stat Assoc 96(456):1348–1360. https://doi.org/10.1198/016214501753382273
32. Huang J, Ma S, Zhang CH (2008) Adaptive Lasso for sparse high-dimensional regression models. Stat Sin 18(4):1603–1618
33. Meinshausen N, Bühlmann P (2006) High-dimensional graphs and variable selection with the Lasso. Ann Stat 34(3):1436–1462
34. Zhao P, Yu B (2006) On model selection consistency of Lasso. J Mach Learn Res 7:2541–2563
35. Zhang CH (2010) Nearly unbiased variable selection under minimax concave penalty. Ann Stat 38(2):894–942. https://doi.org/10.1214/09-AOS729
36. Agresti A (2002) Categorical data analysis, 2nd ed. Wiley. https://doi.org/10.1002/0471249688
37. Javanmard A, Montanari A (2014) Confidence intervals and hypothesis testing for high-dimensional regression. J Mach Learn Res 15(1):2869–2909. https://www.jmlr.org/papers/volume15/javanmard14a/javanmard14a.pdf
38. Bühlmann P, Mandozzi J (2014) High-dimensional variable screening and bias in subsequent inference, with an empirical comparison. Comput Statist. https://doi.org/10.1007/s00180-013-0436-3
39. Fan J, Lv J (2008) Sure independence screening for ultrahigh dimensional feature space. J R Stat Soc: Ser B (Stat Methodol) 70:849–911. https://doi.org/10.1111/j.1467-9868.2008.00674.x
40. Taylor J, Tibshirani R (2018) Post-selection inference for L1-penalized likelihood models. Can J Stat 46(1):41–61. https://doi.org/10.1002/cjs.11313
41. Koo M, Yoo JE (under review) Investigating teachers' team innovativeness using glmmLasso: a machine learning analysis of the teaching and learning international survey 2018. Large-scale Assessments in Education

42. Beemer J, Spoon K, He L, Fan J, Levine RA (2018) "Ensemble learning for estimating individualized treatment effects in studentsuccess studies," Int. J. Artif. Intell. Educ., vol. 28, no. 3, pp. 315–335. https://doi.org/10.1007/s40593-017-0148-x
43. Meier, L. (2022). Package 'grplasso' (version 0.4-7). https://cran.r-project.org/web/packages/grplasso/grplasso.pdf

Predicting Student Performance

Schools Students Performance with Artificial Intelligence Machine Learning: Features Taxonomy, Methods and Evaluation

Alain Hennebelle, Leila Ismail, and Tanya Linden

Abstract Identifying students who might have difficulty in their course of studies ahead of time is crucial. There can be many reasons for performance issues, such as personality, family, social, and/or economic. We advocate that educational systems should use machine learning to predict students' performance based on performance factors. This would allow educational professionals and institutions to put in place a preventive plan to help students towards achievements of their educational goals and success. In this chapter, we propose a student performance prediction method and evaluate its performance. We provide a taxonomy of performance factors that help to gauge students' performance from different perspectives and give insights on the categories and features that have a more significant impact on students' performance. The results of this work can be used by education institutions to put in place a student-centric approach to tackle performance issues before they create long-term effects on student's life. In addition, it will help education policymakers to introduce a tailored approach for the population in specific areas.

Keywords Artificial intelligence · Classification models · Educational data mining · Educational machine learning · Feature selection · Student performance prediction · Taxonomy

L. Ismail
Cloud Computing and Distributed Systems Lab, School of Computing and Information Systems, The University of Melbourne, Parkville, VIC, Australia

A. Hennebelle · L. Ismail · T. Linden
School of Computing and Information Systems, Faculty of Engineering and Information Technology, The University of Melbourne, Parkville, Australia

L. Ismail (✉)
Intelligent Distributed Computing and Systems Lab, Department of Computer Science and Software Engineering, College of Information Technology, United Arab, Emirates University, Abu Dhabi, United Arab Emirates
e-mail: leila@uaeu.ac.ae

M. S. Khine (ed.), *Machine Learning in Educational Sciences*,
https://doi.org/10.1007/978-981-99-9379-6_6

1 Introduction

We live in a society where lifelong learning is part of life. Children start learning from the day they are born and the environment they grow up in supports their mental, emotional, and social development. This development lays the foundations for future success in life, including academic success. Achievements in formal education often pave the way to social and economic success in life [1]. Therefore in many countries, education has always been at the center of governments' attention [2, 3].

Information Systems used in education accumulate massive amounts of data about students, their demographics, educational background, and progress through their studies. Educational data analytics can be a valuable tool to support educators in the early identification of students with demographic factors putting them at risk. It has been established that early intervention can improve developmental and study outcomes [4].

Previous research studied a diverse range of factors that could predict success or impede it. One set of factors is classified as socio-economic status which is predetermined by parents' education and occupation as well as family (or household) income level [5]. Families where parents obtained high education degrees usually provide an environment fostering the need and desire to study and to strive for higher levels of achievement [6].

The rise of online learning during the COVID-19 pandemic, its continuity post-pandemic, the explosion of eLearning resources, and the use of ChatGPT [7] led to the creation of state databases and the establishment of large repositories. In addition, new types of learning environments have been emerging at an unprecedented pace, such as ubiquitous mobile learning, virtual reality, games, and metaverse, in a world that becomes smarter, thanks to the advancement in technologies to support various domains including education [8]. A huge amount of data about students and their learning is being produced in different formats, making it very challenging to analyze manually to get insights about the factors which impact students' performance and take the necessary actions. Artificial Intelligence machine learning technology has been used to predict the benefits of an educational environment, and the academic performance of students to develop a prevention plan for failures [9]. Predicting students' performance also enables educational policymakers and school administration to put in place a plan to improve the educational system, propose strategies to help weak learners, and prevent academic failure, based on factors affecting students' learning outcomes, such as demographic, psychological, environmental, and socio-economic.

Research work on machine learning evaluates the accuracy of students' performance classification models using heterogeneous datasets and evaluation metrics. In addition, those works do not provide an analysis of the features that impact students' performance and the rationale behind the models' performance. In this chapter, we evaluate and compare the performance of two mostly used classification models in the literature, Support Vector Machine (SVM), and Random Forest (RF). This is done in a unified environmental setup using the same dataset [10]. We evaluate the performance of the models in terms of accuracy and F-measure. The machine learning models, generated, after being trained on the dataset, are evaluated using F1-measure and accuracy. F1-measure is the harmonic mean of precision and recall. Accuracy refers to the degree to which the result of a prediction conforms to the correct value.

The key research contributions are as follows:

- We propose a taxonomy of students' performance factors based on their common characteristics and analyze which categories are more significant than others in predicting school students' performance.
- We design, develop, and implement a machine learning based framework for the prediction of students' failures/successes.
- We evaluate the performance of the two most used machine learning algorithms in the literature in terms of accuracy and F-measure in a unified environmental setup, using a real-world student dataset with and without feature selection.

The rest of the chapter is organized as follows. In Sect. 2, we present a taxonomy of factoring impacting students' successes for failures. Section 3 discusses the related work. The classification models used in this study are explained in Sect. 4 in the context of students' performance prediction. The machine learning models under study are explained in Sect. 5. The experimental setup, experiments, and the analysis of the results in terms of accuracy, F-measure are described in Sect. 6. The chapter is concluded in Sect. 7 along with recommendations and future research directions.

2 Taxonomy of Students' Performance Factors

This section presents a taxonomy of students' performance factors (Fig. 1). We classify them into 6 categories (1) demographic, (2) socio-economic, (3) school environment, (4) student education-centric, (5) student social-centric, and (6) student health-centric. This classification aims to analyze which categories of the students' performance factors significantly contribute to the prediction of students' failures or successes.

The majority of past studies discuss gender and age as important factors affecting students' success. In terms of gender, many studies point out that female students often perform better than their male counterparts [11, 12]. The reason is that girls often inspire to achieve better results in their school studies to lay the foundations of success for their future since historically it has been more challenging for females to get ahead in life [13]. However, some studies also point out that family circumstances and ethnic culture have an impact on female students interfering with their success, especially in developing world countries or within poorer groups of the population [14, 15]. Studies that examined age association with pathways to the university observed a negative correlation between mature-age students and intentions to continue to university [16]. The study suggests that if mature-age students

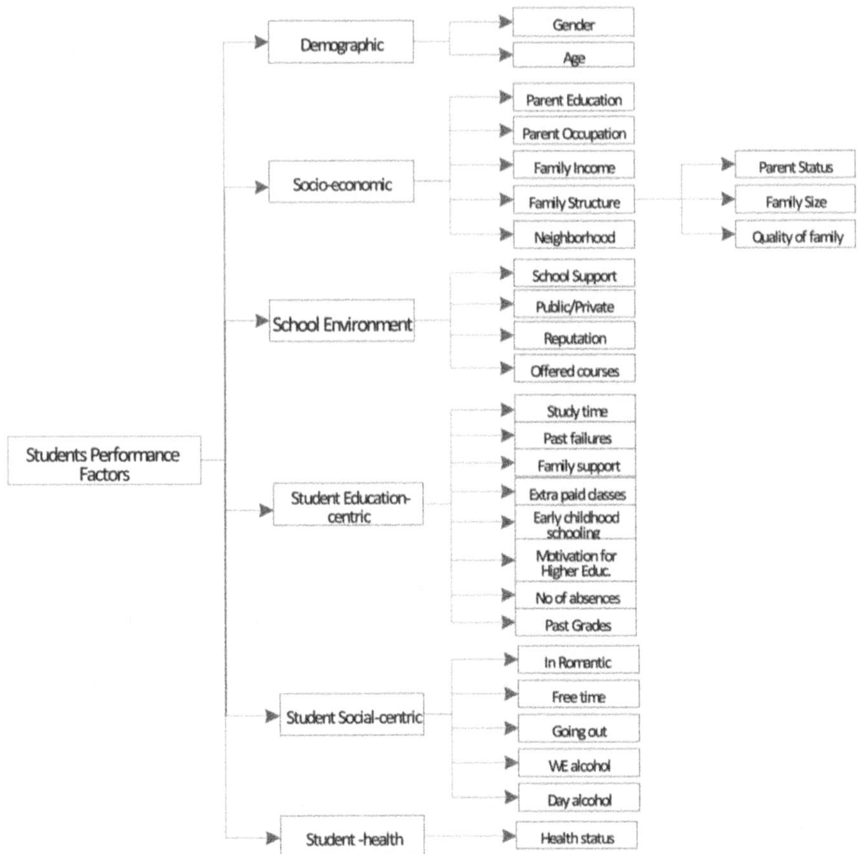

Fig. 1 Taxonomy of student success factors

experienced challenges in their secondary education, they are likely to continue experiencing challenges in the follow-up studies. Other research pointed out the family demands on mature-age students, thus interfering with the success of their studies [17]. However, there are other studies demonstrating that mature-age students are more likely to succeed in the first year of university studies than school leavers [18].

Socioeconomic factors often pre-determine students' success in their studies. These factors include parents' education and occupation [19], family income [20], family structure, and neighborhood as influencing parameters. Parents are important influencers on their children. Parents from middle-class financial backgrounds want their children to achieve more and motivate them as role models, with financial support as well as learning resources [21]. In some cases, parents from lower classes attempt to provide a roadmap to success for their children. They work hard, and they immigrate or send their children to countries with better opportunities so their children can study and have a better life than their parents. But they also apply motivational strategies to keep their children inspired [22]. However, there is a negative correlation between parents' social status and the achievements of their children due to the inability to provide access to learning resources and therefore hindering their children's prospects [23].

Family structure refers to incomplete (single-parent) vs complete families and fragile vs non-fragile families (i.e. families with a high risk of homelessness and families with both parents) [24]. Complete and non-fragile families are considered to provide a stable nourishing environment for children's development [25]. Disadvantaged neighborhoods (e.g. with high unemployment rates and lack of positive role models because a father figure is missing) are more likely to outweigh the family and school influence and would create a contradictive environment where attitudes and behavioral patterns encouraged by schools are not supported in children's surroundings and therefore the children would not consider success in traditional understanding and behaviors leading to it as necessary or useful [26].

Schools can be considered significant contributors to children's development and significantly impact their students' future. A school environment can create a pathway into higher education. It has been demonstrated that type of school (public vs private where fees are high) may affect student's performance and therefore their path into the higher education sector. Private schools are expensive but they often have better-trained teachers and better school facilities and study equipment, therefore providing a nourishing learning environment [27]. Public schools were created to be available for every child regardless of socio-economic background and to provide equal opportunities. In reality, public schools are often understaffed, and underfunded and seek contributions from parents and private sources [28]. This lack of funding could result in deteriorating facilities and out-of-date equipment, leading to teachers' and students' dissatisfaction [29], which in turn affects students' grades. A similar impact is played by the school location (which could be evaluated by the prestigiousness of a suburb and house prices in the area as well as by comparing metropolitan schools vs regional and socio-economic factors). In the USA, students' success is determined by the state and the socio-economic characteristics of the community they live in and where the school is located [16, 30].

3 Related Work

Several works discuss the use of machine learning to predict students' performance. Table 1 presents a comparison. Work in the literature evaluated the performance of machine learning classification models for students' performance. Table 1 shows the work on SVM, and RF, the most used models in the literature. However, they are trained and developed using different datasets containing different features, and evaluated with the means of different performance metrics making an objective comparison difficult. None of these studies compare these models using a public dataset with a large number of observations. Using a private dataset does not enable the research

Table 1 Evaluation of past works on SVM, and RF models

Work	#Features	#Observations	Accuracy	F-measure	Insights on features impact on students' performance	Performance features taxonomy
Mehboob et al. [31]	28	450	✓	×	x	x
Tekin [32]	49	127	×	×	x	x
Almutairi et al. [33]	16	480	✓	✓	x	x
Daud et al. [34]	23	776	×	✓	House expense	x
Rivas et al. [35]	39	32,593	✓	✓	x	x
Ajibade et al. [36]	16	500	✓	✓	x	x
Costa et al. [37]	19	262	×	✓	x	x
	15	161				
Kostopoulos et al. [38]	16	344	✓	×	x	x
Kiu [39]	33	395	×	✓	x	x
Migueis et al. [40]	19	2459	✓	×	x	x
Rimadana et al. [41]	11	125	✓	×	x	x
Lopez et al. [42]	11	114	✓	×	x	x
This study	33	1044	✓	✓	✓	✓

community to reproduce the results. Moreover, these works evaluate the models in terms of accuracy only. We argue that for an imbalanced dataset, it is important to use F-measure as accuracy alone can be misleading.

In this chapter, we compare the models under study in terms of both accuracy and F-measure on a public large dataset. In addition, this study generates insights into the features impacting students' performance and provides a taxonomy.

4 Students Performance Features Exploration

The dataset under study consists of data collected during the 2005–2006 school year from two public schools, in the Alentejo region of Portugal. The students and families answered some questionnaires, and data were completed with the school reports.

Table 2 presents the list of features in the dataset. There is a total of 33 features which encompass the six categories of our taxonomy (Fig. 1). In this study, our main goal is to predict the outcome corresponding to G3 (grade at the end of the school year). To determine the features which are impactful on the classification decision, we identify the correlation between the different features and classification outcome as shown in Fig. 2. The figure shows that the target class G3 is highly correlated with previous grades during the year. Figure 2 shows an important negative correlation between G3 and the number of previous failures during the student's school history, indicating that in general repeating a class will not improve the student's level in future year level. Other negative correlations with academic results include alcohol consumption and age. It confirms that alcohol is a negative indicator for school students as far as the grade is concerned. The age negative correlation is expected as the number of failures will result in an increased age for the student. On the other hand, we can see the quality of family relationships is negatively correlated with alcohol consumption. The mother's and father's education levels (Fedu, Medu) are highly correlated, which tends to show that families are created with parents having the same education level. The higher positive correlation for G3 is the desire to take higher education, which seems that motivation cat as a booster for school achievement. Next, positive correlations are with mother and father education and study time. Study time can somehow be expected, and importance of parents' education pinpoints the influence of the family environment and especially the "knowledge" of education system that experienced parents can dispense to their children, together with a will to go for higher education.

There is a strong correlation between the target class grade (G3) and the previous grades. To generalize the model and capture the relative importance of features related to demographic, socio-economic, school environment, social, and health factors, we develop a machine learning model to predict student performance without considering the previous grades.

To extract the most relevant features for the prediction of students' performance, we use RF regression model and generate "feature importance" (Fig. 3.). The figure shows that the previous grades G1 and G2 are the most important features which determine the students' performance. In addition, the number of absences during the year is the top second factor for the model prediction.

In summary, some features are more important than others for the prediction of students' performance. For instance, in previous grades G1 and G2, the number of absences, student age, and mother education are among the features that should be captured by a machine learning model.

In addition, we extract the importance of the dataset features by applying RF regression after the removal of the previous grades (G1, G2). This is to get insights on the influence of other factors, for instance, socio-economic and social factors (Fig. 4). For instance, Fig. 4 shows that the number of absences, and the number of failures, both numerical features, influence most of the model for the prediction of students' results.

Table 2 List of features of the dataset

Attribute	Description (Domain)
Sex	Student's sex (binary: female or male)
Age	Student's age (numeric: from 15 to 22)
School	Student's school (binary: Gabriel Pereira or Mousinho da Silveira)
Address	Student's home address type (binary: urban or rural)
Pstatus	Parent's cohabitation status (binary: living together or apart)
Medu	Mother's education (numeric: from 0 to 4 a)
Mjob	Mother's job (nominalb
Fedu	Father's education (numeric: from 0 to 4 a)
Fjob	Father's job (nominal b)
Guardian	Student's guardian (nominal: mother, father or other)
Famsize	Family size (binary: ≤3 or >3)
Famrel	Quality of family relationships (numeric: from 1—very bad to 5—excellent)
Reason	Reason to choose this school (nominal: close to home, school reputation, course preference or other)
Traveltime	Home to school travel time (numeric: 1—<15 min., 2—15 to 30 min., 3—30 min. to 1 h or 4—>1h)
Studytime	Weekly study time (numeric: 1—<2 h, 2—2–5 h, 3—5–10 h or 4—>10 h)
Failures	Number of past class failures (numeric: n if 1 ≤ n < 3, else 4)
Schoolsup	Extra educational school support (binary: yes or no)
Famsup	Family educational support (binary: yes or no)
Activities	Extra-curricular activities (binary: yes or no)
Paidclass	Extra paid classes (binary: yes or no)
Internet	Internet access at home (binary: yes or no)
Nursery	Attended nursery school (binary: yes or no)
Higher	Wants to take higher education (binary: yes or no)
Romantic	With a romantic relationship (binary: yes or no)
Freetime	Free time after school (numeric: from 1—very low to 5—very high)
Gout	Going out with friends (numeric: from 1—very low to 5—very high)
Walc	Weekend alcohol consumption (numeric: from 1—very low to 5—very high)
Dalc	Workday alcohol consumption (numeric: from 1—very low to 5—very high)
Health	Current health status (numeric: from 1—very bad to 5—very good)
Absences	Number of school absences (numeric: from 0 to 93)
G1	First period grade (numeric: from 0 to 20)
G2	Second period grade (numeric: from 0 to 20)
G3	Final grade (numeric: from 0 to 20)

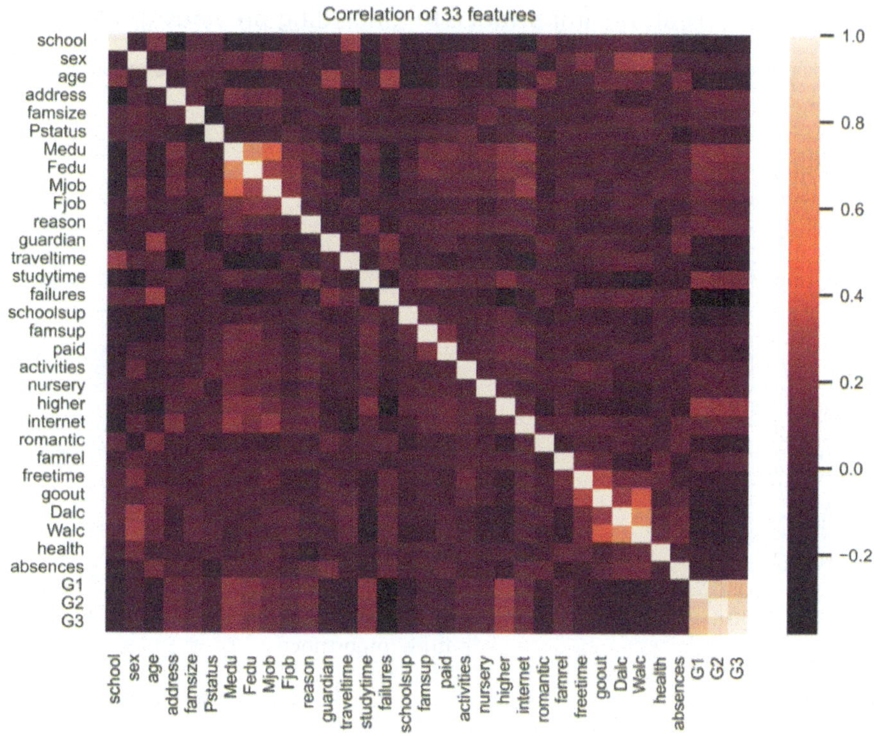

Fig. 2 Correlation of original 33 features

5 Machine Learning Models for Students Performance Prediction

In this section, we explain the machine learning classification models under study for students' performance prediction. We take an example of classifying students' marks into three levels: high, medium, and low. However, other classification labels, such as A, B, C, D, and F can be used.

5.1 Support Vector Machine

The SVM model separates the n-dimensional students' information into high, medium, and low classes. This is by creating a decision boundary, known as a hyperplane, for separation. The hyperplane is generated with the help of the data points in each class that are the closest to the data points in the other class. These data points are known as support vectors. The generation of a hyperplane is an iterative process where the objective is to find the maximum possible margin between the support

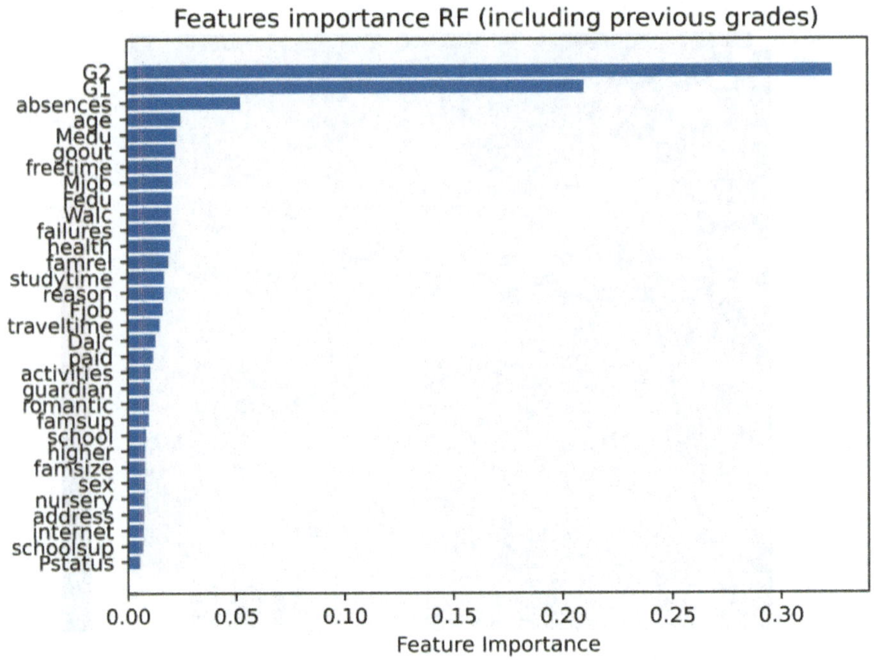

Fig. 3 Relative feature importance (including previous grades) from random forest classifier

vectors of the opposite classes. For a students' performance dataset with a set of F features and $C = \{\text{High, Medium, Low}\}$ classes, the hyperplane that separates these classes can be represented using Eq. (1).

$$\mathrm{wF} + \mathrm{b} = 0$$

$$s.t., wF_i + b > 0, \text{ if } C_i = +1 \; \boldsymbol{and} \; wF_i + b < 0, \text{ if } C_i = -1 \tag{1}$$

where w is normal to the hyperplane and b is the bias.

To find the maximum possible margin for optimal hyperplane, the norm of the margin should be minimized as stated in Eq. (2).

$$\min \frac{1}{2} ||w||^2 s.t., C_i(w.F_i + b) \geq 1, \forall F_i \tag{2}$$

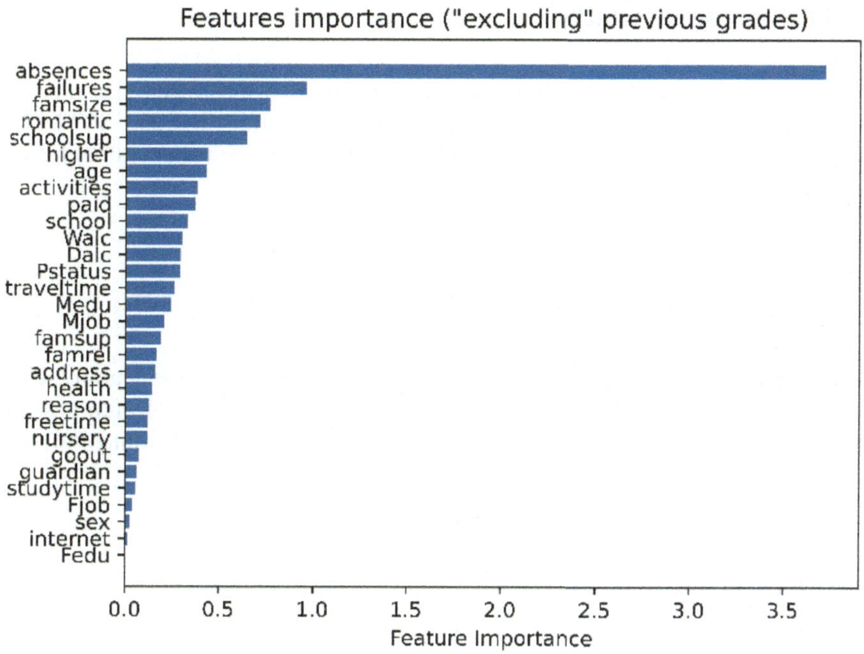

Fig. 4 Relative feature importance (including previous grades) from random forest classifier

5.2 *Random Forest*

RF is an ensemble technique that uses a set of Decision Trees. Each tree is constructed using a randomly selected sample of the dataset [43]. Each Decision Tree predicts the class for a student (high/medium/low) based on the features and voting is performed on the output of each tree. In RF, each DT model will consider only a randomly selected subset of features at each node for splitting the tree and traversing down. The RF model then decides on the high/medium/low class based on the majority of the votes as stated in Eq. (3).

$$C_{\mathrm{RF}}(s) = \text{majority vote}\{C_n(s)\}_1^N \tag{3}$$

where N represents the number of Decision Trees used.

6 Performance Analysis

In this section, we compare the performance of the two most used classification models in the domain of students' performance prediction. RF and SVM. The models are evaluated in terms of Accuracy, and F-measure, with and without feature selection.

6.1 *Experimental Environment*

To evaluate the performance of the models under study with the feature selection algorithm, we use the Information-Gain Attribute Evaluator algorithm as it is found to be the most accurate among others in education machine learning [44]. For implementation, we use the sklearn Python library.

6.2 *Experiments*

Data preprocessing. For the dataset under study, we first convert the categorical features into numerical ones. For this dataset. The separate files for the Mathematics and Portuguese language subjects are combined. We created class labels for the grade to be predicted ('G3'). This is done by mapping the grades to 5 classes based on the Erasmus grade conversion system (*Erasmus Programme*, n.d.), i.e. grade A (16–20), grade B (14–15), grade C (12–13), grade D (10–11) and grade F (0–9).

Model Building and Testing. For model building and testing, we use a tenfold cross-validation method to obtain the training and testing datasets respectively. For the SVM model, we implement the linear kernel. We measure the accuracy, F-measure, for each model. The accuracy is calculated using Eq. (4).

$$\text{Accuracy} = \frac{\text{TP} + \text{TN}}{\text{TP} + \text{FP} + \text{TN} + \text{FN}} \tag{4}$$

where

TP = True Positive = #observations in positive class that are predicted as positive.

TN = True Negative = #observations in negative class that are predicted as negative.

FP = False Positive = #observations in negative class that are predicted as positive.

FN = False Negative = #observations in positive class that are predicted as negative.

The F-measure is calculated using precision and recall as stated in Eq. (5). The values of precision and recall are calculated using Eqs. (6) and (7) respectively.

$$F - \text{measure} = \frac{2(\text{Precision} * \text{Recall})}{\text{Precision} + \text{Recall}} \tag{5}$$

$$\text{Precision} = \frac{\text{TP}}{\text{TP}} + \text{FP} \tag{6}$$

$$\text{Recall} = \frac{\text{TP}}{\text{TP} + \text{FN}} \tag{7}$$

6.3 Experimental Results Analysis

Figure 5 shows the accuracy and F-measure for the dataset without using feature selection. It shows that the RF model has the highest accuracy and F-measure values compared to the other models. This is because RF selects a subset of the features randomly to fit the data while the other models use all the features resulting in over-fitting and consequently inaccurate prediction. When applying the feature selection algorithm on the dataset, the selected 10 features are: the previous two grades, G1 and G2, the number of past class failures, the number of school absences, going out with friends, desire for higher education, study time, father's education, mother's education. With this dataset, we then can see the great influence of the historical results, like previous grades and number of failures in previous years. This importance of previous failures seems to confirm the analysis in Sect. 2, about mature-age students showing less intention to continue to university, because they have lower results, they will not pursue higher education [16]. The next important factor is the health, it is important in the fact that bad health is associated with lower achievement. This can be correlated and explained by an increased number of absences, and less availability for studying.

Figure 6 shows the accuracy and the F-measure with feature selection. It shows that the performance of the models is enhanced by using the selected features. In particular, the accuracy of SVM is increased by 4.64% and the accuracy for RF is increased by 1.54%.

6.4 Study of Data Without Previous Grades

Figure 7 shows that the performance of the models under study decreases significantly when the training is performed on the dataset without the previous grades.

Figures 8 and 9 show the confusion matrices generated by the models under study with and without using previous grades respectively. Figure 9 indicates without the previous grades, RF and SVM models are less accurate than when previous grades are available (Fig. 8). However, Fig. 9 reveals that the prediction is not far from the

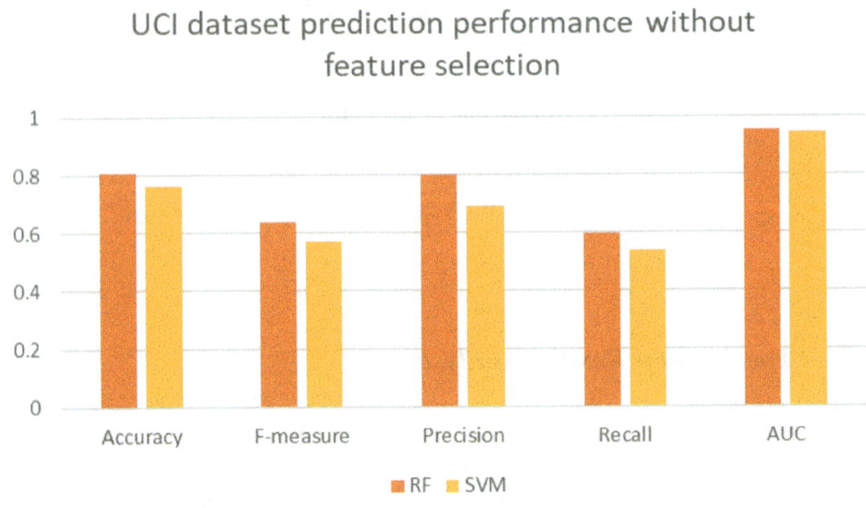

Fig. 5 Accuracy and F-measure of the models for the UCI dataset without feature selection

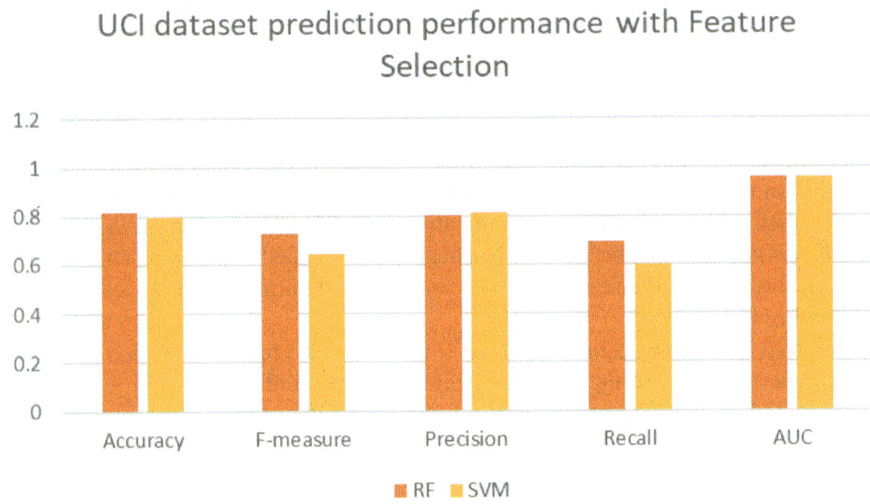

Fig. 6 Accuracy and F-measure of the models for the UCI dataset with feature selection

true values, exhibiting the predictive power of the models for students' performance. In particular, the errors occur in cells where B is chosen instead of C, or C is chosen instead of B, showing that most of the data stays "near" the diagonal of the matrix.

In summary, for the UCI Portuguese dataset having 1044 observations, RF outperforms SVM. Furthermore, the performance of SVM in our experimental environment is the same as in [32]. However, our results are not consistent with the ones obtained by [31, 36, 38, 40]. The discrepancy in the results is due to heterogeneity in the

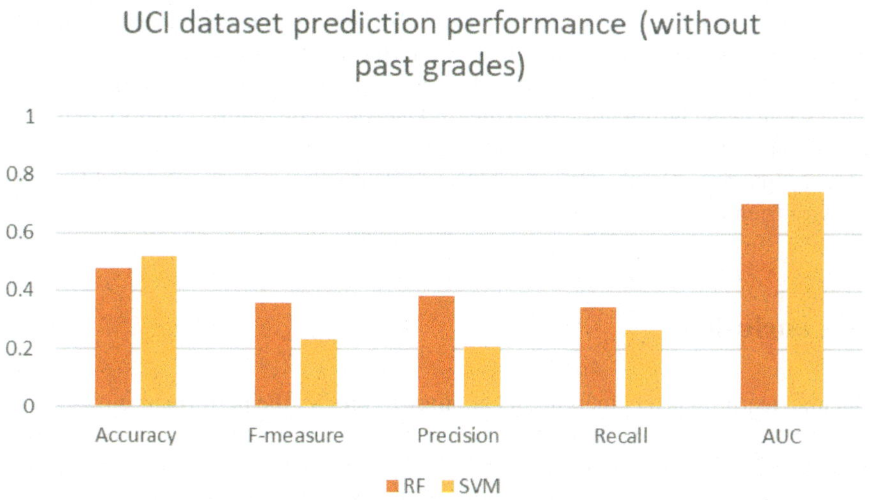

Fig. 7 Accuracy and F-measure of the models for the UCI dataset not using previous grades

Fig. 8 Confusion matrices of prediction models using previous grades features G1 and G2. "0" represents "F" class, "1" is "D", "2" is "C", "3" is "B" and "4" is "A"

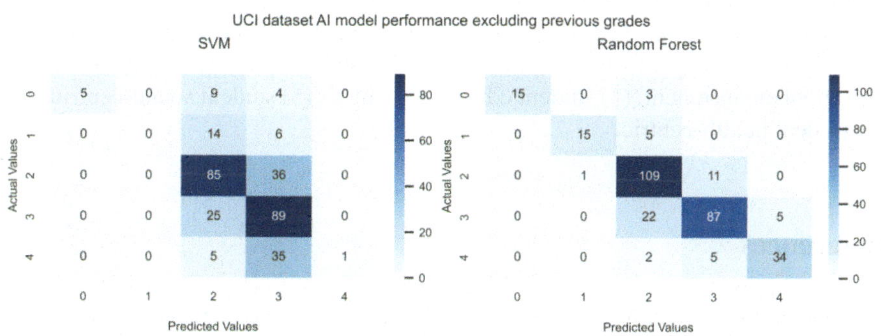

Fig. 9 Confusion matrices of prediction models not using previous grades features G1 and G2. "0" represents "F" class, "1" is "D", "2" is "C", "3" is "B" and "4" is "A"

features used in these works, and they also rely more on grades obtained in previous periods of time. The size of the dataset and the features impact the performance of the models. The study of the model's performance also reveals that prediction is better when using information from the student education-centric category like study time, desire for higher education, and number of past failures, from socio-economic category like parent's education, and from student social-centric category like going out with friends.

7 Conclusions

Machine learning is an emerging technology to predict the performance of students with the main goal of developing strategies by educational professionals to improve the learning environment. Several works in the literature have used various machine learning approaches for the prediction of students' performance. However, these studies used different datasets and evaluation metrics, making it difficult to compare the developed models. In addition, they did not provide insights into the features from socio-economic and social perspectives. In this chapter, we evaluate and compare the performance of the two most used machine learning classification models, i.e. SVM, and RF, to predict students' performance using a dataset with a large number of features impacting the performance. We provide a taxonomy categorizing these features and analyze their impact on student performance. Our taxonomy is based on demographic, socio-economic, school environment, student education-centric, student social-centric, and student health-centric factors. We conduct a comparative analysis of the models under study using a real-life student information dataset. Our experimental results show that RF outperforms SVM. In addition, when selecting a machine learning model, it is essential to consider features such as parents' educational and work background, student study time, number of past class failures, plan for higher education, number of school absences, previous grades, and information about the school. This is because these features improve prediction accuracy. Furthermore, to enable an insightful analysis of the features' impact on student's performance, it is important to provide a mapping between those features and their corresponding categories that we identified: (1) demographic, (2) socio-economic, (3) school environment, (4) student education-centric, (5) student social-centric, and (6) student health-centric.

References

1. Krueger AB, Lindahl M (2001) Education for growth: why and for whom? J Econ Litte 39(4):1101–1136
2. Eric H (2016) Education and the nation's future. Blueprint for America. pp 89–108
3. United Nations (2020) Policy brief: education during COVID-19 and beyond. https://reliefweb.int/report/world/policy-brief-education-during-covid-19-and-beyond-august-2020

4. Ramey SL, Ramey CT (1992) Early educational intervention with disadvantaged children—to what effect? Appl Prev Psychol 1(3):131–140. https://doi.org/10.1016/S0962-1849(05)80134-9
5. Avvisati F (2020) The measure of socio-economic status in PISA: a review and some suggested improvements. Large-Scale AssessS Educ 8(1):8. https://doi.org/10.1186/s40536-020-00086-x
6. Chevalier A, Lanot G (2002) The relative effect of family characteristics and financial situation on educational achievement. Educ Econ 10(2):165–181. https://doi.org/10.1080/09645290210126904
7. OpenAI. (2022). *ChatGPT*. https://chat.openai.com
8. Ismail L, Zhang L (2018). Information innovation technology in smart cities. https://doi.org/10.1007/978-981-10-1741-4
9. Ismail L, Materwala H, Hennebelle A (2021) Comparative analysis of machine learning models for students' performance prediction. Int Conf Adv Digit Sci 149–160
10. Cortez P, Silva AMG (2008) Using data mining to predict secondary school student performance
11. Bugler M, McGeown SP, St Clair-Thompson H (2015) Gender differences in adolescents' academic motivation and classroom behaviour. Educ Psychol 35(5):541–556. https://doi.org/10.1080/01443410.2013.849325
12. Sheard M (2009) Hardiness commitment, gender, and age differentiate university academic performance. Br J Educ Psychol 79(1):189–204. https://doi.org/10.1348/000709908X304406
13. Spencer R, Walsh J, Liang B, Mousseau AMD, Lund TJ (2018) Having it all? a qualitative examination of affluent adolescent girls' perceptions of stress and their quests for success. J Adolesc Res 33(1):3–33. https://doi.org/10.1177/0743558416670990
14. Chavous T, Cogburn CD (2007) Superinvisible women: black girls and women in education. Black Women Gend + Fam 1(2):24–51
15. Ramanaik S, Collumbien M, Prakash R, Howard-Merrill L, Thalinja R, Javalkar P, Murthy S, Cislaghi B, Beattie T, Isac S, Moses S, Heise L, Bhattacharjee P (2018) Education, poverty and "purity" in the context of adolescent girls' secondary school retention and dropout: a qualitative study from Karnataka, southern India. PLoS ONE 13(9):e0202470. https://doi.org/10.1371/journal.pone.0202470
16. Shulruf B, Hattie J, Tumen S (2008) Individual and school factors affecting students' participation and success in higher education. High Educ 56(5):613–632. https://doi.org/10.1007/s10734-008-9114-8
17. Lincoln D, Tindle E (2000) Mature age students in transition : factors contributing to their success in first year. In: 4th Pacific Rim first year in higher education conference (FYHE) : creating futures for a new millennium
18. McKenzie K, Gow K (2004) Exploring the first year academic achievement of school leavers and mature-age students through structural equation modelling. Learn Individ Differ 14(2):107–123. https://doi.org/10.1016/j.lindif.2003.10.002
19. Daleure GM, Albon R, Hinkston K, Ajaif T, McKeown J (2014) Family involvement in Emirati college student education and linkages to high and low achievement in the context of the United Arab Emirates. FIRE: Forum Int Res Educ 1(3)
20. Farooq MS, Chaudhry AH, Shafiq M, Berhanu G (2011) Factors affecting students' quality of academic performance: a case of secondary school level. J Qual Technol Manag 77(2):1–14
21. Irwin S, Elley S (2013) Parents' hopes and expectations for their children's future occupations. Sociol Rev 61(1):111–130. https://doi.org/10.1111/j.1467-954X.2012.02139.x
22. Gutman LM, McLoyd VC (2000) Parents' management of their children's education within the home, at school, and in the community: an examination of African-American families living in poverty. Urban Rev 32(1):1–24. https://doi.org/10.1023/A:1005112300726
23. Eamon MK (2005) Social-demographic, school, neighborhood, and parenting influences on the academic achievement of Latino young adolescents. J Youth Adolesc 34(2):163–174. https://doi.org/10.1007/s10964-005-3214-x
24. O'Malley M, Voight A, Renshaw TL, Eklund K (2015) School climate, family structure, and academic achievement: a study of moderation effects. Sch Psychol Q 30(1):142–157. https://doi.org/10.1037/spq0000076

25. McLanahan S, Garfinkel I (2000) The Fragile Families and Child Wellbeing Study: Questions, Design and a Few Preliminary Results
26. Ainsworth JW (2002) Why does it take a village? the mediation of neighborhood effects on educational achievement. Soc Forces 81(1):117–152. https://doi.org/10.1353/sof.2002.0038
27. Bragg DD, Loeb JW, Gong Y, Deng C-P, Yoo J, Hill JL (2002) Transition from high school to college and work for tech prep participants in eight selected consortia
28. Thompson G, Hogan A, Rahimi M (2019) Private funding in Australian public schools: a problem of equity. Aust Educ Res 46(5):893–910. https://doi.org/10.1007/s13384-019-00319-1
29. Moore CM (2012) The role of school environment in teacher dissatisfaction among U.S. public school teachers. SAGE Open 2(1):215824401243888. https://doi.org/10.1177/2158244012438888
30. Hochschild JL (2003) Social class in public schools. J Soc Issues 59(4):821–840
31. Mehboob B, Muzamal Liaqat R, Abbas N (2017) Student performance prediction and risk analysis by using data mining approach. J Intell Comput 8(2)
32. Tekin A (2014) Early prediction of students' grade point averages at graduation: a data mining approach. Eurasian J Educ Res 54:207–226
33. Almutairi S, Shaiba H, Bezbradica M (2019) Predicting students' academic performance and main behavioral features using data mining techniques. International conference on computing. pp 245–259
34. Daud A, Aljohani NR, Abbasi RA, Lytras MD, Abbas F, Alowibdi JS (2017) Predicting Student Performance using Advanced Learning Analytics. In: Proceedings of the 26th international conference on world wide web companion. pp 415–421
35. Rivas A, Gonzalez-Briones A, Hernandez G, Prieto J, Chamoso P (2020) Artificial neural network analysis of the academic performance of students in virtual learning environments. Neurocomputing 423:713–720
36. Ajibade S-SM, Ahmad NBB, Shamsuddin SM (2019) Educational data mining: enhancement of student performance model using ensemble methods. IOP conference series: materials science and engineering
37. Costa EB, Fonseca B, Santana MA, de Araujo FF, Rego J (2017) Evaluating the effectiveness of educational data mining techniques for early prediction of students' academic failure in introductory programming courses. Comput Hum Behav 73:247–256
38. Kostopoulos G, Lipitakis A-D, Kotsiantis S, Gravvanis G (2017) Predicting student performance in distance higher education using active learning. international conference on engineering applications of neural networks. pp 75–86.
39. Kiu C-C (2018) Data mining analysis on student's academic performance through exploration of student's background and social activities. In: 2018 Fourth international conference on advances in computing, communication and automation (ICACCA)
40. Migueis VL, Freitas A, Garcia PJ, Silva A (2018) Early segmentation of students according to their academic performance: a predictive modelling approach. Decis Support Syst 115:36–51
41. Rimadana MR, Kusumawardani SS, Santosa PI, Erwianda MSF (2019) Predicting student academic performance using machine learning and time management skill data. Int Semin Res Inf Technol Intell Syst (ISRITI) 2019:511–515
42. Lopez MI, Luna JM, Romero C, Ventura S (2012) Classification via clustering for predicting final marks based on student participation in forums. International Educational Data Mining Society
43. Liaw A, Wiener M (2002) Classification and regression by Randomforest. R News 2(3):18–22
44. Ramaswami M, Bhaskaran R (2009) A study on feature selection techniques in educational data mining. J Comput 1(1):7–11
45. Erasmus Programme. (n.d.). https://en.wikipedia.org/wiki/Erasmus_Programme. Accessed 18 Dec 2020

Predicting Response Latencies on Test Questions Based on Features of the Questions

Madelynn Denner, Xiangyu Xu, Teresa M. Ober, Bo Pei, and Ying Cheng

Abstract Estimating the amount of time students spend on a given test question has many benefits. For example, it can be used as a guide on how much time to allocate for an assessment, or as a valuable tool in gauging student effort on a question. The goal of the present study is to identify the characteristics of a written test question (e.g., coverage of topics within the knowledge domain, use of content vocabulary, linguistic features, and question design features) that were most likely to predict median response times for each test question. The dataset consisted of 610 Advanced Placement statistics questions, and the following regression models were constructed: Stepwise Forward and Backward regression, Lasso regression, and a regression tree. Our findings suggest that: (a) variables pertaining to linguistic features, cognitive complexity, and question design all play a significant role in predicting response latency; and (b) these variables collectively explain a substantial amount of variation in response time (~25%).

Keywords Response times · Advanced Placement Statistics · Stepwise regression · Lasso regression · Regression tree

1 Introduction

The amount of time a student takes to answer a test question can provide useful insights for instructional design purposes. For educators, such information may be used to anticipate an approximate amount of time needed to allocate for students to complete an assessment in its entirety. For computer-administered tests, such information can also be used to identify students who engage in "speeded" behaviors and are unlikely to have earnestly attempted an answer [1]. However, estimating response time often requires collecting data from a representative sample of students which is timely and expensive. Modeling associations between features of test questions and

M. Denner · X. Xu · T. M. Ober · B. Pei · Y. Cheng (✉)
University of Notre Dame, 390 Corbett Hall, Nortre Dame, IN, USA
e-mail: ycheng4@nd.edu

M. S. Khine (ed.), *Machine Learning in Educational Sciences*,
https://doi.org/10.1007/978-981-99-9379-6_7

response time could help estimate response time without collecting student data in the future for new questions.

While there has been substantial research on the relationships between item features and item difficulty (e.g. [2]), there has been little work relating item features with response latency, with a couple of exceptions [3–5]. However, the motive to do so is becoming more and more prominent due to the growing popularity of computerized and/or online learning and testing. In the context of a medical licensure exam, [3] used natural language processing as an automated approach to extracting item features and used these features to model response latency. They chose to rely on natural language processing (NLP) for its robust and efficient extraction of item features, which allowed them to obtain a large span of complex and diverse linguistic features of written texts. For example, they were able to extract the synthetic complexity and ambiguity of each word of question text. Baldwin et al. [3] used various machine learning algorithms from the WEKA tool kit in order to fit a model that had an optimal bias and variance tradeoff. A tenfold cross-validation was used to evaluate each model, and the random forest model was found to perform the best.

In order to compare the predictive power of certain types of features, [3] fit separate random forest models on the three types of features derived: linguistic, information retrieval, and embeddings. Information retrieval (IR) features pertain to the relevance of the item text to publications in medicine. A corpus text composed of articles in the MEDLINE database was constructed, along with a query for each item. The query consisted of the stem of both the question text and the correct answer. Each item query is assigned various IR scores, reflecting different aspects of the relevance between the query and the matched abstracts. In total, 16 unique IR features were derived. Word embeddings are vector representations of a word, where each dimension pertains to a different aspect of the semantics of the word. Hence, words similar in meaning will have similar vector representations. Baldwin et al. [3] used the pretrained algorithm, *Word2Vec,* which resulted in vector representations of 300 dimensions [6]. The *Word2Vec* approach learns the standalone semantics of a given word based on its local usage context across all occurrences of the word within the text. The local context is defined by a window of words, which is a parameter of the model. Ultimately, the semantics of the word are represented across the 300-dimensional vector. They found that each of the types of features provides some unique information.

Baldwin et al. [3] also used Lasso regression to determine which linguistic features were the most indicative of response latency. The top five features most positively correlated with response latency were unique word count, presence of an image, number count (normalized by word count), conditional clause count (normalized by word count), and concreteness ratio. The top five features most negatively correlated with response latency were count of additive connective words, number count, count of words not in the 5,000 most common word list, referential pronouns count, and negated word count (normalized by word count).

While Baldwin et al. [3] was a very original paper on using automatically extracted item features to predict response latency, its focus on medical licensure exam and linguistic features (113 features) made it less informative for tests in K-12 settings and tests whose linguistic complexities are limited. In contrast, our analysis in this paper focused on a STEM item bank for high school students and placed a great emphasis on cognitive complexity or knowledge domain of an item. Although we also derived linguistic features, linguistic complexities of the items are not expected to be as high as a medical licensure exam given the intended reading level for K-12 students.

More specifically, we focused on characteristics of test items reflecting content for an Advanced Placement (AP) Statistics course. The AP program provides an opportunity for high school students to take college-level courses. Enrolling in AP-level classes gives high school students the opportunity to earn college credit and satisfy requirements before university enrollment, which can save them money and time in the future. It also allows students to explore different subject areas, giving them a better idea of what they may want to study in college. Finally, earning college credit through the AP program can provide students with more flexibility in college, allowing for them to study abroad, double major, enroll in internships, etc. For these reasons, there has been a rapid growth of AP participation nationwide [7]. On the other hand, the AP program is expected to have growing importance in college admission as more colleges are going test-optional or test-blind, because transcripts and GPA carry an increasing amount of weight in admission decisions, and many high schools give significant weights to grades of AP courses in their calculation of student GPA [8]. For these reasons, it is very important to examine the AP course context. However, few studies have specifically looked into this context [9].

Our focus on the subject of statistics is driven by a rapid expansion of it in the US education system and the labor market. Between 1990 and 2019, the percentage of high school students taking statistics and probability grew over 16 times (1 vs. 16.7%), overtaking calculus (7.2 vs. 15.8%) in 2019 (NCES 2019). In 2022, 216,968 students took the AP statistics exam, and 2,480 colleges accepted AP courses in high school for college credits [10]. In the labor market, there is an expected 31% increase in employment from 2021 to 2031 according to the U. S. Department of Labor Statistics [35]. Thus, we felt that our analysis on AP statistics is particularly important and timely.

The qualities of an item that we chose to focus on included metrics pertaining to linguistic features, cognitive complexity, and question design. We sought to answer the following research questions:

1. How well can we predict the response latency given these item features?
2. Which item features are the most predictive of response latency?
3. Do certain types of features (cognitive domain, linguistic features, and question design) tend to be more predictive of response latency than others?
4. Which modeling method leads to the most accurate and interpretable answers?

All features used as predictors were derived from the content of the test question itself and did not require any student-level data. An advantage of this approach is that we were able to make predictions about the typical response time required for an item without the monetary and time requirements of data collection on students.

2 Method

The features of each test question were summarized by 37 unique feature variables. Four regression models were constructed using these features to model the median response time of an item: Stepwise forward and backward regression, Lasso regression, and regression tree.

3 Data Source

We focused on item-level metrics of test items for AP Statistics. A dataset was created consisting of 37 unique features of 610 question items (see Table 1). Some features reflect the cognitive complexity involved in solving the questions (e.g., *sid count*, which is the number of steps it takes to answer a question), some may capture the linguistic complexity of an item (e.g., *polys*, the number of words with more than three syllables in question text), and others may involve aspects of item design (e.g., *imagecount*, which is the number of images included in a question). We chose to include *imagecount* in our analysis due to the growing impact of visual learning within the academic community. Visualization of learning concepts has been proven to improve students' retention of information, and further promote their creative and analytical thinking skills [11]. Although the addition of visuals may be an effective asset in the learning process, they tend to increase the time spent with the given material, as it takes time to visually process the images. The increasing use of visuals in the teaching setting is mirrored in the testing setting. Thus, it may be beneficial to study the relationship between the inclusion of an image(s) and response time. The comprehensive list of features is explained in further detail in Table 1.

3.1 Response Time

Response time was captured between the point at which a computerized assessment question was presented to the student and submission of an answer (*Minimum* = 2 s, *Median* = 59 s, *IQR*: 36.0–97.4, *Maximum* = 331 s). Between 2016 and 2020, response time data were obtained from students who used a computerized assessment platform for AP Statistics [16]. Response times were collected as part of the process data, which were considered more objective indicators of student engagement and self-regulation [17, 18].

Data were based on answers to 610 questions submitted by high school students in five states throughout the eastern and midwestern United States ($N = 1{,}244$; $M_{\text{age}} = 16.73$; 53.1% female; 1.5% American Indian/Alaskan Native, 20.9% Asian/Asian American, 8.2% Black/African American, 0.2% Hawaiian/Pacific Islander, 68.1% White/European American, 7.1% Hispanic/Latinx, 2.1% Other, 8.3% Multi-racial;

Table 1 List of features of items used in predicting response time

Feature label	Feature description
Knowledge domain	
section_n	Number of unique sections covered by an item
topic_n	Number of unique topics covered by an item
attr_n	Number of unique attributes covered
Topic_1.A-4.B	16 content topics in the AP statistics knowledge domain (binary)
sid count	The number of steps required to answer an item
term_total*	Count of total content terms
term_unique	Count of unique content terms
Linguistic features	
chars*	Number of characters in the question stem
word_chars	Number of alphanumeric characters in the question stem
words	Number of words in the question stem
nonwords	Number of words in the question stem that are non-English language words
sents	Number of sentences in the question stem
sylls*	Number of syllables in question text
polys	Number of words with more than three syllables in question text
re*	Flesch reading ease score of question text [12]
ari	Automatic readability index score of question text [13]
gl	Flesch Kincaid grade level score of question text [12]
smog	Measure of Gobbledygook score of question text [14]
cl	The Coleman Liau index score of question text [15]
Question design	
type_bf1	Blank field (1) or Multiple-choice (0)
image_count	Count of images in text
answers_n	Number of answer options

Note * Indicates the feature was removed as a predictor due to multicollinearity

10.5% free/reduced price lunch eligible). Median response time to an item was used to indicate the typical response time required by an item. Before conducting regression analyses, median response latencies were log-transformed as the dependent variable (see Fig. 1). This transformation helped to resolve the positive skew of the raw response time data. The resulting log-response time was roughly normal, except for a small amount of very short response times.

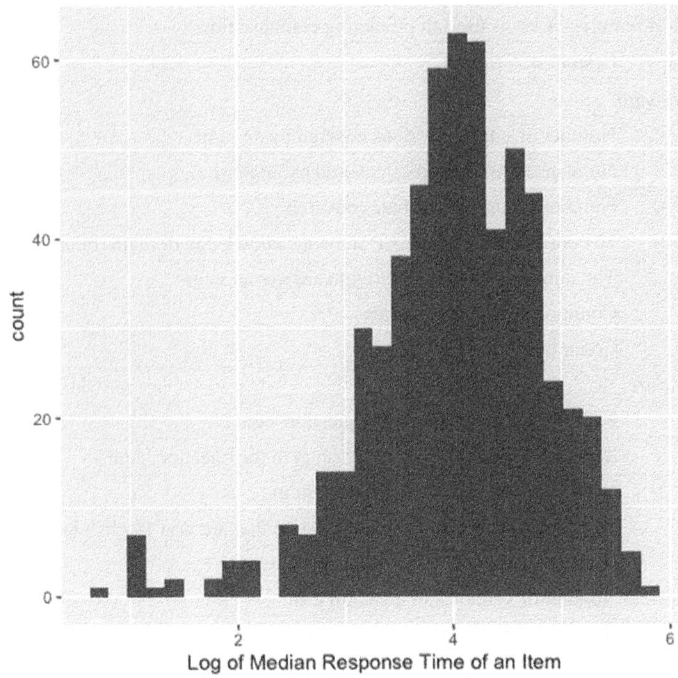

Fig. 1 Distribution of the log of the median response time across all items

3.2 Cognitive Complexity of an Item

An item can require complex cognitive processing in many different ways. The first type of metric we investigated was the content coverage of an item, i.e., the coverage of different topics of an item within the AP Statistics knowledge domain.

Each item is classified as covering a certain number of learning attributes (e.g., "*Interpreting slope*"), according to the AP Statistics test framework [33], College Board, 2010. There are a total of 157 unique learning attributes, which are organized into 16 different topics (e.g., "*Exploring bivariate data*"), which were subsumed by 4 unique sections (e.g., "*Exploring Data*"). Both the number of unique sections, attributes, and topics that a question addressed, as well as the specific topics covered (which is a dummy coded categorical variable) were included as predictor variables in this analysis. We chose to examine the content coverage at the topic level as opposed to the attribute level to allow for greater interpretability and simplicity. In the item bank, an item covered at most 2 unique sections, 3 unique topics, and 6 unique attributes. The number of steps required to complete an item ranged from 1 to 9, with a median of 3.

The second type of metric that indicates an item's cognitive complexity was derived based on the content vocabulary, i.e., the number of terms involved in an item that are often used in the scope of a high school AP Statistics course. The list of terms was compiled from a variety of resources, including a glossary of the 4th edition of the *Practice of Statistics* textbook [19] and multiple study guides compiled by AP statistics teachers. From there, the Hunspell package in R was used to correct any misspelled words, stem any plural form of a word, and compound any terms that required it [20]. Next, a Document Term Matrix (DTM) was constructed and words that appeared on the vocabulary list were filtered. The count of content terms was calculated by summing up the number of occurrences of any word on the content vocabulary terms list, as they were found in the question text. The count of unique terms was also calculated so that if the same term appeared multiple times in the question it was counted only once. The 610 items in the question bank each included anywhere from 1 to 13 unique vocabulary words (the average number of content terms included was 4.6).

3.3 Linguistic Features

Linguistic features of the question text pertained to either the length or the readability of the text. The length was quantified through a variety of different measurements, including the word, character, sentence, and syllable count. Readability—which is a measure of the familiarity, complexity, and interpretability of a question text [21]—was measured through five different metrics: Flesch Reading Ease, Flesch-Kincaid Grade Level [12], Automated Readability Index [13], Simple Measure of Gobbledygook [14], and Coleman-Liau [15]. The score of each metric on each item was derived using the *sylcount* package in R [22]. The maximum alphanumeric character count of an item was 283, with a mean of 115 and median of 104. The maximum word count of an item was 127, with a mean of 41 and median of 46. The 610 items had anywhere between 1 and 29 sentences, with an average of 7.9. Finally, the number of words with more than 3 syllables in an item ranged anywhere from 1 to 32, with mean of 8.6 and median of 8. See Table 2 for descriptive statistics of the different readability tests.

Table 2 Descriptive statistics for the readability indices

Readability index	Minimum	1st quantile	Median	3rd quantile	Maximum
ari	1.000	9.000	12.000	14.000	22.000
gl	−1.535	4.000	5.220	6.681	16.462
smog	3.129	7.492	8.842	9.643	14.192
cl	−19.575	4.210	6.270	8.133	18.129

3.4 Question Design

Finally, a few features pertaining to the design of the question were examined. These additional features included item type (blank field or multiple choice), the number of answer options, and the number of images that appeared in the question text. Among the 610 items, there were 154 blank field questions and the remaining 465 were multiple choice. For multiple choice items, there were anywhere from 2 to 8 answer options provided. Each item included at minimum 0 and at most 3 images.

4 Correlation Analysis

Before performing regression, a bivariate correlation analysis was completed among the 37 predictor variables. The Pearson correlation coefficients were calculated between all pairs of predictor variables, as visualized in Fig. 2. As to be expected, the total count of content terms and unique count of content terms were highly correlated. Typically, the frequency of a unique content vocabulary word is not impactful on the overall cognitive complexity of a question. For example, while the inclusion of the term *confidence interval* is a valid indicator of the cognitive complexity of the item, the effect is the same regardless of how many times the term is included. Therefore, the total count of content terms was removed for the remainder of this investigation, and only the unique count of content terms was kept. In addition, there were high levels of multicollinearity among the variables pertaining to readability and question length. Hence, the character count (denoted as *chars*), syllable count (denoted as *sylls*), and the flesch reading ease score (denoted as *re*) were all removed. In total, four features were removed from further analyses.

5 Traditional Regression Analysis

Four regression-based approaches were compared to identify the features that best predict the median response time taken to answer a test question. Stepwise regression with forward and backward selection, Lasso (Least Absolute Shrinkage and Selection Operator) regression, and a regression tree analysis were performed. The training ($N = 427$) and testing ($N = 183$) datasets were randomly split for the purpose of cross-validation. We then evaluated the four models according to their Mean Squared Error (MSE) and coefficient of determination (R^2).

Stepwise regression was chosen due to its ease of interpretation with a large number of features. The approach works by removing or adding a single feature on one iteration, and testing for the statistical significance after each iteration [23]. A feature is deemed statistically significant by calculating the t-statistic from the coefficient in the current model [24]. Then, the t statistic is squared to give either

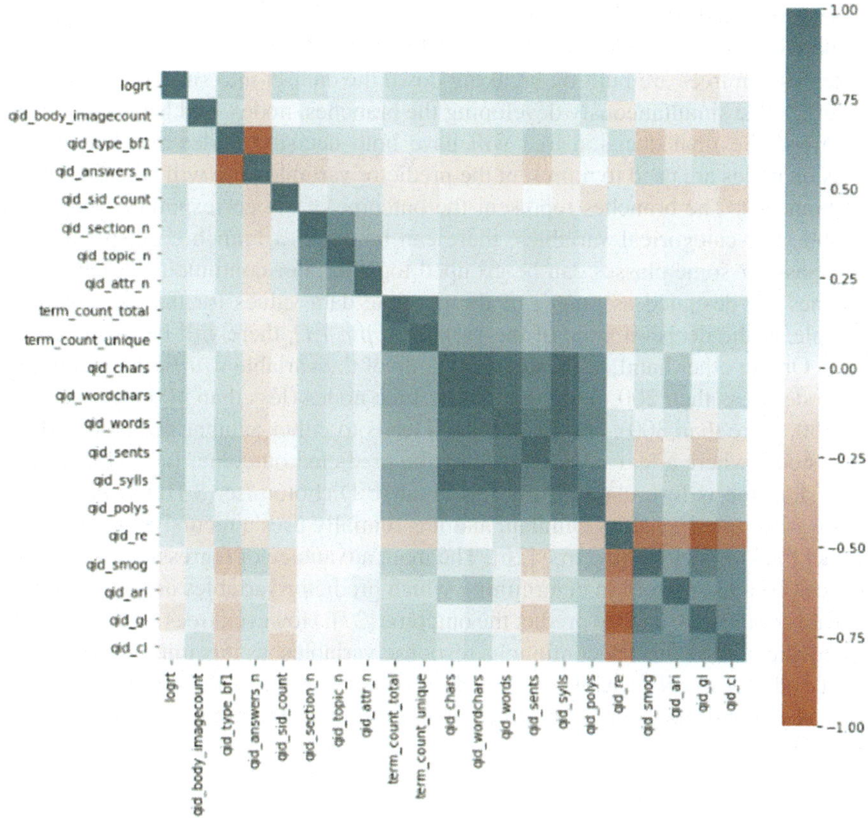

Fig. 2 Correlation heatmap among the 37 predictor variables

its "F-to-add" statistic or its "F-to-remove" statistic. The forward selection method starts with an empty model and adds features one-by-one according to the highest "F-to-add" statistic of the features not included in the current model. Backward selection starts with the full model and removes features one by one according to the lowest "F-to-remove" statistic of the features included in the current model. This procedure stops when all the remaining "F-to-add" and "F-to-remove" statistics are below a threshold of significance.

6 Machine Learning Methods

Lasso regression is a widely used penalized regression approach for feature selection. In ordinary least squares, a large number of features may result in overfitting. To solve this problem, lasso includes a penalty term that requires the sum of absolute values of coefficients to be less than a constant [25]. Therefore, some of the coefficients can shrink to 0 and the complexity of the model is reduced.

Regression trees are built by breaking down the dataset into smaller and smaller subsets, while simultaneously developing the branches, nodes, and leaves of a decision tree. The final decision tree will have both decision nodes and leaf nodes. Decision nodes are used to represent the predictor variables and will contain at least two branches. The branches represent the outcome of the corresponding predictor variable. For categorical variables, there can be either a branch corresponding to each class, or some classes can be grouped together. For continuous variables, the branches are designed as ranges of the possible data values for this variable. For example, at the decision node of the variable *topic 1.D*, there will be two nodes: 1 and 0. On the other hand, at the decision node of the variable *words*, there could be two nodes (less than 200, more than 200), three nodes (less than 100, between 100 and 300, more than 300), etc. Each branch leads to either another decision node or a leaf node, where a leaf node represents the predicted outcome. In R, the optimal regression tree is formulated using the Iterative Dichotomiser 3 (ID.3) algorithm. ID.3 was invented by Ross Quinlan, and it essentially uses a recursive process with no backtracking to grow the tree [26]. The main advantage of regression trees is that they can be a useful tool in determining which predictor variables are important, as well as how they are used to predict the outcome [27]. However, trees tend to produce inadequate predictions for continuous response variables, as it is impossible to have a leaf node for every possible predicted outcome. Therefore, the regression tree in this analysis will be used as an indicator in examining the predictive power of the feature variables.

7 Results

The stepwise regression model with backward selection was considered the best model given that it had the lowest testing MSE and only a slightly lower R^2 value than Lasso (see Table 3). It was also more interpretable than Lasso with only 8 variables kept as predictors (Table 4), whereas Lasso kept 25 predictors (Table 6). In contrast, forward selection and the regression tree seem to overfit the training dataset (see Table 3). Balancing accuracy and interpretability, the stepwise regression model with backward selection was the best. Roughly 25% of the variation of median response time can be explained by the item features.

Given the backward selection model appeared best fitting and most intepretable, we focus specifically on the results of this model. Based on the results of the backward selection model, eight-item feature variables were kept as significant predictors of (log-transformed) median response time, which included five features reflecting cognitive complexity (*sid_count* and four topic variables), one linguistic feature (*word_chars*), and two features pertaining to item design: *type_bf1* (question type: fill-in-the-blank or multiple choice), and *qid_answers_n* (number of answer options).

Table 3 Model fit for four analytic techniques

	Training MSE	Testing MSE	R^2
Forward selection linear model	0.5365	0.6736	0.2446
Backward selection linear model	0.5325	0.5256	0.2502
Lasso regression	0.5370	0.5475	0.2531
Regression tree	0.4687	0.5383	0.1836

Table 4 Features and coefficients for the backward selection linear model

Feature	Coefficient	Standard error
(Intercept)	2.703	0.226
type_bf1	0.655	0.190
sid_count	0.115	0.030
answers_n	0.139	0.047
word_chars	0.003	0.0005
X1.D.1	−0.303	0.108
X1.E.1	−0.538	0.233
X2.B1	−0.516	0.153
X2.C1	−0.517	0.190

The four topic variables are: *1.D* ("Exploring Bivariate Data"), *1.E* ("Exploring Categorical Data"), *2.B* ("Planning and Conducting Surveys"), and *2.C* ("Planning and Conducting Experiments"). These findings imply that all three aspects of an item—cognitive complexity, linguistic features, and question design—are important for predicting response time. Notably, the features kept as significant had considerable overlap between the backward selection model (see Table 4) and the forward selection model (see Table 5).

The coefficients of the predictor variables *answers_n*, *sid count*, and *word_chars* were estimated as positive. This implies that students will take longer on items that

Table 5 Features and coefficients for the forward selection linear model

Feature	Coefficient	Standard error
(Intercept)	3.331	0.107
polys	−0.028	0.012
sid_count	0.105	0.030
word_chars	0.005	0.001
X1.D.1	−0.251	0.110
X1.E.1	−0.628	0.230
X2.B1	−0.564	0.153
X2.C1	−0.452	0.191
X3.C1	0.234	0.139

Table 6 Features and coefficients for the Lasso regression model

Feature	Coefficient
(Intercept)	3.046
type_bf1	0.449
image_count	−0.059
answers_n	0.102
sid_count	0.095
section_n	0.214
topic_n	−0.098
attr_n	−0.004
word_chars	0.003
polys	−0.006
gl	−0.046
ari	0.030
term_unique	0.017
X1.A	−0.030
X1.B	0.039
X1.D	−0.191
X1.E	−0.451
X2.B	−0.439
X2.C	−0.391
X2.E	−0.074
X3.A	0.143
X3.B	0.295
X3.C	0.231
X3.D	−0.064
X4.A	0.163
X4.B	−0.068

pose more answer options, require multiple steps to complete, and are lengthier. Furthermore, the coefficient for *question type* (multiple choice vs fill-in-a-blank) was estimated as positive, suggesting students take longer to respond to blank-field questions than multiple choice. All of these findings were to be expected, as they are consistent with the typical test-taking behavior of students. For topics pertaining to content covered earlier in the semester (e.g., $X.1$, $X.2$), which should be easier for students and take them less time to complete, there was a negative coefficient. This suggests that questions associated with these topics were less time-consuming to complete.

According to the variable importance scores obtained from the regression tree (see Fig. 3), the top five item features that are the most predictive of median response time are word count, alphanumeric character count, number of sentences, number of

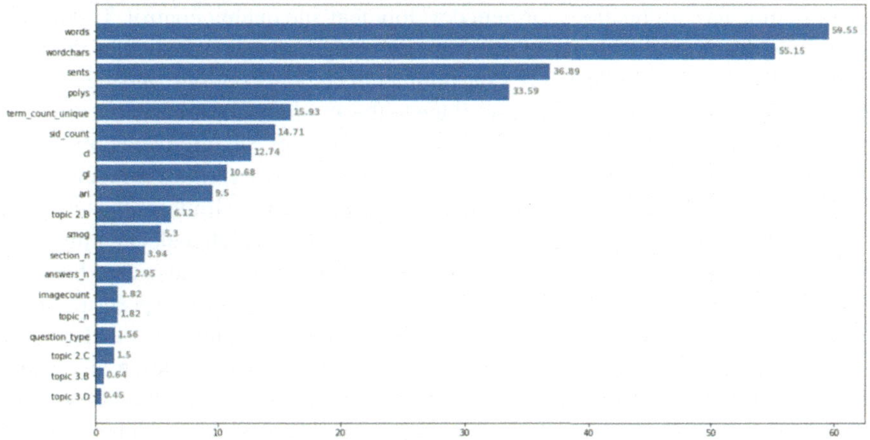

Fig. 3 Variable importance scores of the regression tree

words with more than 3 syllables, and the unique count of AP stats content terms. All five of these variables fell into the linguistic feature category, with all but the unique count of content terms pertaining to question length. Overall, the linguistic feature variables obtained the highest variable importance (VI) scores, as the Coleman-Liau index, grade level, automatic readability, and smog test scores all received VI scores above 5. Therefore, it is likely that the remaining three regression techniques did not select all of these variables due to high levels of multicollinearity among them.

8 Conclusions and Discussion

In this chapter, we wanted to establish the relationship between item characteristics and the time required to complete an item. Stepwise forward selection, stepwise backward selection, Lasso regression, and a regression tree were each fit to a training and testing set of AP Statistics test questions. The stepwise regression model using backward selection was found to be the best model given its balance of model performance and interpretability. The model selected eight item features, which spanned across all three types of item characteristics: cognitive complexity, linguistic features, and question design. Together these features could account for a substantial amount (~25%) of the variation in response latency of items. According to Cohen [28], this is considered a large effect size in the social sciences. On the other hand, there is room for improvement and possibly large language models (LLMs) could be applied to help.

There are several important implications of these findings. First, understanding which features of a written test question are likely to contribute to variation in response time has implications for item development and quality control. Test developers and educators alike may benefit from knowing how features such as the item type (blank field vs. multiple choice), the number of response options, and the number of steps in the answer solutions all contribute to the amount of time needed to process the question and attempt to solve it. Second, these findings can improve applications of response times in measurement, for example, using response times for identifying aberrant response behaviors [29] or setting an appropriate time limit for the test [30]. Third, the present investigation focuses on response time, though understanding how the item features also predict the proportion of respondents who answered the item correctly could also be highly informative, especially considering speed-accuracy trade-offs on cognitively demanding tasks [31]. Finally, for any tests that are not strictly a power test, the time-consumingness of items has implications for exam fairness. An item that requires a high reading level and hence becomes very time-consuming on a math test may cause fairness issues because respondents differ in their language backgrounds.

Acknowledgements This work was supported by the National Science Foundation CAREER award (Grant #DRL-1350787) and Institute of Education Sciences (Grant #R305A180269) to Dr. Ying Cheng. The funding supported the development of the AP-CAT (Advanced Placement-Computerized Adaptive Testing) and iDAP (Intelligent Diagnostic Assessment Platform) projects. It was through these computerized learning platforms in which the item bank was obtained, and the response time data were collected.

References

1. Wise SL (2017) Rapid-guessing behavior: its identification, interpretation, and implications. Educ Meas Issues Pract 36(4):52–61
2. Scheuneman J, Gerritz K, Embretson S (1991) Effects of prose complexity on achievement test item difficulty. ETS Res Rep Ser 1991(2):i–53
3. Baldwin P, Yaneva V, Mee J, Clauser BE, Ha LA (2021) Using natural language processing to predict item response times and improve test construction. J Educ Meas 58(1):4–30. https://doi.org/10.1111/jedm.12264
4. Smith RW (2000) An exploratory analysis of item parameters and characteristics that influence item level response time. [Doctoral dissertation, University of Nebraska-Lincoln]. ProQuest Dissertations Publishing
5. Swanson DB, Case SM, Ripkey DR, Clauser BE, Holtman MC (2001) Relationships among Item characteristics, examine characteristics, and response times on USMLE Step 1. Acad Med 76(10):S114–S116
6. Mikolov T, Chen K, Corrado G, Dean J (2013) efficient estimation of word representations in vector space. International conference on learning representations
7. College Board (2019) Student participation and performance in Advanced Placement Rise in tandem. https://www.collegeboard.org/releases/2018/student-participation-and-performance-in-ap-rise-in-tandem
8. Rosinger KO, Sarita Ford K, Choi J (2021) The role of selective college admissions criteria in interrupting or reproducing racial and economic inequities. J High Educ 92(1):31–55. https://doi.org/10.1080/00221546.2020.1795504

9. Suzuki H, Hong M, Ober T, Cheng Y (2022) Prediction of differential performance between advanced placement exam scores and class grades using machine learning. Front Educ. https://doi.org/10.3389/feduc.2022.1007779
10. College Board (2022) (rep.) Program summary report. https://apcentral.collegeboard.org/media/pdf/program-summary-report-2022.pdf.
11. Aisami RS (2015) Learning styles and visual literacy for learning and performance. Procedia Soc Behav Sci. 538–545. https://core.ac.uk/download/pdf/82167009.pdf
12. Kincaid JP, Fishburne RP Jr, Rogers RL, Chissom BS (1975) Derivation of new readability formulas (automated readability index, fog count and flesch reading ease formula) for navy enlisted personnel. Technical report, Naval Technical Training Command Millington TN Research Branch
13. Senter R, Smith EA (1967) Automated readability index. Technical report, Cincinnati Univ OH
14. McLaughlin GH (1969) Smog grading-a new readability formula. J Read 12(8):639–646
15. Coleman M, Liau TL (1975) A computer readability formula designed for machine scoring. J Appl Psychol 60(2):283
16. Brodersen A, Carter MF, Liu C, Cheng Y (2021) Collection of process data in web-based assessment systems and its applications to validating non-cognitive constructs. In: Jiao H, Lissitz R (eds) Enhancing effective instruction and learning using assessment data. Information Age Publisher, Charlotte, NC
17. Ober TM, Hong M, Rebouças D, Carter M, Liu C, Cheng Y (2021) Linking self-report and process data to performance across different assessment types. Comput Educ. https://doi.org/10.1016/j.compedu.2021.104188
18. Whitney BM, Cheng Y, Brodersen AS, Hong MR (2019) The Survey of student engagement in statistics: initial development and validation. J Psychoeduc Assess 37:553–565. https://doi.org/10.1177/0734282918769983
19. Starnes DS, Yates D, Moore DS (2010) The practice of statistics. Macmillan
20. Ooms J (2022) hunspell: High-Performance Stemmer, Tokenizer, and Spell Checker. https://docs.ropensci.org/hunspell/
21. DuBay WH (2004) The principles of readability. Impact Information, Costa Mesa
22. Schmidt D (2017) Sylcount: syllable counting and readability measurements. R. R package version. pp 2–5. https://cran.r-project.org/web/packages/sylcount/index.html
23. Zhang Z (2016) Variable selection with stepwise and best subset approaches. Ann Transl Med 4(7):136
24. Taylor J, Tibshirani R (2015) Statistical learning and selective inference. Proc Natl Acad Sci. https://doi.org/10.1073/pnas.1507583112
25. Tibshirani R (1996) Regression shrinkage and selection via the lasso. J Roy Stat Soc 58(1):267–288
26. Quinlan JR (1986) Induction of decision trees. Mach Learn 1:81–106. https://doi.org/10.1007/BF00116251
27. Loh WY (2011) Classification and Regression Trees. Wiley Interdiscip Rev: Data Min Knowl Discov 1:14–23. https://doi.org/10.1002/widm.8
28. Cohen J (1992) A power primer. Psychol Bull 112:155–159. https://doi.org/10.1037/0033-2909.112.1.155
29. Shao C, Cheng Y (2022) Application of change point analysis of response time data to detect test speededness. Educ Psychol Measur 82(5):1031–1062
30. Van der Linden WJ (2011) Setting time limits on tests. Appl Psychol Meas 35(3):183–199
31. Liesefeld HR, Janczyk M (2019) Combining speed and accuracy to control for speed-accuracy trade-offs. Behav Res Methods 51(1):40–60
32. AP statistics. AP Statistics—AP Students|College Board. (n.d.)
33. College Board (2010) AP statistics course description
34. Digest of Education Statistics (2021) National Center for Education Statistics (NCES) Home Page, a part of the U.S. Department of Education. (n.d.)

35. Office of Occupational Statistics and Employment Projections, Mathematicians and statisticians : Occupational Outlook Handbook (2023). U.S. Bureau of Labor Statistics, Washington DC

Predicting Student Attrition in University Courses

László Bognár

Abstract Educational institutions are actively engaged in extensive initiatives to mitigate student dropout rates. In addition to various strategies, the integration of machine learning (ML) models stands out as a prominent tool. These models play a crucial role in pinpointing students who are at risk of discontinuing their studies and offer tailored interventions to assist them in catching up with their academic coursework. Developing effective educational machine learning models entails consideration of specific features. Typically, the aim is to leverage a large sample size and incorporate an extensive set of predictors for optimal model construction. While the criterion for a substantial sample size is often met in higher education or degree-level analyses, identifying, and incorporating effective predictors pose challenges. In the realm of education, the prospect of building a reliable model is enhanced when predictors are tailored to subjects or even specific curricula. This customization is justified by the inherent differences in content, student enrollment, and requirements across various university courses. However, a critical challenge emerges as the use of course-specific predictors proliferates—it leads to a subdivision of existing observations into numerous parts. Consequently, this segmentation diminishes the size of the sample available for model training, posing a potential limitation to the robustness of the machine learning model. Predicting within a semester also poses special problems, since the observations that follow each other in time are not independent of each other. This chapter is about building ML models of university courses to better handle the prediction of student attrition.

Keywords Student dropout predictors · Causes of student dropout · Educational machine learning models · Performance metrics for dropout prediction · Factors influencing student dropout

L. Bognár (✉)
University of Dunaújváros, Dunaújváros, Hungary
e-mail: bognarl@uniduna.hu; drbognar@gmail.com

M. S. Khine (ed.), *Machine Learning in Educational Sciences*,
https://doi.org/10.1007/978-981-99-9379-6_8

1 Addressing Student Drop-Out in Universities: Understanding the Underlying Causes and Implementing Multifaceted Solutions

Student drop-out is a significant problem for universities because it leads to decreased retention rates, decreased revenue, and a negative impact on the university's reputation. In addition, dropouts often leave with significant debt, which can affect their financial well-being and increase the risk of defaulting on loans. The high dropout rate is a multifaceted issue with many underlying causes that universities must address to improve student retention.

One of the primary reasons for student drop-out is financial difficulty. A report by the National Student Clearinghouse Research Center found that students who receive Pell Grants, which are awarded based on financial need, have a higher drop-out rate than students who do not receive these grants [29]. Financial difficulty can also be a barrier to students' ability to complete their coursework, as they may need to work long hours to support themselves or their families, leaving less time for studying.

Another significant factor in student drop-out is academic preparedness. Students who are not academically prepared for college-level work may struggle in their coursework and become discouraged [24]. This source provides statistics on high school and college dropout rates in the United States and notes that academic preparedness is a key factor in college dropout rates. It underscores the significance of offering assistance to students facing academic challenges, particularly those at risk of discontinuing their studies.

The preservation of mental health and well-being emerges as pivotal in the context of student retention. Individuals grappling with mental health challenges like anxiety, depression, or substance abuse may encounter difficulties in effectively handling their academic obligations and other responsibilities. According to research conducted by the American College Health Association, a substantial 60% of college students acknowledged experiencing overwhelming anxiety, while 40% reported instances of depression within the preceding 12 months [4]. Such mental health concerns can exert a profound influence on students' capacity to successfully navigate their coursework, potentially culminating in dropout scenarios.

The COVID-19 pandemic has also had a significant impact on student retention. Many universities have had to switch to online learning, which can be challenging for students who do not have access to reliable technology or a suitable study environment. In addition, students may be struggling with the emotional and financial impacts of the pandemic, making it difficult to focus on their studies [18].

To address the issue of student drop-out, universities must implement a multifaceted approach that addresses the underlying causes. This may include providing financial assistance and support to students who are struggling financially, offering academic support and resources to students who are not academically prepared, and providing mental health services and support to students who are struggling with mental health issues. In addition, universities must adapt to the changing circumstances brought on by the COVID-19 pandemic and find innovative ways to support students through these challenging times.

2 How Can Machine Learning Help in this Dropout Problem for Both the Universities and the Students?

Machine learning can help address the problem of student drop-out in several ways, both for universities and for individual students. For universities, machine learning can be used to identify at-risk students early and provide targeted support to help them succeed. By analyzing large amounts of data such as academic performance, attendance, financial aid status, and other factors, machine learning algorithms can identify patterns that may indicate a student is struggling. Once at-risk students have been identified, universities can provide targeted interventions to help them stay on track. These interventions may include academic support such as tutoring or mentoring, financial assistance, or mental health services. By intervening early and providing support, universities can help prevent students from dropping out and improve overall retention rates.

For individual students, machine learning can provide personalized recommendations to help them succeed. For example, a machine learning model could analyze a student's academic performance and recommend courses or study strategies that are likely to be most effective for that student's learning style. Additionally, machine learning could provide personalized financial planning advice to help students better manage their finances and reduce the risk of dropping out due to financial difficulty.

3 Categories of Factors that Influence Dropout

Numerous studies have examined the factors influencing student dropout, categorizing them based on various dimensions and criteria. In a synthesis of 44 studies, Kehm et al. [17] identified nine dimensions: "(a) study conditions, (b) academic integration at university, (c) social integration at university, (d) personal efforts and motivations for studying, (e) information and admission requirements, (f) prior academic achievement in school, (g) personal characteristics of the student, (h) sociodemographic background of the student, and (i) external conditions". Similarly, Alyahyan and Düştegör [3] conducted a comprehensive analysis of 89 studies and classified factors into five categories: "(a) prior academic achievement, (b) student demographics, (c) e-learning activity, (d) psychological attributes, and (e) environments". Furthermore, Arias & Dehon (2008) identified three categories of factors: "(a) personal characteristics, (b) high-school path characteristics, and (c) socioeconomic factors".

The elements contributing to student dropout manifest a multifaceted nature, originating from intricate and interconnected causes. Scholars such as Arias and Dehon [5], Behr et al. [7], Burnett [10], Kehm et al. [17], and Rastrollo-Guerrero et al. [27] have underscored the diverse array of factors playing a role in this phenomenon.

4 Models with Time-Invariant and Time-Variant Predictors

In the construction of machine learning models for prediction, predictors, also referred to as indicators, are established by considering the aforementioned factors. These predictors represent quantified characteristics associated with the factors and are commonly normalized to a standardized range of -1 to 1, as outlined in studies such as [1, 11, 13].

Machine learning models can be used to predict student dropout based on various predictors. Two types of machine learning models are commonly used to predict student dropout: models with time-invariant predictors and models with time-variant predictors.

Models with time-invariant predictors:

Time-invariant predictors, which remain constant throughout the learning process and are generally known at the outset, play a crucial role in estimating the likelihood of failure before learning commences. However, determining their values is often a nuanced and intricate task. While certain predictors like gender, age, race, nationality, class type, type of program, number of years repeated in high school, type of school, and academic measures such as prior academic achievement GPA and CGPA are relatively straightforward to define, others present greater complexity.

Factors like student interest, motivation, self-confidence, beliefs, and attitudes fall into this category of more challenging-to-define predictors. Additionally, collecting data for certain predictors may pose difficulties, whether due to sensitivity, complexity, unattainability, or legal constraints such as those imposed by GDPR. Examples of such intricate predictors encompass socioeconomic elements (household structure, parents' educational level, occupational activity, income) or life circumstances (health, employment, family responsibilities). The intricate nature of these predictors underscores the complexity involved in comprehensive data collection for predictive modeling in education.

Models with time-variant predictors:

Factors associated with student performance emerge as critical predictors of attrition, as highlighted in various studies [6, 14, 15, 19, 21, 22, 27, 33]. Predictors that dynamically change based on student performance during the study are termed time-varying or time-variant predictors.

Student performance can be gauged in several ways. While straightforward indicators like the percentage of successfully completed courses are employed, more detailed predictors, such as semester GPA, calculated from grades and credit hours, are often utilized. Insufficient credits to complete a semester can be indicative of a potential risk of dropping out.

Indicators reflecting student performance can dynamically change throughout the learning process based on the student's activities. These values, often derived from data extracted from Learning Management System (LMS) log files, are associated with specific learning resources or activities [34]. Learning resources encompass files, folders, pages, glossaries, and URL links, while learning activities involve tasks such as task sharing, self-assessment, communication, orientation, assignments, tests, forums, chats, etc. The combined term for these elements is learning components. Predictors based on components include metrics like the number of materials viewed, frequency of using individual curriculum elements, self-assessment quiz activities (including the number of attempts and the maximum grade achieved on attempts), and student engagement in social networks, among others [34].

In addition to the time-invariant and time-variant models, the use of hybrid models, which use both type of predictors, is also common to improve prediction accuracy. By incorporating multiple sources of data, hybrid models can provide a more comprehensive understanding of complex phenomena like student attrition. Furthermore, hybrid models can help to address some of the limitations of time-invariant and time-variant models, such as the lack of sensitivity to changes in student behavior over time.

Overall, the use of hybrid models shows great promise for improving the accuracy of student attrition prediction and could have significant implications for educational institutions seeking to improve student retention rates.

5 Models at Different Levels

Degree-level model:

This model uses data at the level of the entire degree program to predict student dropout. Predictors might include demographic information such as gender, race/ethnicity, and age, as well as prior academic performance such as high school grades, SAT/ACT scores, and previous college coursework. Other potential predictors could include financial aid status, whether the student is a first-generation college student, and whether the student is enrolled full-time or part-time.

Year-level model:

This model uses data at the level of each academic year to predict student dropout. Predictors might include the number of credits a student is taking, the number of courses a student is enrolled in, and the student's GPA for that year. Other potential predictors could include whether the student changed their major, whether they studied abroad, and whether they participated in extracurricular activities.

Course-level model:

This model uses data at the level of each individual course to predict student dropout. Predictors might include the student's grades in the course, their attendance record, and whether they participated in class discussions or completed homework assignments. Other potential predictors could include the difficulty level of the course, the instructor's teaching style, and the format of the course (e.g., online vs. in-person).

Exam-level model:

This model uses data at the level of each individual exam to predict student dropout. Predictors may include the student's score in previous examinations, their study habits leading up to the exam (e.g., how much time they spent studying, whether they studied alone or with others), and their motivation for taking the exam (e.g., whether they were required to take it for a particular course or program). Other potential predictors could include the format of the exam (e.g., multiple choice vs. essay) and the level of difficulty of the exam.

It's worth noting that these models are not mutually exclusive and could be used in combination to provide a more comprehensive picture of factors that contribute to student dropout. For example, a model that incorporates predictors at multiple levels (e.g., demographic information, academic performance, course grades) may be more effective in predicting student dropout than a model that relies on predictors at only one level. Additionally, these models may be refined over time as more data becomes available and as the factors that contribute to student dropout change.

6 Specialties of Educational Machine Learning Models

Machine learning models for student dropout prediction are specialized models that are designed to predict the likelihood of a student dropping out of school based on various factors such as academic performance, social and economic background, and personal characteristics. Compared to models in other fields, such as finance or healthcare, machine learning models for student dropout prediction have several unique specialties.

Feature selection: Feature selection is the procedure of identifying the most pertinent features for predicting the output of a model. In the context of student dropout prediction, feature selection is crucial because it can help to identify the most important factors that contribute to dropout, such as academic performance and socioeconomic status. Various feature selection techniques, including correlation-based feature selection (CFS), principal component analysis (PCA), and numerous others, have been employed in machine learning models dedicated to predicting student dropout [12, 16].

Longitudinal data analysis: Longitudinal data analysis is a statistical method that is used to analyze data collected over time. In the context of student dropout prediction, longitudinal data analysis is important because it can help to identify trends and patterns in student behavior that may lead to dropout [28, 30].

Domain-specific knowledge: Domain-specific knowledge refers to knowledge about the particular domain being studied. In student dropout prediction, domain-specific knowledge might include knowledge of the education system, the factors that contribute to student dropout, and the interventions that can be used to prevent dropout. Machine learning models for student dropout prediction may incorporate domain-specific knowledge through the use of expert knowledge or by incorporating data from multiple sources, such as student records and surveys.

Model transparency, interpretability: Model transparency refers to the ability to understand how a model arrived at its predictions. Model transparency is important because it can help educators and policymakers to understand why a particular student is at risk of dropping out and what interventions might be effective in preventing dropout. Machine learning models for student dropout prediction may incorporate model transparency through the use of interpretable models such as decision trees and random forests. Interpretability techniques such as feature importance analysis and partial dependence plots have been used in machine learning models for student dropout prediction [23].

Imbalanced data handling: Imbalanced data handling involves techniques designed for datasets where the number of instances in one class significantly differs from that in another class. This is particularly important in scenarios where the number of students who drop out is typically much smaller than the number of students who do not drop out. In the context of machine learning models for student dropout prediction, various imbalanced data handling techniques, such as oversampling and undersampling, have been utilized [20].

Model evaluation: Model evaluation refers to techniques for evaluating the performance of a machine learning model. It can help to determine the accuracy and reliability of the model's predictions. The evaluation of machine learning models for student dropout prediction must take into account the specificity and sensitivity of the model, as well as its overall predictive power. This requires the use of metrics such as precision, recall, fallout, and F1 score in addition to traditional evaluation metrics like accuracy and AUC-ROC [26].

Transfer learning: Transfer learning refers to the process of using knowledge gained from one domain, such as healthcare or finance, to improve performance in another domain, in the present case to improve the accuracy of student dropout prediction models. Transfer learning techniques such as deep learning and convolutional neural networks have been used in machine learning models for student dropout prediction. There are cases when transfer learning framework outperforms traditional machine learning models, such as logistic regression and decision trees, in terms of accuracy and F1 score [31].

Contextual factors: Contextual factors refer to the environmental and social factors that may influence student behavior and academic performance. Contextual factors might include the student's family background, socioeconomic status, and school environment [2].

Continuous monitoring and feedback: Continuous monitoring and feedback refer to the process of monitoring student behavior and academic performance over time and providing feedback to students, educators, and parents. Machine learning models for student dropout prediction may incorporate continuous monitoring and feedback through the use of predictive analytics and automated notifications [32].

7 The Predictive Power of Models

The predictive power of machine learning models is contingent upon various factors. The effectiveness of these models is evaluated based on their ability to "accurately" forecast the target, the anticipated value they are designed to predict. The target can encompass factors such as exam success or failure (Y = 0 or Y = 1) or the grade/score attained in the exam (Y = A, B, C, D, F, or Y = Score). Detailed descriptions of the measures used to assess predictive power are provided in subsequent chapters.

In a general sense, "accuracy" refers to the proportion of students within the model whose predicted outcomes align with the actual results. Computing this ratio involves more than a simple count of students. Given that data can be collected or sampled from students across multiple courses, and even at various intervals during a single course (as explained in the "Model Building" chapter), it is preferable to calculate the proportion of samples in which the predicted (Y Pred) and actual (Y Act) target values coincide (see Fig. 1).

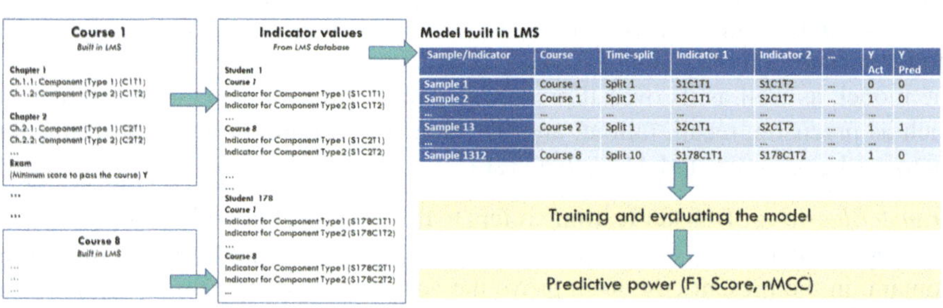

Fig. 1 The model's predictive ability is assessed based on the extent of agreement between the predicted and actual Y target values for the samples

8 Model Building Actors, Competences, and Exploitation of Results

Successful model development necessitates collaborative efforts from professionals specializing in education, digital curriculum design, applied statistics, and computer science. Drawing upon the necessary expertise, the key contributors to machine learning (ML) model construction encompass roles such as teacher, course designer, software engineer, LMS system administrator, and educational data analyst (Fig. 2).

Teacher

The main responsibility of the teacher in constructing the model involves establishing the curriculum, designing assignments and tasks, creating engaging student activities, selecting appropriate tools to convey the curriculum, and scheduling performance assessments. The teacher should possess expertise in e-learning instructional technology, a comprehensive understanding of the essential features and components of the applied Learning Management System (LMS), and familiarity with concepts like predictive models, predictive capability, and model usability. Leveraging insights derived from predictive models, the teacher actively intervenes in

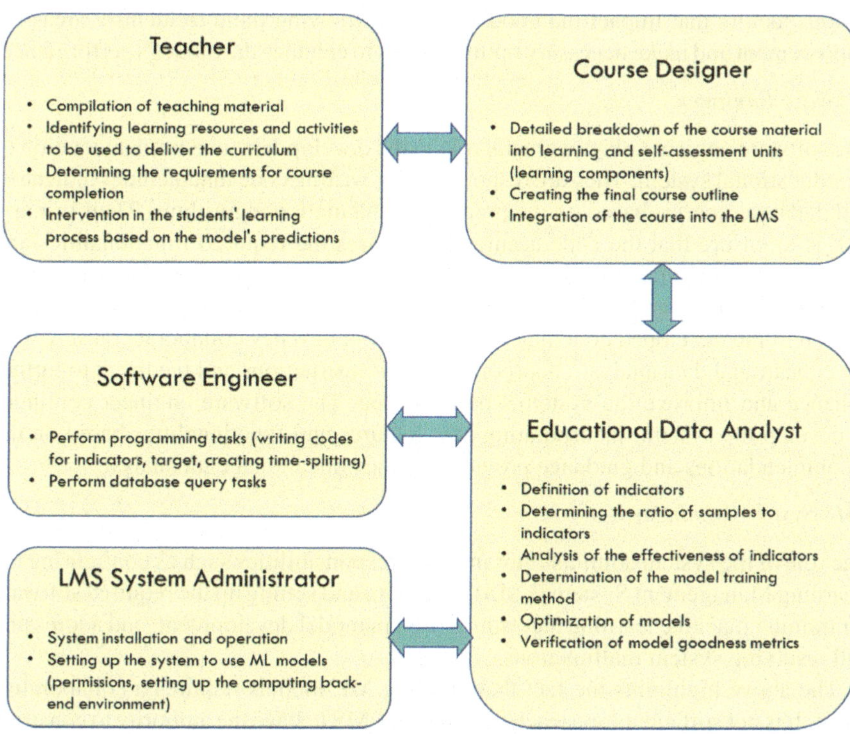

Fig. 2 The actors in ML model building and their tasks

the learning process, providing additional teaching support to students who may be lagging behind.

Course Designer

The course designer possesses a comprehensive understanding of the teacher's role and possesses expertise in developing digital learning materials. They work closely with the teacher and educational data analyst to ensure a collaborative approach.

The course designer identifies the diverse factors influencing the predictive capability of the models and adjusts the course outline accordingly to align with the intended predictive model. They are responsible for building the course within the LMS, leveraging its features and functionalities to deliver an effective and engaging learning experience.

Educational Data Analyst

An educational data analyst is a proficient expert who specializes in constructing, analyzing, and optimizing models within an educational context. They possess a deep understanding of the capabilities offered by the LMS and can effectively utilize its features for data analysis purposes. The data analyst is skilled in interpreting goodness-of-fit metrics of models, which provide insights into how well the model aligns with the available data. Additionally, they are knowledgeable about the various factors that impact the goodness-of-fit, allowing them to identify areas for improvement and make necessary optimizations to enhance the model's performance.

Software Engineer

The software engineer plays a crucial role in the development and implementation of the educational system. They are responsible for writing code that includes indicators and database queries based on the predefined calculation algorithms. Their primary task is to ensure that the code accurately captures the required functionalities and meets the system's specifications. The software engineer conducts thorough testing of the code to identify and rectify any potential issues or bugs, ensuring the smooth integration of the developed code into the overall system. They collaborate closely with the educational data analyst, incorporating their instructions and feedback to further enhance and improve the system's performance. The software engineer continues to develop the system, incorporating new features and functionalities based on the recommendations and guidance provided by the educational data analyst.

LMS System Administrator

The role of the system administrator involves responsibilities such as configuring the Learning Management System (LMS), installing and setting up the required software for training machine learning algorithms used in model development, and addressing and resolving system malfunctions.

The above highlights the fact that building ML models requires a collaborative effort. It is not sufficient for a teacher within an LMS to have the authority to construct such models. Building efficient models with high effectiveness relies on having the right expertise and specialized knowledge. Treating the system as a black box without proper understanding can lead to limitations and obstacles. Inaccurate models can result in false alarms, causing significant inconvenience.

To mitigate these challenges, it is crucial to involve professionals with the necessary skills, such as educational data analysts and software engineers. They bring expertise in building, analyzing, and optimizing ML models for educational systems. Through a collaborative approach and leveraging the skills of multiple individuals, the likelihood of constructing reliable and efficient ML models increases significantly.

9 Model Building

9.1 Embedded Machine Learning Models in an LMS

Incorporating a machine learning model within an LMS is a widely adopted practice due to the requirement of gathering extensive data about each student and their learning activities, particularly those closely tied to their performance. The system's underlying logic and process are depicted in Fig. 3, providing a visual representation of the model's functioning and its integration into the LMS.

The system records logs of students' activities, which serve as the primary source for tracking their progress. These activity logs enable the identification and calculation of indicators for various types of learning activities, customized for each student in different learning phases. These indicators hold significant importance in the construction of machine learning models. By utilizing these models, it becomes possible to identify students who are experiencing performance gaps and are at risk of not successfully completing the course.

Figure 4 provides an overview of the components involved in building machine learning models within an LMS, as well as the interconnections and relationships between these components.

The aim of constructing a model is to predict the value of the target variable (Y) for students who are currently participating in a course at an intermediate point of the ongoing session. It is beneficial to design the course in a way that allows

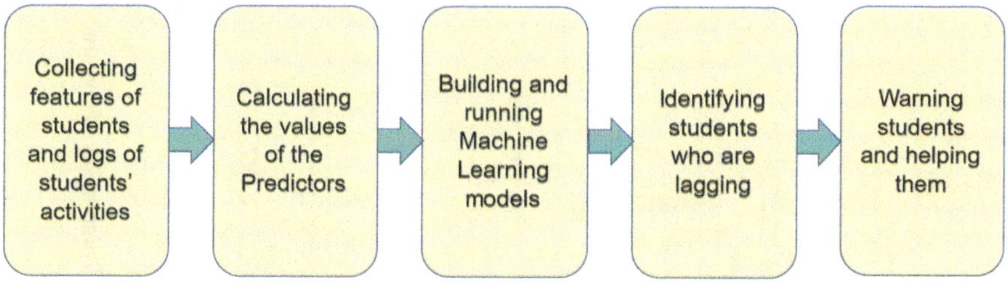

Fig. 3 Workflow for constructing a machine learning model within an LMS

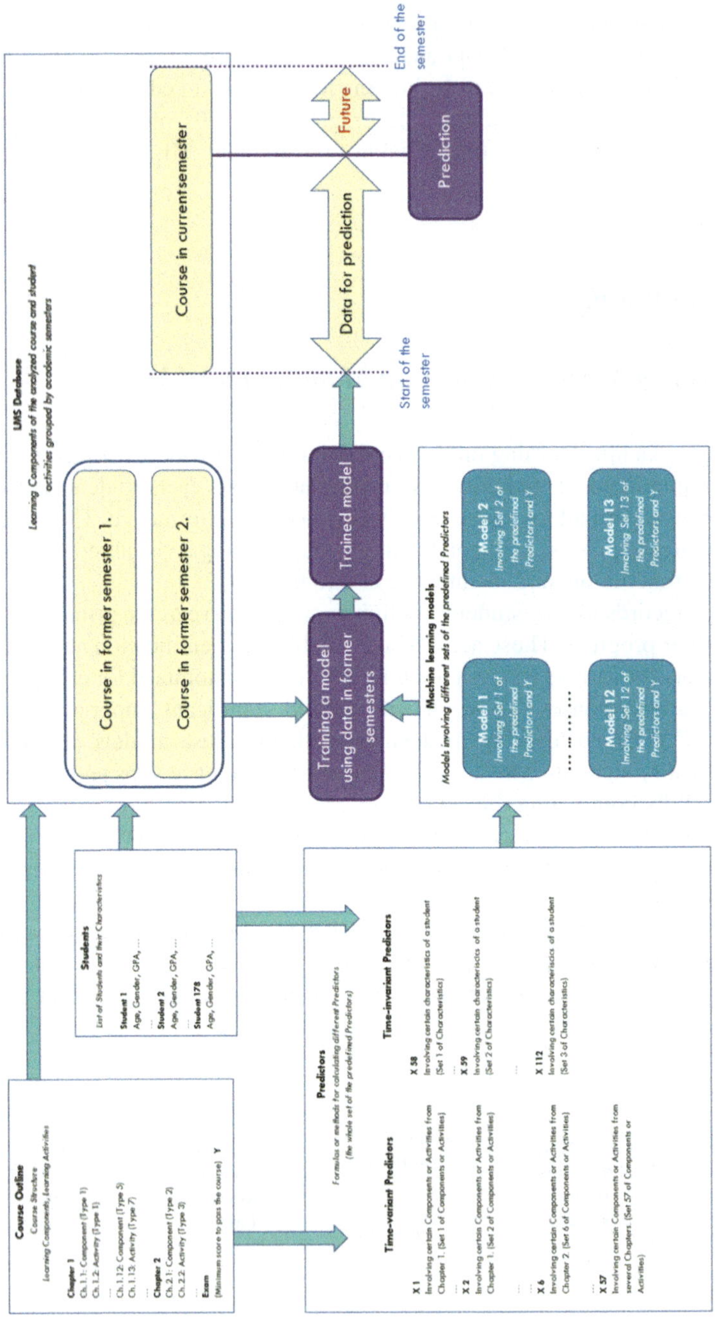

Fig. 4 Components of constructing a machine learning model in an LMS

assigning indicators to each learning component, which exhibit a strong correlation with the target variable. The course design process produces a course outline that is optimized for the machine learning (ML) model, enabling it to make the most accurate predictions.

To train a machine learning model, data from courses with identical outlines, conducted in preceding academic semesters, are employed. Leveraging the activities and achievements of students in these past courses, indicator values are computed, and a dataset of samples is constructed. Each sample includes indicator values for the course components along with the corresponding actual target value (Y) for that student.

A diverse set of indicators can be established according to a specific course outline. Within this pool of potential indicators, various models can be created depending on the inclusion of specific indicators. Following the training of multiple models, the model demonstrating the highest predictive accuracy is chosen. This selected trained model is subsequently utilized to forecast the success of students in the current course, employing samples from students currently enrolled in the ongoing course.

By leveraging this approach, the model-building process maximizes the accuracy of predictions, aiding in foreseeing the future performance of students in the current course.

9.2 Levels of Indicators

The value of a time-variant predictor (indicator) is influenced by the interactions within its corresponding learning component, which can occur at multiple levels. These levels refer to the scope of calculation for an indicator associated with a specific type of component. This computation can encompass all student activities across the entire Learning Management System (LMS) site, a specific course, or an individual chapter within a course. Consequently, indicators can be defined at the site level, course level, and chapter level (as depicted in Fig. 5).

A *site-level indicator* encapsulates a student's activity across all the courses they have undertaken, yielding a singular value for a specific component. For example, in the case of tests distributed across various courses with diverse content or multiple tests within a course and chapter, the site-level indicator for a test captures the overall student activity across all tests in all courses.

In contrast, a *course-level indicator* assigns the same value to a specific course, reflecting the student's activity solely within that particular course.

Moreover, a *chapter-level indicator* measures the activity within individual chapters of a course. Consequently, for a given component, there are as many indicators as there are chapters in the course. These chapter-level indicators offer insights into the student's engagement within each chapter.

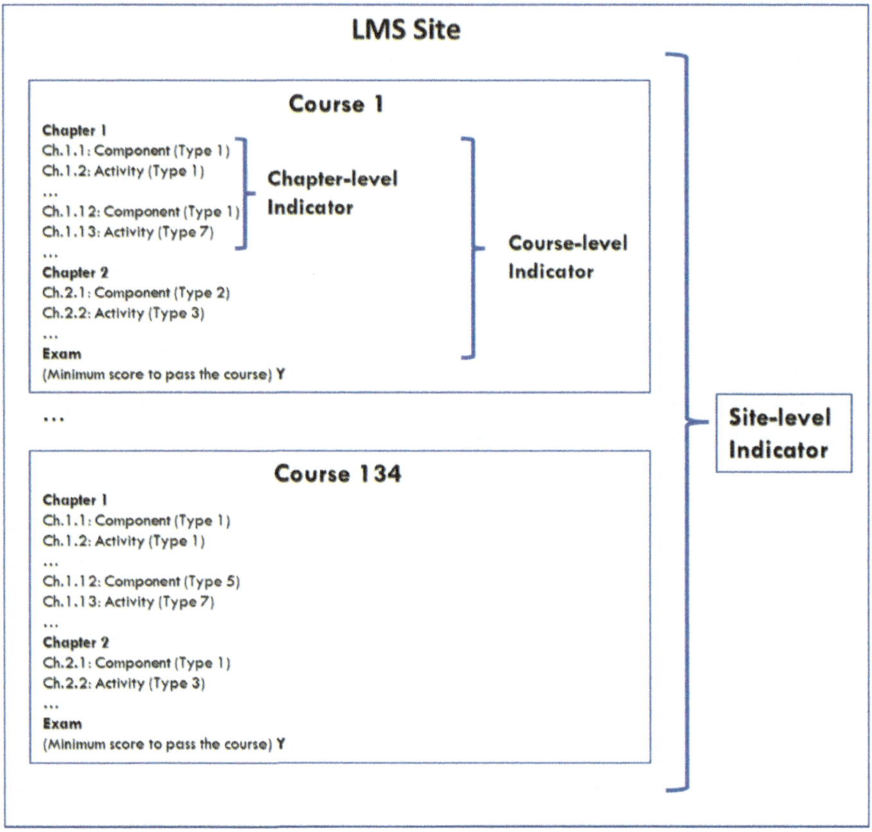

Fig. 5 Interpretation of indicator levels

9.3 Training-Prediction Schemes

A very important part of building a machine learning model is determining how the model is taught, namely, what time intervals does the algorithm use to teach the model and then perform predictions.

For the first training-prediction variant, we use the same duration of previous courses to collect training data as we use the same duration of observed data in the currently running course to predict. Neither the duration of the training data nor the duration of the currently running course is subdivided (Fig. 6).

Figure 7 illustrates the second potential approach to training and predicting. In the right-hand segment of the diagram, the objective is to predict the success of students beyond the 7th time division in the "Current semester". To make this prediction, we leverage the activities conducted during the time period indicated by the arrow labeled "Data for prediction" in the figure. In a 10-time division study, the period

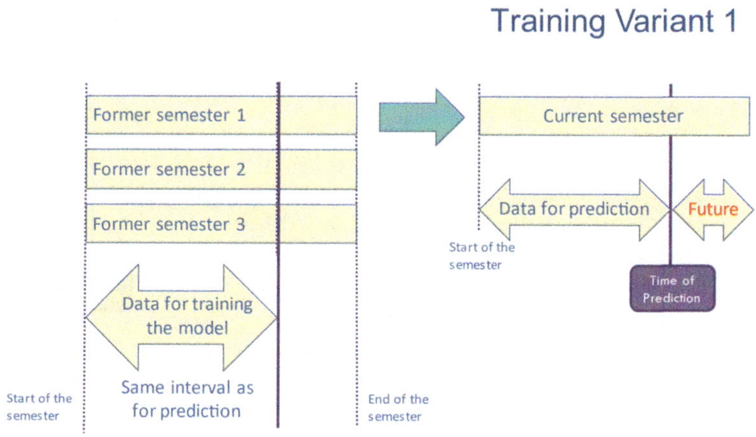

Fig. 6 Model is trained up using the former semester data without splitting the time interval

after the 7th time division signifies the future for the students and has not occurred yet.

For model training, we can use data from similar courses undertaken in previous semesters and years. In this model, data on student activities are collected separately for each subinterval of time. Notably, in this specialized model, student activities in subsequent time periods are not entirely independent of one another, and a chronological order exists between these activities. By considering the chronological relationship between student activities, we can ensure that the training process captures the temporal dynamics of their learning journey, leading to a more accurate and contextually relevant model.

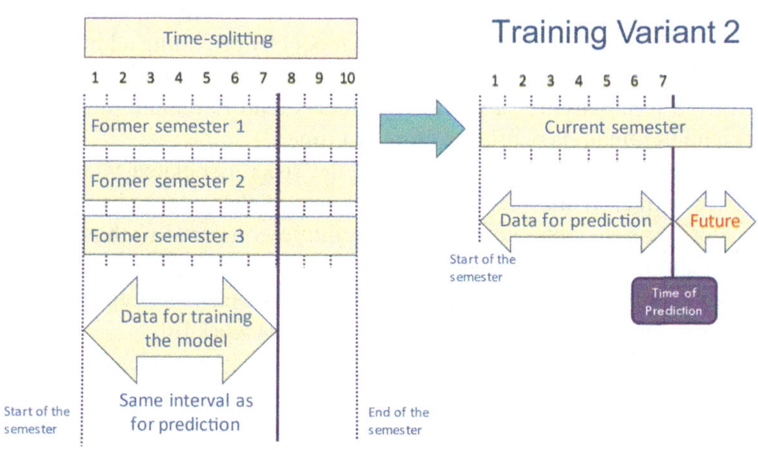

Fig. 7 Model is trained up to the "time of prediction" with subdividing the time intervals

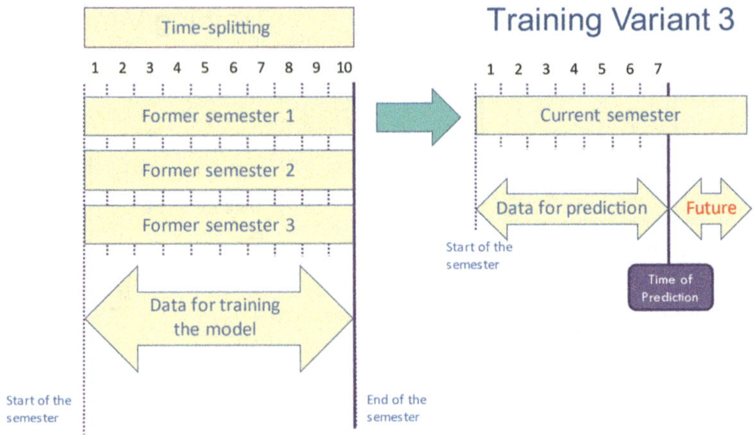

Fig. 8 Model is trained up to the end of semester with subdividing the time intervals

The third variant of the training-forecasting scheme follows a more traditional approach similar to non-educational machine learning models. In this scenario, all data encompassing every activity performed by all students in prior courses, spanning from the commencement to the conclusion of the course, are employed to train the model. Please refer to Fig. 8 for a visual representation of this scheme.

Although not entirely logically sound, the teaching method described above has demonstrated improved performance indicators and the ability to generate more reliable forecasts. This outcome can be attributed to a common characteristic of the techniques employed in training machine learning models. When faced with a scarcity of training samples, it is possible to artificially increase the sample count by creating slightly modified copies of existing samples. This technique is known as data augmentation. In practice, incorporating activities from a specific student that occurred after the forecast time enhances the model by providing additional training samples. These supplementary samples are unlikely to contradict the predictions obtained from past activities, as diligent students tend to remain diligent and lazy students tend to remain lazy after the forecast time.

However, it is essential to acknowledge the risks associated with this model-building technique. While there is a risk of bias if there are substantial changes in the majority of students' activities in the future compared to the forecast time, our investigations consistently demonstrated that this training-forecasting approach consistently produced superior results across various cases. A more comprehensive exploration of this topic can be found in our previous work [8].

9.4 *Predictor Matrices and Target Vectors in the Different Schemes*

The scenario described in Variant 1 represented the simplest model, featuring a uniform time interval. In this case, the predictor matrix X and target vector Y associated with each student's activities in the different learning components can be described by the simple scheme shown in Fig. 9.

If the model is later employed for predictions within a course timeline, as in Variants 2 and 3, the entire course duration needs to be segmented at specific time points. Figure 10 illustrates the structure of the X matrix and Y vector generated through this methodology.

When the semester is segmented into c equal parts, the number of samples, represented by the rows of the X matrix, becomes c times n for n students. Each student's indicator value is recorded in each time division or time split. The vector Y, containing the target values, is repeated for each student in each time division.

To incorporate the time division information, the indicator columns are expanded with time division indicators, signifying the specific time intervals of interest. In

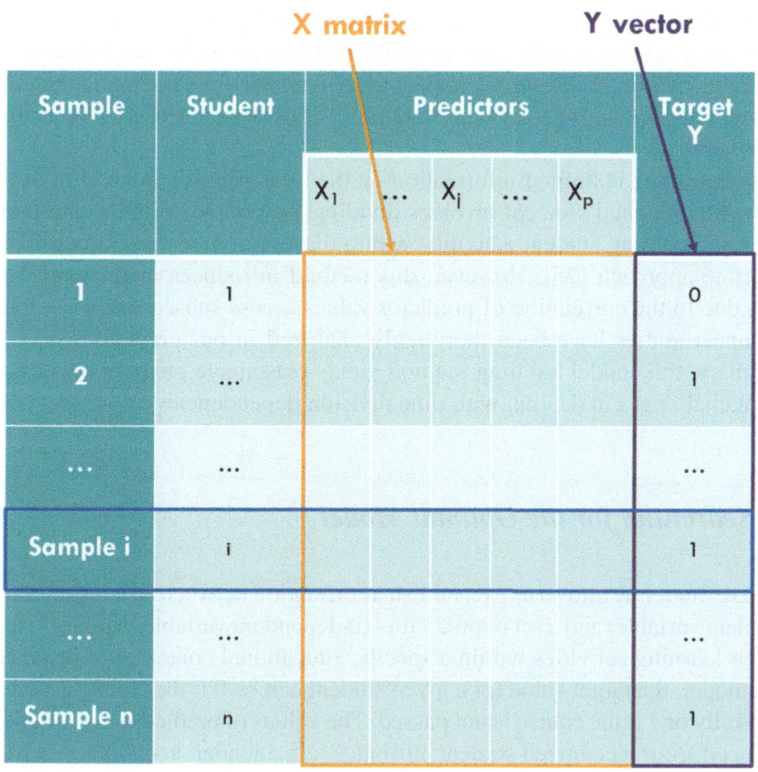

Fig. 9 The X predictor matrix and the Y target vector for the Variant 1 training-prediction scheme

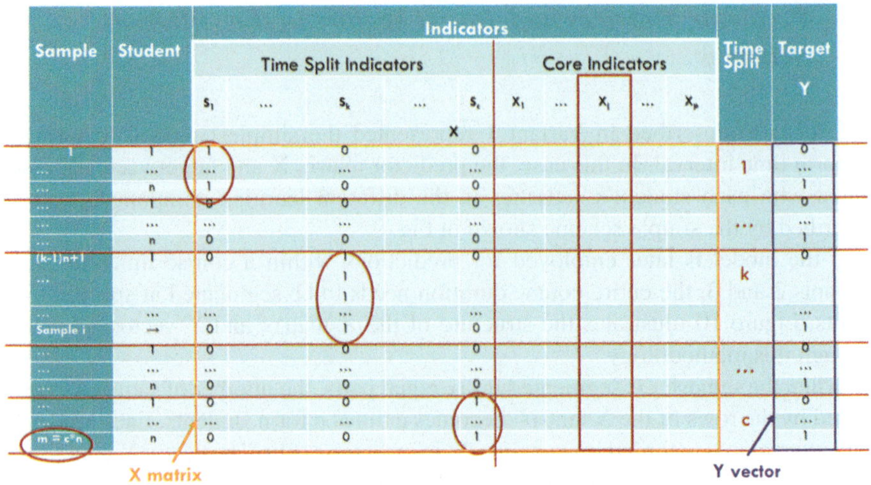

Fig. 10 The X predictor matrix and the Y target vector for the variant 2 and 3 training-prediction scheme

columns S1 to SC, a value of 1 is observed only for the rows corresponding to the calculation of activity indicators in the respective time divisions, while the other rows have a value of zero. There are various options available for calculating the indicator values within a given time division, including considering activities preceding the time division.

Although there is limited information on the time-splitting method in the literature, the fundamental concept involves dividing the course duration into parts and separately recording student activities within these time intervals, is considered an appropriate approach [25]. However, this method introduces mathematical uncertainties due to the correlation of predictor values across subsequent time divisions. These uncertainties have been thoroughly analyzed in our previous works [8, 9]. In summary, this model building method yields reasonable predictions, despite the inherent challenges in dealing with time division dependencies.

9.5 *Searching for the Optimal Model*

Predictive modeling allows us to establish a correlation between an educational target (dependent variable) and a set of predictors (independent variables) linked to learners and their learning activities within a specific educational context. For instance, in a binary model, the target value for a given student can be 0 if the course is completed successfully or 1 if the course is not passed. The values of predictors for a student can be deduced from recognized student attributes (e.g., gender, age, grades in previous courses) or extracted from activity logs stored in an LMS throughout the course.

If comprehensive records of students' learning activities exist, predictors derived from activity logs are frequently employed. In the realm of Supervised Learning (SL), algorithms strive to discover a modeling function denoted as f that captures the relationship between the predictors and the target. Following this, the established function can be employed for predictive purposes.

In mathematical terms, this relationship can be represented as a hypothesis in the general form:

$$Y = f(X_1, \ldots X_j, \ldots, X_p) + \varepsilon$$

Here, Y represents the target variable, f denotes an unknown function of the predictors X_1, ..., X_p (typically a fixed function with unknown parameters), and ε represents a random error term that is independent of the predictors and has a mean of zero.

The SL algorithms aim to obtain the optimal $\hat{f}$ estimate for the unknown function f by utilizing the available data comprising predictors and the target variable. The term "optimal" is defined based on minimizing the cost function Q, typically calculated as the sum of squared difference between the actual target values (Y) and the predicted values ($\hat{y}$) using all samples in the summation.

$$Q = Sum\,[(Y - \hat{Y})^2] = \text{minimum}$$

The cost function quantifies the overall prediction error, and the goal is to find the function that minimizes this error. By iteratively adjusting the parameters of the model, the SL algorithms optimize the function f to achieve the best possible predictions. The specific algorithm employed will determine the approach used to minimize the cost function and estimate the optimal parameters for f.

In practice, we use a certain subset of the existing data called *Training set* and get a certain $\hat{f}$. If we use a different *Training set*, we are very likely to get a different $\hat{f}$. As we keep changing *Training sets*, we get different outputs for $\hat{f}$. The amount by which it varies as we change the *Training sets* is called *Variance*. This model variance refers to the error that occurs due to the model's sensitivity to the noise or fluctuations in the training data. This can lead to overfitting, where the model fits too closely to the training data and fails to generalize to new, unseen data, resulting in poor performance.

To estimate the true function f using various methods such as linear or logistic regression, we often employ a simplified function between the predictors and the target. However, in most real-life scenarios, the true relationship is more intricate. The simplifying assumptions introduce *Bias* to a model. *Bias*, in this context, refers to the error stemming from oversimplifying the underlying patterns in the data. This oversimplification can lead to underfitting, where the model fails to capture the complexity of the relationships between the input features and the target variable, resulting in subpar performance. The more inaccurate the assumptions regarding the true relationship, the higher the *Bias*, and vice versa.

In general, a model will exhibit some error when tested on validation or test data. It can be demonstrated mathematically that both *Bias* and *Variance* only contribute to a model's error. We aim for low error, necessitating the minimization of both *Bias* and *Variance*. However, achieving this is not entirely feasible; there exists a trade-off between *Bias* and *Variance*.

In the context of predicting student dropout, a model with high bias might overlook important predictors of dropout, such as family background or previous academic performance, while a model with high variance might overfit to specific patterns in the training data, such as the demographics of the students or the distribution of grades, that do not generalize to new data. Achieving a balance between bias and variance is crucial for building an accurate and robust predictive model.

10 Supervised Learning Algorithms

Several supervised learning algorithms are commonly used for predicting student dropout. Here are some frequently used algorithms along with the circumstances where they are preferred:

Logistic Regression:

Logistic regression is a widely used algorithm for binary classification problems like student dropout prediction. It is preferred when the relationship between the input features and the target variable is assumed to be linear or when interpretability of the model is important.

Decision Trees:

Decision trees provide transparent and interpretable results. The generated tree structure allows for easy understanding of the decision-making process, making it useful for educational institutions and stakeholders to gain insights into the factors influencing student dropout. Decision trees can reveal the relative importance of different features or variables in predicting student dropout. By examining the splits and hierarchy of the tree, one can identify the key factors contributing to dropout risk.

They can capture nonlinear relationships between predictors and the target variable. In predicting student dropout, there might be complex interactions and nonlinear dependencies among various factors. Decision trees can effectively model such relationships without relying on assumptions of linearity. Decision trees can handle a mix of categorical and numerical data.

Random Forest:

Random Forest is an ensemble learning method that combines multiple decision trees to make predictions. It is preferred when there is a complex relationship between the input features and the target variable and when handling a large number of features. Random Forests can handle both numerical and categorical features and can provide feature importance information.

Support Vector Machines (SVM):

SVM is a powerful algorithm for both linear and nonlinear classification problems. It is preferred when there is a clear margin of separation between the classes and when dealing with high-dimensional data. SVM can also handle both numerical and categorical features effectively.

Gradient Boosting Methods (e.g., XGBoost, LightGBM):

Gradient boosting methods are popular for building highly accurate models. They create an ensemble of weak prediction models and gradually improve them. These methods are preferred when high predictive performance is desired and when handling large datasets with complex relationships. Gradient boosting methods can handle both numerical and categorical features and automatically handle missing data.

Neural Networks (e.g., Multilayer Perceptron, Recurrent Neural Networks):

Neural networks are versatile models that can capture complex patterns in the data. They are preferred when dealing with large-scale datasets and when there are nonlinear relationships between the input features and the target variable. Neural networks can handle both numerical and categorical features, as well as sequential data (e.g., temporal patterns in student dropout).

The choice of algorithm depends on various factors, including the nature of the dataset, the complexity of the relationships, the size of the dataset, and the desired interpretability or predictive performance. It is often recommended to experiment with multiple algorithms and compare their performance using appropriate evaluation metrics to determine the most suitable model for a given student dropout prediction task.

11 Performance Metrics and Their Use for Checking Model Bias and Variance

To quantify the models' performance, different widely accepted metrics computed from the confusion matrix are used (Fig. 11).

In the present context, the elements of the confusion matrix:

- *Number of true positives (TP): when for a sample the prediction is 1 (the student fails to pass the course) when the actual value of the target is 1 (failed),*
- *Number of false positives (FP): when for a sample the prediction is 1 (the student fails to pass the course) when the actual value of the target is 0 (passed),*
- *Number of true negatives (TN): when for a sample the prediction is 0 (the student passes the course) when the actual value of the target is 0 (passed),*
- *Number of false negatives (FN): when for a sample the prediction is 0 (the student passes the course) when the actual value of the target is 1 (failed).*

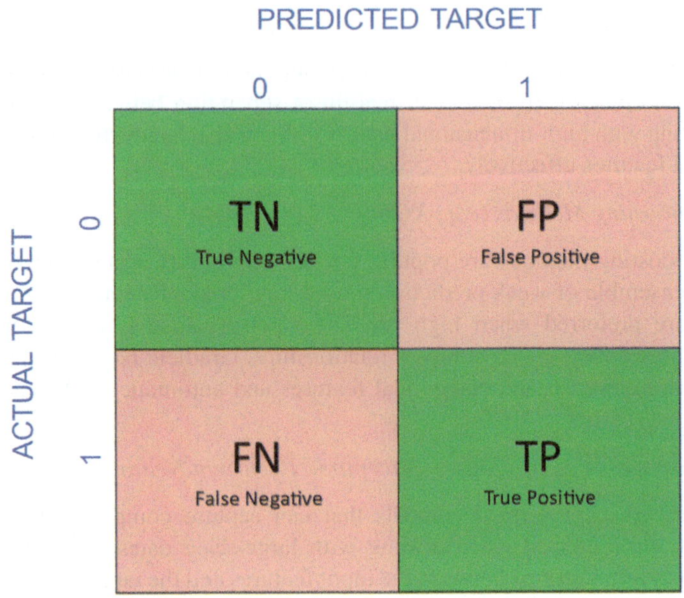

Fig. 11 The confusion matrix

The most used performance metrics:

- *Accuracy:* the ratio of correctly predicted samples (True Positives + True negatives) to the total number of samples.

$$Accuracy = (TP + TN)/n$$

- *Error:* subtracting the *Accuracy* value from 1.

$$Error = 1 - Accuracy$$

- *Recall:* measures the fraction of correctly classified failed students among all students who failed.

$$Recall = TP / (TP + FN)$$

- *Fallout:* measures the fraction of students incorrectly classified as failed among all students who passed the course.

$$Fallout = PF / (FP + TN)$$

- *Precision:* measures the fraction of correctly classified failed students among all students predicted as failed.

$$Precision = TP / (TP + FP)$$

- *F1 Score:* combines *Precision* and *Recall* into a single value between 0 and 1. It reflects the model's goodness of fit even in the case of unbalanced data.

$$F1\,Score = 2 * (Precision * Recall) / (Precision + Recall)$$

- *nMCC (Normalized Matthew Correlation Coefficient):* is another complex measure that ranges from 0 to 1. It is capable of assessing the model's goodness of fit, even with unbalanced data.

$$nMCC = (TP * TN - FP * FN)/sqrt((TP + FP) * (TP + FN) * (TN + FP) * (TN + FN))$$

These metrics can be used to assess both model bias and model variance in the following ways:

Model Bias:

- *Accuracy* and *Error*: These metrics can provide an indication of model bias. A high error rate and low accuracy suggest that the model may be biased and is underfitting the data, failing to capture the underlying patterns and relationships.
- *Recall*: If the model consistently misclassifies failed students (low *Recall*), it indicates a bias towards incorrectly predicting the outcome. This suggests that the model is not adequately capturing the characteristics associated with student dropout.

Model Variance:

- *Fallout*: A high *Fallout* rate suggests that the model is incorrectly classifying non-dropout students as dropouts. This indicates a high variance as the model is sensitive to noise or fluctuations in the data, leading to overfitting.
- *Precision*: If the *Precision* is low, it indicates that the model is classifying many non-dropout students as dropouts. This suggests that the model has high variance and is overfitting to the training data.
- *F1 Score and nMCC*: These metrics can also provide insights into both bias and variance. If the *F1 score* and *nMCC* are low, it suggests a balance issue between bias and variance. A low *F1 score* indicates poor overall performance, which can be caused by both underfitting (high bias) and overfitting (high variance).

In summary, by examining the *Accuracy*, *Error*, *Recall*, *Fallout*, *Precision*, *F1 score*, and *nMCC*, you can gain insights into both model bias (underfitting) and model variance (overfitting). It is important to strike a balance between bias and variance to build a well-performing and robust model for predicting student dropout.

12 The Use of Performance Metrics in Model Training, Validation, and Testing Phase

The metrics can be utilized in the model training, validation, and testing phases as follows:

Model Training:

- During model training, you can use the training data to fit the model and optimize its parameters.
- To monitor the model's performance and make decisions on model adjustments, you can calculate metrics such as *Accuracy*, *Recall*, *Precision*, and *F1 score* on the training data.
- These metrics help you assess the initial performance of the model and identify areas that require improvement.

Model Validation:

- After training, it's important to evaluate the model's generalization ability on unseen data. This is done through model validation using a separate validation dataset.
- Calculate the same metrics (*Accuracy, Recall, Precision, F1 score*) on the validation dataset to assess how well the model performs on new, unseen data.
- These metrics provide insights into the model's performance and help in tuning hyperparameters or adjusting the model architecture to optimize its performance.

Model Testing:

- Once you have selected the best model based on validation results, it's crucial to assess its performance on a completely independent and unbiased test dataset.
- Apply the trained model to the test dataset and calculate the metrics (*Accuracy, Recall, Precision, F1 score*) to evaluate its performance.
- The test dataset provides an unbiased estimate of the model's performance in real-world scenarios and helps determine its readiness for deployment.

By utilizing these metrics in each phase, you can iteratively train, validate, and test the model, making necessary adjustments to improve its performance. It is essential to ensure that the evaluation is conducted on different datasets to avoid overfitting and obtain reliable performance estimates.

13 Learning Curve of the Model

An alternative method for assessing the bias and variance of a model involves plotting its Learning Curve. This curve illustrates the evolving trend of the error rate as the number of samples used to train the model continuously increases, with an independent validation set employed for validation after each training session.

By plotting the error rate against the sample size for both the training and validation sets, the Learning Curves can be visualized. These Learning Curves not only reveal the actual bias and variance of the model but also provide insights into potential avenues for model improvement (e.g., whether more samples or adjusting the number of indicators would be beneficial).

The Learning Curve is a useful tool for detecting high bias or high variance in a model. By analyzing the Learning Curve, you can identify signs of bias or variance based on the following observations:

High Bias (Underfitting):

- Training and Validation *Error rate*: If the training and validation *Error rate* curves converge to a relatively high value and remain close together, it indicates that the model has high bias.
- Performance Plateau: The Learning Curve shows stagnant or limited improvement in performance (e.g., accuracy, F1 score) as the training dataset size increases. This indicates that the model is unable to capture the complexity of the data and is underfitting.

High Variance (Overfitting):

- Training and Validation *Error rate* Gap: If there is a significant gap between the training and validation *Error rate* curves, with the training *Error rate* being much lower, it suggests high variance. This indicates that the model is fitting the training data too closely but failing to generalize to new, unseen data.
- Overfitting Indicators: The Learning Curve shows that the training performance (e.g., accuracy) keeps improving while the validation performance plateaus or starts to degrade. This suggests that the model is memorizing the training data and is overly sensitive to its noise or fluctuations.

By closely examining the Learning Curve and making adjustments to the model, you can strike a balance between bias and variance to achieve optimal performance.

In Figs. 12 and 13, two different models' Learning Curves can be seen as an example for comparison.

The proximity of the training and validation curves in Fig. 12 signifies low *Variance*, and the low level of the *Validation error rate* suggests low *Bias*. The curves also indicate that collecting more *Samples* would not enhance this model, as both the *Training* and *Validation Error Rate* curves have reached their "horizontal imaginary asymptote." To further decrease *Bias*, additional predictors should be incorporated. Figure 13 illustrates a model with high bias and high variance.

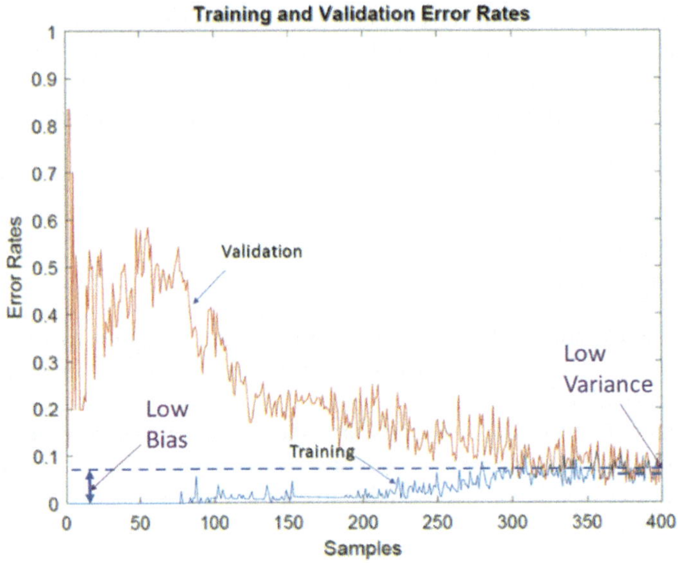

Fig. 12 The learning curve of a low bias low variance model

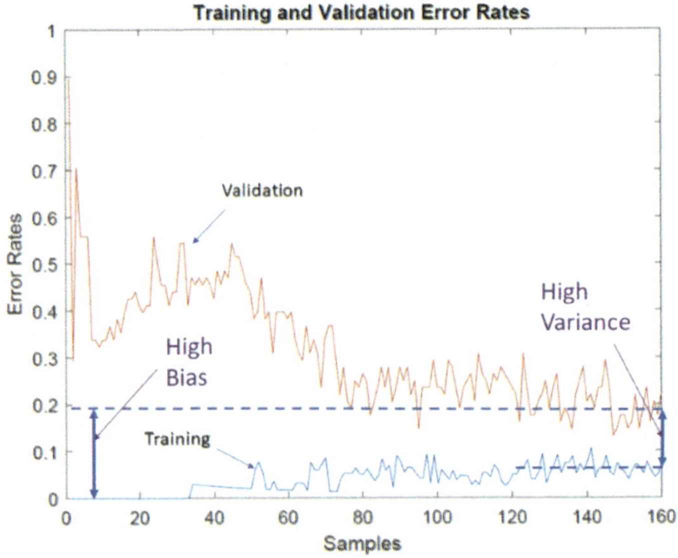

Fig. 13 The learning curve of a high bias high variance model

14 Conclusion

In conclusion, addressing student dropout in universities requires a deep understanding of the underlying causes and the implementation of multifaceted solutions. Machine learning can play a crucial role in tackling this issue by providing insights and predictive models that help both universities and students.

There are different categories of factors that influence dropout, including models with time-invariant predictors, and models with time-variant predictors. These models can be built at different levels, such as degree level, year level, course level, and exam level, to capture the specific dynamics of student performance.

Educational machine learning models have their own specialties, such as incorporating indicators derived from student activities and learning components. These models can be embedded within a Learning Management System (LMS) to gather extensive data and track student progress, enabling the identification of at-risk students.

The predictive power of these models relies on training them with data from previously completed courses and utilizing indicators that exhibit a strong correlation with the target variable. By selecting the model with the highest predictive power, it becomes possible to forecast the success of students in the current course.

The training-prediction schemes used in model building involve different approaches to time intervals and data utilization. These schemes consider the chronological order of student activities and aim to capture the temporal dynamics of their learning journey.

When building models, it is important to search for the optimal model by minimizing the error and balancing the bias and variance. Bias refers to the error caused by oversimplification, while variance refers to the error due to sensitivity to training data. Achieving a balance between bias and variance is crucial for building accurate and robust predictive models.

Several supervised learning algorithms, such as logistic regression, decision trees, random forest, support vector machines, gradient boosting methods, and neural networks, can be utilized for predicting student dropout based on the specific circumstances and characteristics of the dataset.

Performance metrics, including accuracy, precision, recall, and F1 score, can be used to evaluate the models and check for bias and variance. These metrics are computed from the confusion matrix, which captures the true positives, false positives, true negatives, and false negatives.

An alternative approach to examining the *Bias* and *Variance* of a model is to illustrate its Learning Curve.

In conclusion, addressing student dropout requires a comprehensive understanding of the factors involved and the implementation of effective predictive models. By leveraging machine learning and considering the balance between bias and variance, universities can take proactive measures to support students and improve retention rates.

References

1. Abu Saa A, Al-Emran M, Shaalan K (2019) Factors affecting students' performance in higher education: a systematic review of predictive data mining techniques. Technol Knowl Learn 24:567–598. https://doi.org/10.1007/s10758-019-09408-7
2. Aina C, Baici E, Casalone G, Pastore F (2022) The determinants of university dropout: a review of the socio-economic literature. Socioecon Plann Sci 79:101102. https://doi.org/10.1016/j.seps.2021.101102
3. Alyahyan E, Düştegör D (2020) Predicting academic success in higher education: literature review and best practices. Int J Educ Technol High Educ 17:1–21. https://doi.org/10.1186/s41239-020-0177-7
4. American College Health Association (2021) American college health association-national college health assessment III: undergraduate student reference group executive summary spring 2021. https://www.acha.org/documents/ncha/NCHA-III_SPRING-2021_UNDERGRADUATE_REFERENCE_GROUP_EXECUTIVE_SUMMARY_updated.pdf
5. Arias Ortiz E, Dehon C (2008) What are the factors of success at university? A case study in Belgium. CESifo Econ Stud 54(2):121–148. https://academic.oup.com/cesifo/article-abstract/54/2/121/454068
6. Asif R, Merceron A, Abbas S, Ghani N (2017) Analyzing undergraduate students' performance using educational data mining. Comput Educ 113:177–194. https://www.sciencedirect.com/science/article/abs/pii/S0360131517301124
7. Behr A, Giese M, Teguim Kamdjou HD, Theune K (2021) Motives for dropping out from higher education—An analysis of bachelor's degree students in Germany. Eur J Educ 56(2):325–343. https://doi.org/10.1111/ejed.12433
8. Bognár L, Fauszt T, Nagy B (2021a) Machine learning model building techniques for small and medium-sized university courses. Int J Artif Intell 19(2):20–43. https://www.scopus.com/record/display.uri?eid=2-s2.0-85123345005&origin=inward&txGid=45ea1626166793b7f5df57ba6c4d3cdd
9. Bognár L, Fauszt T, Nagy GZ (2021b) Analysis of conditions for reliable predictions by Moodle machine learning models. Int J Emerg Technol Learn 16(6):106–121. https://online-journals.org/index.php/i-jet/article/view/18347
10. Burnett K (2001) Interaction and student retention, success and satisfaction in web-based learning. https://eric.ed.gov/?id=ED459798
11. Dharmawan T, Ginardi H, Munif A (2018) Dropout detection using non-academic data. In: 2018 4th International conference on science and technology (ICST). IEEE, pp 1–4. https://ieeexplore.ieee.org/document/8528619
12. Enaro AO, Chakraborty S (2020) Feature selection algorithms for predicting students academic performance using data mining techniques. Int J Sci Technol Res 9(04):3622. https://www.ijstr.org/final-print/apr2020/Feature-Selection-Algorithms-For-Predicting-Students-Academic-Performance-Using-Data-Mining-Techniques.pdf
13. Er E (2012) Identifying at-risk students using machine learning techniques: a case study with IS 100. Int J Mach Learn Comput 2(4):476. http://www.ijmlc.org/show-32-132-1.html
14. Ghorbani R, Ghousi R (2020) Comparing different resampling methods in predicting Students' performance using machine learning techniques. IEEE Access 8:67899–67911. https://ieeexplore.ieee.org/document/9062549
15. Hussain M, Zhu W, Zhang W, Abidi SMR, Ali S (2019) Using machine learning to predict student difficulties from learning session data. Artif Intell Rev 52(1):381–407. https://doi.org/10.1007/s10462-018-9620-8
16. Jalota C, Agrawal R, Jalota C (2021) Feature selection algorithms and student academic performance: a study. In: International conference on innovative computing and communications. Springer, pp 269–276. https://doi.org/10.1007/978-981-15-5113-0_23
17. Kehm BM, Larsen MR, Sommersel HB (2019). Student dropout from universities in Europe: a review of empirical literature. Hung Educ Res J 9(2):147–164. https://akjournals.com/view/journals/063/9/2/article-p147.xml

18. Kuhfeld M, Soland J, Tarasawa B, Johnson A, Ruzek E, Liu J (2020) Projecting the potential impact of COVID-19 school closures on academic achievement. Educ Res 49(8):549–565. https://doi.org/10.3102/0013189X20965918
19. Marbouti F, Diefes-Dux HA, Madhavan K (2016) Models for early prediction of at-risk students in a course using standards-based grading. Comput Educ 103:1–15. https://www.sciencedirect.com/science/article/pii/S0360131516301634
20. Mduma N (2023) Data balancing techniques for predicting student dropout using machine learning. Data 8(3):49. https://doi.org/10.3390/data8030049
21. Mengash HA (2020) Using data mining techniques to predict student performance to support decision making in university admission systems. IEEE Access 8:55462–55470. https://ieeexplore.ieee.org/document/9042216
22. Miguéis VL, Freitas A, Garcia PJ, Silva A (2018) Early segmentation of students according to their academic performance: a predictive modelling approach. Decis Support Syst 115:36–51. https://www.sciencedirect.com/science/article/abs/pii/S0167923618301428
23. Nagy M, Molontay R (2023) Interpretable dropout prediction: towards XAI-based personalized intervention. Int J Artif Intell Educ (Advance online publication). https://doi.org/10.1007/s40593-023-00331-8
24. National Center for Education Statistics (2019) Fast facts: dropout rates. https://nces.ed.gov/fastfacts/display.asp?id=16
25. Olivé DM, Huynh DQ, Reynolds M, Dougiamas M, Wiese D (2020) A supervised learning framework: using assessment to identify students at risk of dropping out of a MOOC. J Comput High Educ 32(1):9–26
26. Pelanek R (2015) Metrics for evaluation of student models. J Educ Data Mining 7(2):83–95. https://jedm.educationaldatamining.org/index.php/JEDM/article/view/JEDM087
27. Rastrollo-Guerrero JL, Gómez-Pulido JA, Durán-Domínguez A (2020). Analyzing and predicting students' performance by means of machine learning: a review. Appl Sci 10(3):1042. https://www.mdpi.com/2076-3417/10/3/1042
28. Respondek L, Seufert T, Hamm JM, Nett UE (2020) Linking changes in perceived academic control to university dropout and university grades: a longitudinal approach. J Educ Psychol 112(5): 987–1002. https://psycnet.apa.org/fulltext/2019-40705-001.html
29. Shapiro D, Dundar A, Huie F, Wakhungu PK, Bhimdiwala A (2017) Completing college: a national view of student attainment rates. National Student Clearinghouse Research Center. https://nscresearchcenter.org/wp-content/uploads/SignatureReport12.pdf
30. Song Z, Sung SH, Park DM, Park BK (2023) All-year dropout prediction modeling and analysis for university students. Appl Sci 13(2):1143. MDPI AG. https://doi.org/10.3390/app13021143
31. Tsiakmaki M, Kostopoulos GK, Kotsiantis S, Ragos O (2020) Transfer learning from deep neural networks for predicting student performance. Appl Sci 10(6):2145. https://doi.org/10.3390/app10062145
32. Tsiakmaki M, Kostopoulos G, Kotsiantis S, Ragos O (2020) Implementing AutoML in educational data mining for prediction tasks. Appl Sci 10(1):90. https://doi.org/10.3390/app10010090
33. Wang AY, Newlin MH (2002) Predictors of performance in the virtual classroom: identifying and helping at-risk cyber-students. J (Technol Horiz Educ) 29(10):21
34. Zhang H, Almeroth K (2010) Moodog: tracking student activity in online course management systems. J Interact Learn Res 21(3):407–429

Improving Students' Achievement Prediction in Blended Learning Environments with Integrated Machine Learning Methods

Yangyang Luo and Yiran Cui

Abstract Blended learning has been adopted in universities for over a decade. With the advantage of combining online and offline learning contexts, blended learning enables a teacher to design separate instructional models in one curriculum. However, student data from two learning contexts also bring challenges for teachers. Especially, it is difficult to apply machine learning methods to improve students' learning efficiency. Moreover, there are more obstacles to interpreting results from machine learning methods and transforming them into student-supporting strategies. This chapter discusses a case study using machine learning methods to analyze learning data from both online and offline contexts. The study initially measures students' learning styles, collects students' online learning behavior data and offline classroom voice data, and uses an integrated machine learning method to recognize students' data patterns. Periodic achievement tests are arranged in teaching practice to select the students whose academic performance increment is larger than the average increment of the whole class, and the implication of data patterns of selected students is used to modification of teaching strategies. The results indicate that after at least three semesters of iteration, about 79% of students in the class follow the recommended learning pattern. Compared to collaborative learning, this method is more suitable for discussion-based learning.

Keywords Blended learning · Learning achievement prediction · Machine learning · Integrated methods

Y. Luo
Institute of Higher Education, Lanzhou University, Lanzhou, Gansu, China
e-mail: luoyy@lzu.edu.cn

Y. Cui (✉)
School of International Education, Shandong University, Jinan, Shandong, China
e-mail: cuiyirann@163.com

M. S. Khine (ed.), *Machine Learning in Educational Sciences*,
https://doi.org/10.1007/978-981-99-9379-6_9

1 Introduction

In recent years, artificial intelligence (AI) technology has deeply integrated into various industries, bringing significant changes to the way society produces and the structure of the industry [10]. Technology tools like Chat Generative Pre-trained Transformer (ChatGPT) have been transitioning the role of AI from replacing repetitive work to assisting human thinking [12]. However, the integration of AI into daily activities in education is still progressing slowly due to the conservatism of the education industry [25]. As the main academic communities promoting the integration of AI technology into various aspects of educational activities, Learning Analytics and Knowledge (LAK) and Educational Data Mining (EDM) have shown in research that trustworthy AI technologies are becoming crucial tools to advance learning in complex learning environments [26].

Using machine learning techniques to predict learners' future achievement is the research field of AI integrated into education. When researchers conduct a learning achievement prediction study, the main processes include collecting data on various student learning activities, applying machine learning techniques to identify learning patterns, predicting learning outcomes, and providing suggestions for learning strategies [3, 14, 28]. Moreover, identifying meaningful patterns in learning behavior is the most challenging part of this process. Previous studies have indicated that meaningful learning behaviors often depend on the learning scenarios [4, 31]. The more comprehensive the data collected on learning behavior, the more important the prediction of learning outcomes to adjust learning strategies. Compared to fully online or fully face-to-face learning, blended learning environments are more complex. Therefore, meaningful learning behaviors occur in both online and face-to-face scenarios, which brings challenges in using machine learning techniques to analyze learning behaviors and provide suggestions for learning strategies [20, 21].

Firstly, there are challenges in data collection. The blended learning setting means learners need to participate in Internet learning scenarios and Face-to-Face learning scenarios. Activities in Internet learning settings can be recorded by a virtual learning management system (VLM), while activities in face-to-face learning environments usually are recorded only through cameras [23]. Secondly, there are challenges in selecting machine learning algorithms to process various data types. The learning activity data generated in blended learning include various data types, such as digital data, textual data, audio data, video data, and sequential data [2]. Currently, there is no single algorithm that can handle all these types of data. Thirdly, there are also the challenges in identifying meaningful learning activities. There is no research to support the notion that any particular learning behavior or type of behavior significantly influences learning outcomes in every learning scenario. It becomes even more challenging to identify meaningful learning behaviors from both online and face-to-face learning scenarios [11, 24]. Existing research only analyzes online scenarios in blended learning and converts various types of data into digital data for analysis [1, 9]. Some studies have shown that online data alone are sufficient to predict student performance when the diversity and frequency of online learning activities reach a certain level [5, 8]. However, such studies are only applicable to certain courses. As Luo et al. (2022) pointed out, predicting student learning outcomes based on data from online learning activity alone is only applicable to certain types of courses but can't be applied generally to all blended learning scenarios [17]. Some existing researchers have applied machine learning algorithms to analyze various types of data from sources in blended learning and have successfully predicted student grades in educational research [7, 6]. However, there is still very limited research aiming to guide learning strategies with machine learning in blended learning.

This study aims to track and collect data from students in both online and face-to-face learning environments, covering two blended courses over four semesters. The integrated machine learning algorithm is used to analyze students' learning behavior patterns, predict and provide feedback on changes in their learning outcomes, and help them improve academic achievement.

2 Data Collection

This study collaborated with two instruction teams to track and collect data from two courses over four consecutive semesters. The basic information about the two courses is as follows.

English Course (EC): This course was an English class for the whole school and was taught by a team of four instructors. It was a compulsory course. Students were assigned to different classes according to their prerequisite test scores. Before the start of each semester, the instructional team reviewed all learning units and optimized the content, activities, tests, and evaluation criteria based on student's performance in the previous semester. Before online learning began, the instructional team used the Index of Learning Styles Questionnaire (shown in Appendix A) to investigate potential student group characteristics [27]. The questionnaire items such as (1) I found the improvement in myself after I tried it out, or pondered over it; (2) I prefer to think realistic, or innovative; (3) I like to use a picture, or words to think about my activities. In the learning management system (LMS), the instructional team designed the learning objectives for each unit and provided reading materials, interactive videos, and tests to assess learning progress in a specific order. During the face-to-face sessions, the instructional team identified problems in learning based on participation in online activities and test results. They adopted a pedagogical approach with repeated lectures and discussions to address the problems encountered during online learning.

Chemistry Professional Course (CPC): This course was offered to senior undergraduate students in the college, and taught by a team of two instructors. Before the semester began, the instructional team used the Index of Learning Styles Questionnaire (shown in Appendix A) to investigate potential student group characteristics [27]. The team grouped students according to their grades in prerequisite major courses and implemented a combination of theoretical and laboratory instruction. During online learning, the instructional team designed reading materials and tests for each theory learning unit and provided reading materials regarding prerequisite knowledge and demonstration videos related to the experimental operations of each laboratory unit. During face-to-face learning, the instructional team explained the key points and the difficulty of learning tasks based on students' online learning results and asked students to present in groups how to apply the theory to the chemistry experimental practices procedure. Then, the teacher guided each group with chemistry experimental practices based on the quality of the group's previous presentation. Detailed information such as students' numbers and groups is shown in Table 1.

According to previous research, the frequency and duration of students' engagement in learning activities during online learning have a significant impact on their performance [19]. In face-to-face learning, the content and the order of student discussion can also influence student performance [18]. Therefore, we collected online learning data of students through learning management system (LMS) logs, which were transferred into the frequency of student engagement indicators. As shown in Table 2, indicators included the count of student access to the curriculum page, the length of time students remained active on the curriculum page, the count of students reading text material, the count of students watching video material, time length of students watching video material, the count of students posting ideas, the count of responses to others' ideas, the count of submissions on tests, the count of questions asked of the teacher, and the count of notes taken. In offline learning scenarios, we collected voice data from students and teachers through voice recorders. To prevent mutual interference with the voices of teacher and student groups, we dispatched voice recorders to each student group. Therefore, the offline learning data included the discussion voice data with time stamps of students and teachers.

Table 1 Information on the two courses

Course name	Number of students enrolled	Number of classes	Group numbers in each class	Semester	Number of preliminary tests
English Course (EC)	180	4	12	1	5
	164	4	11	2	5
	205	5	11	3	5
	168	4	11	4	5
Chemistry Professional Course (CPC)	54	2	9	1	5
	48	2	8	2	5
	60	2	10	3	5
	72	3	10	4	5

Table 2 Information on data types

Learning setting	Data code	Name of data	Data type
Online learning	Activity1(AC1)	Course page visits(times)	INT
	Activity2(AC2)	Total time of learning activities(min)	INT
	Activity3(AC3)	Number of times of reading text materials(times)	INT
	Activity4(AC4)	Number of times of watching video materials(times)	INT
	Activity5(AC5)	Total time of watching video materials(min)	INT
	Activity6(AC6)	Count of ideas publish	INT
	Activity7(AC7)	Count of responses to others' ideas	INT
	Activity8(AC8)	Count of submitted in a test	INT
	Activity9 (AC9)	Count of asking questions to the teacher	INT
	Activity10 (AC10)	Count of taking notes	INT
Face-to-face learning	Voice1 (VO1)	Voice of teacher leading	WAV
	Voice2 (VO2)	Voice of student discussion	WAV
Student learning style	SLS(SLS)	Result of student learning style investigation	INT
Student Score	SC(SC)	Result of student assessment	INT

Note INT means "integer", which is a coding variable type and to define only the whole numbers data. WAV is the sound file format name, which refers to the file storing digital audio (waveform) data

3 Research Method and Results

To recognize the students' voice patterns in the face-to-face learning context, we used Long Short-Term Memory (LSTM) optimized recurrent neural network (and its variants) to identify the content features of student discussions and the keywords of teacher-led voice. Based on the results of learning style tests conducted at the beginning of each semester, students were pre-categorized into different types, and stratified random sampling was performed based on the proportional representation of each student type. In the face-to-face learning environment, the teacher determines the topics to be discussed by the student and provides requirements before the discussions. Results generated by the machine learning models were used to analyze the completeness of student discussion content and the degree of agreement with the teacher's guiding keywords to assess the quality of student discussions [16]. The analysis of student discussion audio was used as a parameter for predicting student performance in offline learning activities, particularly the scores on periodic tests. To identify effective patterns of student learning activities, we divided the intervals based on the number of times students participated in the periodic tests. What's more, we also collected parameters of online learning behavior, parameters of offline learning behavior, and scores from the periodic tests from students five times per semester.

3.1 Data Preprocessing and Model Debugging

Initially, we collected audio from both teachers and students in each session and mapped every section of voice to a speaker. Secondly, we converted the speech data of each speaker into text data with temporal labels and transformed the text data into word vectors. There are two methods used in word vector transformation commonly. One-hot representation, which uses vector V to represent a word, and there is only one position of vector V is 1, and the reset position consists of 0. Distribution representation can represent the original meaning of words and capture the semantic distances between similar words. However, the one-hot representation method is not able to present the similarity between words, and Distribution representation solves this problem by converting words into real-valued vectors, resolving the curse of dimensionality due to the increasing number of words in a sentence, calculating the associative properties between words in the text, construct hierarchical dimensions that capture meanings, and improve the accuracy of vector semantics [29]. Then we chose the distribution representation as the method for word vector transformation.

Thirdly, we performed Chinese word segmentation and labeled the data. The word segmentation tool used was the THU Lexical Analyzer for Chinese. Compared to popular Chinese text segmentors (ICTCLAS(2015) and Jieba), the F-Measure value of single student discussion content (535.6 KB on average) used by THU Lexical Analyzer is 0.912, which has the best performance [22]. We decomposed a complete semantic sentence into initiation words (I), center words (C), stop words (S), and punctuation (P), where initiation words and stop words are the words closest to the beginning and end of the sentence, respectively. In this study, after automatically labeling all the text words, an issue of class imbalance in word categories arose. The number of words labeled as C was much larger than the number of words labeled as other categories. Therefore, we addressed this issue in two ways. (1) Sentences that only contained C-labeled words without I-labeled or S-labeled words were directly marked as incomplete content and excluded from model training. (2) We assessed the distance between C-labeled words used by students and teachers. Sentences in the student discussion content that contained consecutive C-labeled words unrelated to the teacher's word vectors were directly labeled as low-quality content. This approach helped reduce the disparity in the number of C-labeled words compared to other types of labeled words. Their distribution results are shown in Figs. 1 and 2.

Fourthly, we applied LSTM implemented in Python Keras RNN (Recurrent Neural Network) API to unsupervised learning knowledge hidden in text data. The purpose of the fourth step is to recognize the semantic completeness of student discussion and the semantic matching degree, which is an indicator of how much student discussion content is aligned with teacher guidance content. When we adjust the RNN model, the Keras tuner tools implement the hyperparameter tuning, which includes the size of the hidden layer, number of hidden layers, number of epochs, size of batches, and number of time steps. According to Liashchynskyi and Liashchynskyi [15], our text-time serials data are more than four dimensions in complexity compared to the gridsearchCV method. Therefore, even if randomsearchCV costs more time, it

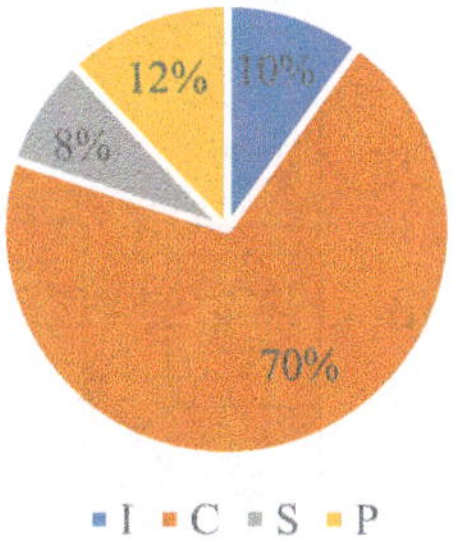

Fig. 1 Word label percentage before balance operation

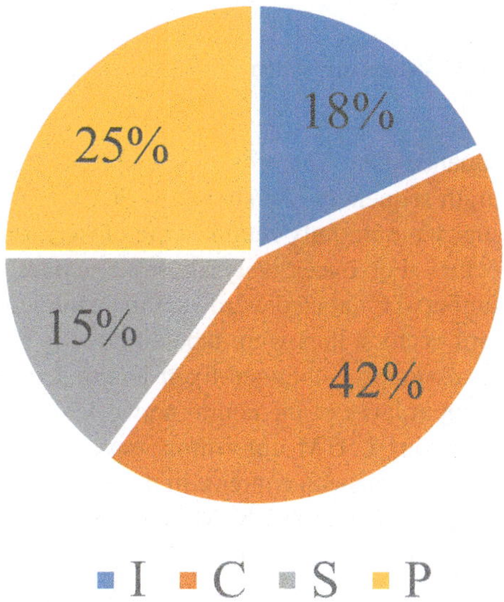

Fig. 2 Word label percentage after balance operation

is the best choice for hyperparameter tuning of our LSTM model. The range and hyperparameter type when we use randomsearchCV tuning are shown below.

Size of hidden layer: [20–2000]
Number of hidden layers: [6–200]
Number of epochs: [1–300]
Size of batches: [20–100]
Number of time steps: [3–300].

3.2 Semantic Completeness Analysis

The student discussion audio data sequences collected in this study have variable lengths. However, LSTM and its related variants can only handle fixed-length

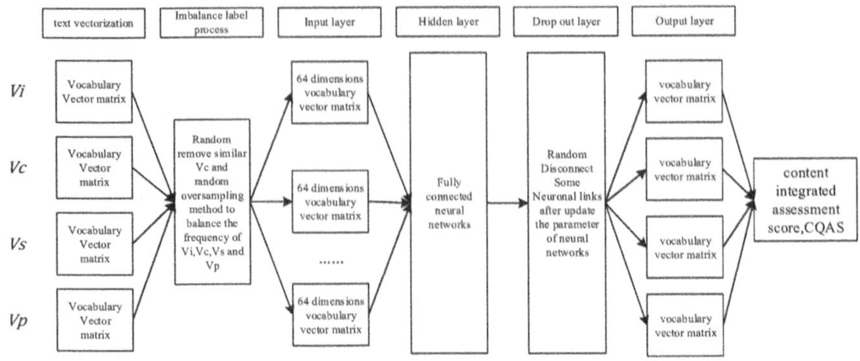

Fig. 3 Student discussion content analysis model implemented by RNN

sequence data. Therefore, we employed a random retention method to handle over-length sequences. As shown in Fig. 3, when individual discussion content was too long, we randomly retained the closely related content within the C-labeled word vectors. For excessively short sequences, we used oversampling to increase the number of C-labeled word vectors in the sequence. If the data points lacked C-labeled word vectors, they were labeled directly in the low-quality discussion. Through the processing, we obtained fixed-length sequences of 64 dimensions, which served as the input layer of the neural network algorithm.

Various LSTM algorithms have been used in previous studies on processing student discussion content. In this paper, we selected the commonly used BI-LSTM, Tree-LSTM, Graph-LSTM, and double LSTM algorithms to compare and analyze the different algorithmic impacts on semantic completeness results.

Furthermore, we added a Dropout layer before the output layer of the model to prevent overfitting and reduce under-fitting during model training. This Dropout layer disconnected a portion of the input layer based on the set probability each time the model updated its parameters. To ensure consistency between the word vector dimension and the number of label categories in the output, the vector dimension of the added Dropout layer was set to be the same as that of the output layer. We used the soft-max activation function and extracted the resulting probabilities to evaluate the completeness of the student discussion.

Figure 4 represents the performance results obtained through cross-validation of the various algorithms in model debugging, showing the training accuracy varying with the parameter number of epochs. The results indicated that the model built using Bi-LSTM achieves the best performance. This study also examined the results of the model's accuracy precision with other hyper-parameters during the debugging process. These results consistently showed that the model constructed with Bi-LSTM is best suited to analyzing the completeness of student discussion included in this study.

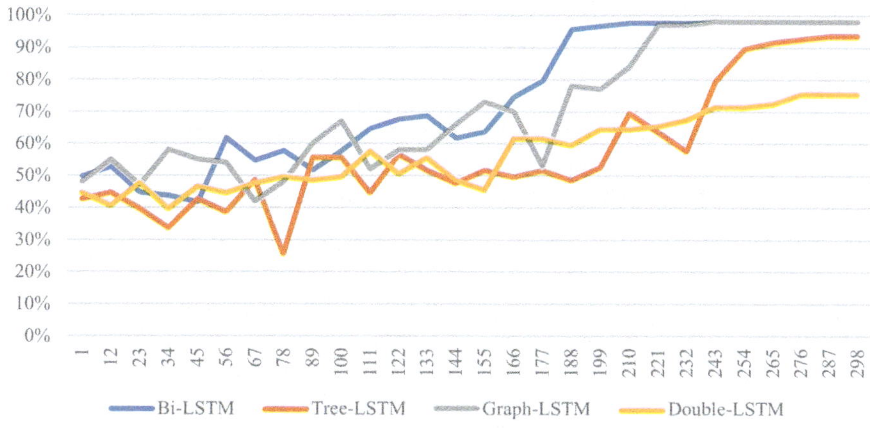

Fig. 4 Training accuracy varies with the number of epochs tuning

3.3 *Semantic Matching Degree Analysis*

To examine the degree of matching between student discussion content and key phrases in the teacher guidance content, the text of the teacher guidance content was encoded into the four-label format described in Sect. 3.1. The distances between the C-labeled word vector of the teacher guidance content and the student discussion content were calculated. The key information in the student discussion content that best matched the context of the teacher guidance content was extracted. The vectors were accumulated using a time series and the weight of the vector matrix was adjusted at each iteration by a loss function. Finally, the probability of matching the content of each student's discussion with the teacher's guidance content was calculated.

To select the best machine learning algorithm for evaluating the matching degree of student discussion content, this study considered the TF-IDF (term frequency–inverse document frequency) algorithm, which is widely used for statistical analysis based on word frequency matching. Additionally, the LSTM algorithm, which is a single-cycle neural network algorithm used for processing single-turn dialogues, and its improved versions, MV-LSTM (Multi-View LSTM) and Match-LSTM, were compared and analyzed. The evaluation metrics Mean Reciprocal Rank (MRR) and Mean Average Precision (MAP) were selected. MRR originally referred to the ability of a neural network model to find more conspicuous items for users, emphasizing positional relationships and sequentially. In this study, MRR refers to the positional matching degree of key phrases in the student's utterances found by the neural network model with those in the teacher guidance content (as shown in Eq. 1). In Eq. 1, N represents the number of matches between key phrases in the student's utterances and those in the teacher guidance content, and Pi represents whether the i-th key phrase corresponds to a match.

$$MRR = \frac{1}{N}\sum_{i=1}^{N}\frac{1}{p_i} \tag{1}$$

MAP represents Mean Average Precision, which originally refers to the average accuracy of retrieving each topic in all relevant documents by the model. However, in this study, MAP refers to the average accuracy of matching each student's discussion to the content of the teacher's guidance. It is represented as Eq. 2.

$$MAP = \frac{1}{S}\sum_{i=1}^{S} AP_i \tag{2}$$

In Eq. 2, S represents the number of students, and AP_i represents the accuracy rate of the i-th student matching the teacher guidance content in their discussion. It can be represented as Eq. 3.

$$AP_i = \frac{1}{N}\sum_{i=1}^{N} if(content_s == content_t) \times \frac{1}{p_i} \tag{3}$$

In Eq. 3, N represents the number of student discussion content keywords, $content_s$ represents the student discussion content keywords, $content_t$ represents the teacher guidance content keywords, and p represents the location of i-th keywords. To compare the performance of each model, classroom discussion data from the first semester of two curriculums were divided into five parts, aligned with the regular achievement tests, and named the sub-datasets as periodical achievement tests 1–5. Then, we input the sub-datasets into each model one by one and selected the best model according to the results of MRR and MAP.

According to Table 3, the MRR and MAP values of MV-LSTM outperformed other models in the datasets of periodic outcome test 1, test 2, and test 3. Although the MRR value of MV-LSTM was 0.01 lower than that of LSTM, the MAP value of MV-LSTM was 0.04 and 0.03 higher than that of LSTM, respectively. Therefore, the MV-LSTM semantic matching degree analysis model was most suitable for our study.

3.4 Recognize Student Learning Pattern

The results of student discussion semantically integrated assessment (SCIA) and student content matching assessment (SCMA) represent the student offline learning data during each periodical achievement test. Combining the SCIA, the SCMA with the online learning behavior data AC1 to AC10, students' learning style test data SLS, and the student learning activities data from both learning settings and the characteristics of students can be collected. The variance of periodical achievement test

Table 3 Semantic matching degree analysis model performance

	Periodical achievement test 1		Periodical achievement test 2		Periodical achievement test 3		Periodical achievement test 4		Periodical achievement test 5	
	MRR	MAP	MRR	MAP	MRR	MAP	MRR	MAP	MRR	MAP
TF-IDF	0.48	0.35	0.6	0.53	0.6	0.56	0.49	0.44	0.54	0.52
LSTM	0.53	0.47	0.52	0.44	0.56	0.49	**0.61**	0.55	**0.65**	0.59
MV-LSTM	**0.59**	**0.55**	**0.64**	**0.57**	**0.65**	**0.58**	0.6	**0.59**	0.64	**0.62**
Match-LSTM	0.52	0.47	0.6	0.57	0.59	0.51	0.59	0.47	0.58	0.46

score ΔSC represents the student learning performance during each learning stage, which means $\Delta SC = SC_i - SC_{i-1}$. As the stage achievement test score is limited to 0–100, two types of students will disturb our predictive model. The first type of students is those who maintain high scores throughout the semester and have a small absolute value of ΔSC. Another type of student is those who maintain low scores throughout the semester and have small absolute values of ΔSC. For the first type of students ($SC >= 90 \& \Delta|SC| < 5$), their learning activities data will be labeled as highly recommended. For the second type of students ($SC <= 60 \& \Delta|SC| < 5$), their learning activities data will be labeled as strongly rejected. For other students, we labeled their score variance by comparing the student score variance and the mean of class score variance. The transfer rule is shown below: if $\Delta SC_i > 1\sigma$ of $\Delta SC_{(mean)}$, then recommended; if $\Delta SC_i > 2\sigma$ of $\Delta SC_{(mean)}$, then highly recommended; if $\Delta SC_i > -1\sigma$ of $\Delta SC_{(mean)}$, then not recommended; if $\Delta SC_i > -2\sigma$ of $\Delta SC_{(mean)}$, then strongly rejected. The rules show that if the ΔSC of a student is over one standard deviation of all classes, it will be labeled as recommended.

In the current study, we have chosen some machine learning algorithms widely used in previous studies to find students' online and offline learning behavior patterns corresponding to different labels, which include SVM (Support Vector Machines), LR (logistic regression), Naïve Bayes, K-Nearest Neighbors (k-NN), and Random Forest [13, 28, 30]. Before data input, the students' dataset was stratified by random sampling in alignment with the students' SLS distribution. We then obtained five sub-datasets of student learning activities for cross-validation. At the stage of model evaluation, we used average precision, average recall, macro-f1, and micro-f1 as indicators to assess the prediction results of the multi-classification model (as shown in Tables 4 and 5).

3.5 Prediction Results Feedback to the Instruction Process

The predictive model constructed using the Random Forest algorithm identifies the importance of various indicators in predicting student learning behavior. In our study, the predicted results of stage assessing score growth for each student were provided as feedback. Additionally, all highly recommended student learning behavior indicators corresponding to score growth and all strongly rejected student learning behavior indicators corresponding to score growth were disclosed. Finally, the percentage of students in the class who followed the recommended student learning behavior pattern was calculated, and the mean score growth for all students between the first and last stage assessment tests of the semester was calculated.

As shown in Fig. 5, by the end of the first semester, 40% of the students participating in the English course followed the learning behavior patterns that corresponded to the "highly recommended" label. The average score for the class increased by 5 points (on a hundred-mark scale). By the time the course reached its third semester, 79% of the class followed the learning behavior patterns corresponding to

Table 4 Prediction result evaluation of each machine learning algorithm in four semesters of the EC curriculum

EC	Semester1			
	k-NN	NB	RF	SVM
AP	0.652	0.685	**0.821**	0.752
AR	0.675	0.691	**0.835**	0.783
Marco-F1	0.663	0.688	**0.828**	0.767
Micro-F1	0.695	0.713	**0.841**	0.791
	Semester 2			
	k-NN	NB	RF	SVM
AP	0.632	0.655	**0.812**	0.701
AR	0.638	0.659	**0.818**	0.712
Marco-F1	0.635	0.657	**0.814**	0.706
Micro-F1	0.641	0.671	**0.833**	0.723
	Semester 3			
	k-NN	NB	RF	SVM
AP	0.612	0.681	**0.835**	0.781
AR	0.627	0.703	**0.848**	0.801
Marco-F1	0.620	0.692	**0.841**	0.791
Micro-F1	0.644	0.720	**0.865**	0.813
	Semester 4			
	k-NN	NB	RF	SVM
AP	0.585	0.701	**0.833**	0.765
AR	0.601	0.725	**0.852**	0.779
Marco-F1	0.593	0.713	**0.842**	0.772
Micro-F1	0.608	0.732	**0.855**	0.797

the "highly recommended" label, and the class average increased by 15.4 points (on a hundred-mark scale) by the end of the fourth semester.

As shown in Fig. 6, at the end of the first semester, 35% of the students participating in the chemistry professional course followed the learning behavior patterns corresponding to the "highly recommended" label. Although there were fluctuations in the percentage of students following our recommended learning behavior patterns during the course iterations, by the end of the fourth semester, 60% of the students followed the recommended learning behavior patterns. In Addition, the average score for the class increased by 10.7 points (on a hundred-mark scale) compared to the beginning of the fourth semester. Therefore, it indicates that our method of analyzing students' online and offline learning processes, predicting their performance on phase assessments, and providing feedback can effectively enhance their learning outcomes in discussion-based and collaborative learning.

Table 5 Prediction result evaluation of each machine learning algorithm in four semesters of the CPC curriculum

CPC	Semester 1			
	k-NN	NB	RF	SVM
AP	0.503	0.62	**0.774**	0.709
AR	0.523	0.633	**0.781**	0.737
Marco-F1	0.513	0.626	**0.777**	0.723
Micro-F1	0.531	0.631	**0.789**	0.732
	Semester 2			
	k-NN	NB	RF	SVM
AP	0.484	0.591	**0.784**	0.681
AR	0.491	0.613	**0.796**	0.693
Marco-F1	0.487	0.602	**0.790**	0.687
Micro-F1	0.496	0.612	**0.801**	0.791
	Semester 3			
	k-NN	NB	RF	SVM
AP	0.467	0.559	**0.767**	0.671
AR	0.475	0.574	**0.781**	0.682
Marco-F1	0.471	0.566	**0.774**	0.676
Micro-F1	0.492	0.573	**0.796**	0.681
	Semester 4			
	k-NN	NB	RF	SVM
AP	0.585	0.607	**0.814**	0.771
AR	0.592	0.611	**0.821**	0.794
Marco-F1	0.588	0.609	**0.817**	0.782
Micro-F1	0.613	0.616	**0.834**	0.801

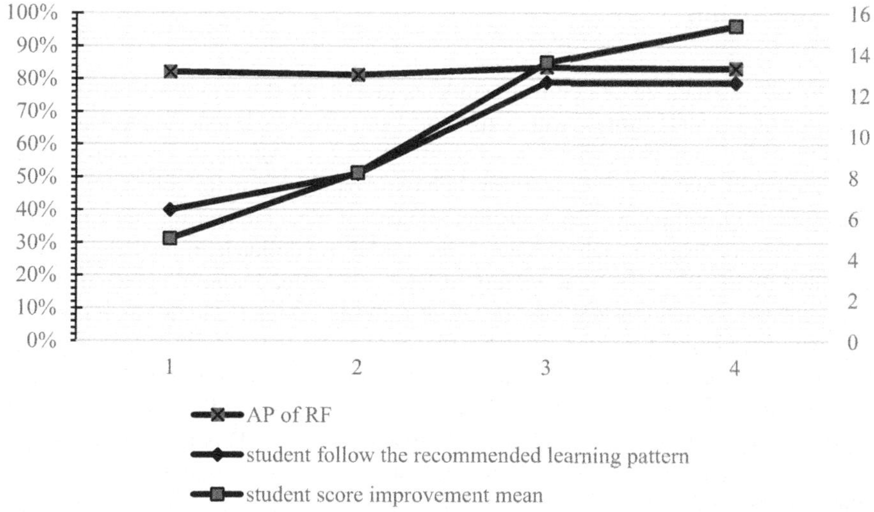

Fig. 5 Student learning pattern, score improvement, and prediction accuracy variant line chart in English Course (EC)

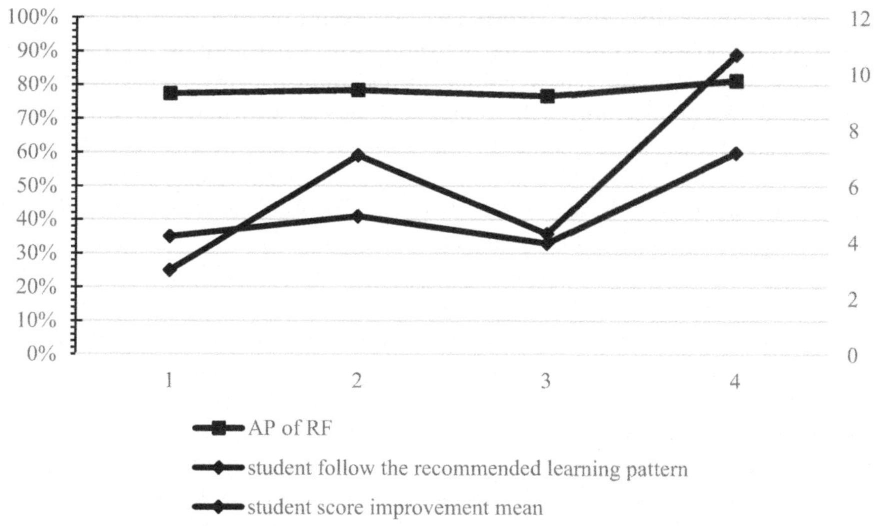

Fig. 6 EC student learning pattern, score improvement, and prediction accuracy variant line chart in Chemistry Professional Course (CPC)

4 Conclusion

How to analyze data generated from students in a blended learning scenario, which combines online and offline learning, has long been an obstacle to leveraging machine learning techniques to promote learner effectiveness. Although previous studies have indicated that learners' learning behavior patterns remain relatively consistent whether in online or offline settings, the use of machine learning techniques to predict learners' learning achievements based solely on online learning behavior data is limited by the frequency and variety of learners' online learning engagement. To comprehensively analyze learners' behavior patterns in blended learning, it is necessary to find a machine learning algorithm that can handle all types of learning behavior data from both online and offline sources. In this study, we collected students' online learning behavior data and offline discussion voice data and conducted embedding analysis using RNN and Random Forest algorithms. This method allows for a more complete analysis of the entire learning process in blended learning and enables tracing of changes in students' learning behavior patterns when their performance changes, thus better-guiding students to adjust their online and offline learning strategies.

The results of this study showed that the above method worked well in both the English Course (EC) and Chemistry Professional Course (CPC). For example, when this method was iteratively used in the English Course until the third semester, 79% of the students followed the recommended learning pattern, and the mean score increased by 15.4 points (on a hundred-mark scale) at the end of the fourth semester. In the Chemistry Professional Course, 60% of the students followed the recommended learning pattern by the fourth semester, resulting in a mean score increase of 10.7 points (on a hundred-mark scale). There are two possible reasons why the method was less effective in the Chemistry Professional Course than in the English Course. Firstly, the Chemistry Professional Course requires more foundational knowledge and practical skills, which are largely determined before students engage in the course. As a result, students with a lower starting point may find it difficult to improve their performance by following the behavior patterns of higher-performing students. Secondly, compared to course discussions in the English Course, the discussions in Chemistry Professional Course only reflect students' knowledge mastery, and it is not possible to rely solely on the data collected in this study to determine whether students can effectively apply the knowledge and skills acquired in practical experiments.

5 Limitations

The results of this study are based on data from only four semesters in two courses, and further validation is required to generalize the results to all discussion-based and collaborative learning. In addition, there was a significant difference in individual students' acceptance of suggestions for adjusting their learning behavior patterns due to the different numbers of students enrolled in the two courses each semester. Finally, although the study found that using machine learning techniques achieved relatively accurate predictions of changes in learning achievement (up to 83.5%), the effectiveness of adjusting learning behavior patterns based on the feedback provided using these predictions remained limited in further improving students' performance. Future research is likely to focus on how to effectively utilize prediction results to improve student's learning achievements.

Appendix A

Index of Learning Styles Questionnaire

Richard M. Felder

Barbara A. Soloman

North Carolina State University

Privacy policy:

Your response data and learning style profile are not stored or sent to anyone other than you. They cannot be recovered once you have submitted the completed form and received the results.

Directions:

Please provide us with your first and last name, which will be printed on the report of results that will be returned to you. (Note the Privacy Policy above if you are concerned about confidentiality.) You may only choose one answer for each question, and you must answer all questions before you can submit the form. If both answers to a question seem to apply to you, choose the one that applies more frequently throughout all your courses. When you have selected answers to all 44 questions, click on the "Submit" button at the end of the form.

Full Name:

1. I understand something better after I
   - try it out.
   - think it through.
2. I would rather be considered
   - realistic.
   - innovative.
3. When I think about what I did yesterday, I am most likely to get
   - a picture.

- words.

4. I tend to
   - understand details of a subject but may be fuzzy about its overall structure.
   - understand the overall structure but may be fuzzy about details.
5. When I am learning something new, it helps me to
   - talk about it.
   - think about it.
6. If I were a teacher, I would rather teach a course
   - that deals with facts and real-life situations.
   - that deals with ideas and theories.
7. I prefer to get new information in
   - pictures, diagrams, graphs, or maps.
   - written directions or verbal information.
8. Once I understand
   - all the parts, I understand the whole thing.
   - the whole thing, I see how the parts fit.
9. In a study group working on difficult material, I am more likely to
   - jump in and contribute ideas.
   - sit back and listen.
10. I find it easier
    - to learn facts.
    - to learn concepts.
11. In a book with lots of pictures and charts, I am likely to
    - look over the pictures and charts carefully.
    - focus on the written text.
12. When I solve math problems.
    - I usually work my way to the solutions one step at a time.
    - I often just see the solutions but then have to struggle to figure out the steps to get to them.
13. In classes I have taken.
    - I have usually gotten to know many of the students.
    - I have rarely gotten to know many of the students.
14. In reading nonfiction, I prefer.
    - something that teaches me new facts or tells me how to do something.

- something that gives me new ideas to think about.

15. I like teachers
    - who put a lot of diagrams on the board.
    - who spend a lot of time explaining.
16. When I'm analyzing a story or a novel.
    - I think of the incidents and try to put them together to figure out the themes.
    - I just know what the themes are when I finish reading and then I have to go back and find the incidents that demonstrate them.
17. When I start a homework problem, I am more likely to.
    - start working on the solution immediately.
    - try to fully understand the problem first.
18. I prefer the idea of
    - certainty.
    - theory.
19. I remember best
    - what I see.
    - what I hear.
20. It is more important to me that an instructor
    - lay out the material in clear sequential steps.
    - give me an overall picture and relate the material to other subjects.
21. I prefer to study
    - in a study group.
    - alone.
22. I am more likely to be considered
    - careful about the details of my work.
    - creative about how to do my work.
23. When I get directions to a new place, I prefer
    - a map.
    - written directions.
24. I learn
    - at a fairly regular pace. If I study hard, I'll "get it."
    - in fits and starts. I'll be totally confused and then suddenly it all "clicks."
25. I would rather first
    - try things out.

- think about how I'm going to do it.

26. When I am reading for enjoyment, I like writers to
    - clearly say what they mean.
    - say things in creative, interesting ways.
27. When I see a diagram or sketch in class, I am most likely to remember
    - the picture.
    - what the instructor said about it.
28. When considering a body of information, I am more likely to
    - focus on details and miss the big picture.
    - try to understand the big picture before getting into the details.
29. I more easily remember
    - something I have done.
    - something I have thought a lot about.
30. When I have to perform a task, I prefer to
    - master one way of doing it.
    - come up with new ways of doing it.
31. When someone is showing me data, I prefer
    - charts or graphs.
    - text summarizing the results.
32. When writing a paper, I am more likely to
    - work on (think about or write) the beginning of the paper and progress forward.
    - work on (think about or write) different parts of the paper and then order them.
33. When I have to work on a group project, I first want to
    - have "group brainstorming" where everyone contributes ideas.
    - brainstorm individually and then come together as a group to compare ideas.
34. I consider it higher praise to call someone
    - sensible.
    - imaginative.
35. When I meet people at a party, I am more likely to remember
    - what they looked like.
    - what they said about themselves.
36. When I am learning a new subject, I prefer to

- stay focused on that subject, learning as much about it as I can.
- try to make connections between that subject and related subjects.

37. I am more likely to be considered
    - outgoing.
    - reserved.
38. I prefer courses that emphasize
    - concrete material (facts, data).
    - abstract material (concepts, theories).
39. For entertainment, I would rather
    - watch television.
    - read a book.
40. Some teachers start their lectures with an outline of what they will cover. Such outlines are
    - somewhat helpful to me.
    - very helpful to me.
41. The idea of doing homework in groups, with one grade for the entire group,
    - appeals to me.
    - does not appeal to me.
42. When I am doing long calculations,
    - I tend to repeat all my steps and check my work carefully.
    - I find checking my work tiresome and have to force myself to do it.
43. I tend to picture places I have been
    - easily and fairly accurately.
    - with difficulty and without much detail.
44. When solving problems in a group, I would be more likely to
    - think of the steps in the solutions process.
    - think of possible consequences or applications of the solution in a wide range of areas.

References

1. Akram A, Fu C, Li Y, Javed MY, Lin R, Jiang Y, Tang Y (2019) Predicting students' academic procrastination in blended learning course using homework submission data. IEEE Access 7:102487–102498

2. Al-Samarraie H, Saeed N (2018) A systematic review of cloud computing tools for collaborative learning: opportunities and challenges to the blended-learning environment. Comput Educ 124:77–91
3. Albreiki B, Zaki N, Alashwal H (2021) A systematic literature review of student performance prediction using machine learning techniques. Educ Sci 11(9):552
4. Ashburn EA, Floden RE (eds) (2006) Meaningful learning using technology: what educators need to know and do. Teachers College Press
5. Buschetto Macarini LA, Cechinel C, Batista Machado MF, Faria Culmant Ramos V, Munoz R (2019) Predicting students success in blended learning—evaluating different interactions inside learning management systems. Appl Sci 9(24):5523
6. Chango W, Cerezo R, Romero C (2021) Multi-source and multimodal data fusion for predicting academic performance in blended learning university courses. Comput Electr Eng 89:106908
7. Chango W, Cerezo R, Romero C (2019) Predicting academic performance of university students from multi-sources data in blended learning. In: Proceedings of the second international conference on data science, E-learning and information systems, pp 1–5
8. Conijn R, Van den Beemt A, Cuijpers P (2018) Predicting student performance in a blended MOOC. J Comput Assist Learn 34(5):615–628
9. Conijn R, Snijders C, Kleingeld A, Matzat U (2016) Predicting student performance from LMS data: a comparison of 17 blended courses using Moodle LMS. IEEE Trans Learn Technol 10(1):17–29
10. Francesc P, Miguel S, Axel R, Paula V (2019). Artificial intelligence in education: challenges and opportunities for sustainable development. Unesdoc.unesco.org. https://unesdoc.unesco.org/ark:/48223/pf0000366994
11. Han F, Ellis RA (2019) Identifying consistent patterns of quality learning discussions in blended learning. Internet High Educ 40:12–19
12. Hariri W (2023) Unlocking the potential of ChatGPT: a comprehensive exploration of its applications, advantages, limitations, and future directions in natural language processing. Preprint at arXiv:2304.02017
13. Hu YH, Lo CL, Shih SP (2014) Developing early warning systems to predict students' online learning performance. Comput Hum Behav 36:469–478
14. Khan A, Ghosh SK (2021) Student performance analysis and prediction in classroom learning: a review of educational data mining studies. Educ Inf Technol 26:205–240
15. Liashchynskyi P, Liashchynskyi P (2019). Grid search, random search, genetic algorithm: a big comparison for NAS. Preprint at arXiv:1912.06059
16. Lugini L, Litman D (2019) Predicting specificity in classroom discussion. Preprint at arXiv: 1909.01462
17. Luo Y, Han X, Zhang C (2022) Prediction of learning outcomes with a machine learning algorithm based on online learning behavior data in blended courses. Asia Pac Educ Rev 1–19
18. López-Pérez MV, Pérez-López MC, Rodríguez-Ariza L, Argente-Linares E (2013) The influence of the use of technology on student outcomes in a blended learning context. Educ Tech Res Dev 61:625–638
19. Mwalumbwe I, Mtebe JS (2017) Using learning analytics to predict students' performance in Moodle learning management system: a case of Mbeya University of Science and Technology. Electr J Inf Syst Dev Ctries 79(1):1–13
20. Nakayama M, Mutsuura K, Yamamoto H (2017) The possibility of predicting learning performance using features of note taking activities and instructions in a blended learning environment. Int J Educ Technol High Educ 14(1):1–14
21. Nguyen VA (2017) The impact of online learning activities on student learning outcome in blended learning course. J Inf Knowl Manag 16(04):1750040
22. Peng H, Cambria E, Hussain A (2017) A review of sentiment analysis research in Chinese language. Cogn Comput 9:423–435
23. Picciano AG (2014) Big data and learning analytics in blended learning environments: benefits and concerns. IJIMAI 2(7):35–43

24. Rasheed RA, Kamsin A, Abdullah NA (2020) Challenges in the online component of blended learning: a systematic review. Comput Educ 144:103701
25. Romanenko IB, Puyu YV, Romanenko YM, Romanenko LY (2021) Digitalization of education: conservatism and innovative development. In: Knowledge in the information society: joint conferences XII communicative strategies of the information society and XX professional culture of the specialist of the future 12. Springer International Publishing, pp 22–29
26. Siemens G, Baker RSD (2012) Learning analytics and educational data mining: towards communication and collaboration. In: Proceedings of the 2nd international conference on learning analytics and Knowledge, pp 252–254
27. Soloman B, Felder R (2005) Index of learning styles questionnaire. https://www.webtools.ncsu.edu/learningstyles/. Accessed 17 May 2023
28. Tomasevic N, Gvozdenovic N, Vranes S (2020) An overview and comparison of supervised data mining techniques for student exam performance prediction. Comput Educ 143:103676
29. Turian J, Ratinov L, Bengio Y (2010) Word representations: a simple and general method for semi-supervised learning. In: Proceedings of the 48th annual meeting of the association for computational linguistics, pp 384–394
30. Waheed H, Hassan SU, Aljohani NR, Hardman J, Alelyani S, Nawaz R (2020) Predicting academic performance of students from VLE big data using deep learning models. Comput Hum Behav 104:106189
31. Weller HG, Repman J, Rooze GE (1994) The relationship of learning, behavior, and cognitive style in hypermedia-based instruction: implications for design of HBI. Comput Sch 10(3–4):401–418

Enhancing Predictive Performance in Identifying At-Risk Students: Integration of Topological Features, Node Embeddings in Machine Learning Models

Balqis Albreiki, Zahiriddin Rustamov, Jaloliddin Rustamov, and Nazar Zaki

Abstract This study proposes an innovative approach to the early identification of at-risk students in higher education settings by augmenting traditional machine learning classifiers with topological features and node embeddings derived from graph-based representations of student data from programming courses at the United Arab Emirates University. Utilizing a dataset comprising student demographic characteristics and course performance results, we construct an adjacency matrix using cosine and Canberra distance metrics, which is then thresholded to form a binary, unweighted graph. Topological features and node embeddings are extracted from this graph, representing the relationships and interactions between students. The extracted features are combined with the original dataset to train various machine learning classifiers, aiming to enhance their predictive accuracy. Experimental results demonstrate a significant improvement in prediction performance, with an increase of almost 10% in accuracy when node embeddings are incorporated. The most notable improvement was observed when using a Multi-Layer Perceptron classifier with the original dataset supplemented with both topological features and node embeddings, achieving 93.5% accuracy. Our findings highlight the potential of graph-based methods to enrich the feature set used by machine learning models, thereby enhancing their capacity to identify at-risk students early. Future work will focus on refining the feature extraction process, exploring other graph methods, and incorporating additional types of data. This study lays the foundation for more comprehensive and effective early-warning systems in higher education, aiming to enhance student support services and improve overall educational outcomes.

Keywords At-risk students · Higher education · Machine learning · Topological features · Node embeddings · Graph representation · Predictive accuracy

---

B. Albreiki · Z. Rustamov · J. Rustamov · N. Zaki (✉)
Department of Computer Science and Software Engineering, College of Information Technology, United Arab Emirates University, Al Ain, UAE
e-mail: Nzaki@uaeu.ac.ae

M. S. Khine (ed.), *Machine Learning in Educational Sciences*,
https://doi.org/10.1007/978-981-99-9379-6_10

1 Introduction

Academic excellence proves essential for continued education sustainability while serving as an accurate predictor for overall student determination. Research reveals that students with better academic performances tend to experience fewer challenges remaining enrolled compared with those who don't perform well academically [13]. Nonetheless, dropout rates continue to present global challenges affecting different countries, including Indonesia, Australia, the UK as well as the US. The UK has seen a surge of non-continuation rates among first-degree entrants projected at over thirty percent by their fourth year; Australia displays lower numbers at over seven percent for full-time entrants; twenty-five percent makes up America's relatively higher dropout rate among its population studying full-time courses during an academic year alone. Further to highlight, only 45% of students who enrol in college graduate within five years, according to the National Collegiate Retention and Persistence report published in 2012. As indicated by recent data, a concerning trend in Indonesia's higher education system is the sizable percentage (28.2%) of students who discontinued their studies in 2017 [1, 10, 11, 13, 15, 17].

According to a review of past studies, it was found that the majority of recent studies focused on the prediction of end-of-term success. However, by the end of the term, it may be too late to intervene and support struggling students, resulting in higher rates of attrition and lower academic achievement [14]. According to Azmat and Iriberri [5], students showed an improvement in grades when they were provided with feedback about their relative performance, suggesting that early feedback is beneficial for students. Similarly, Fritz [7] findings propose that providing students with information on their grades and performance earlier in the semester can encourage them to pursue academic support.

The traditional approach for schools was to depend on teachers' feedback and heuristic rules that relied on metrics such as GPA, absence rates, and tardiness to pinpoint students who might be at risk. While human judgment and heuristic rules can sometimes be correct, they are only general guidelines, remain unchanged, are expensive to maintain, and often contain errors. Moreover, the heuristic rules that prove effective in recognising at-risk students for a specific group of students within a particular school district might not apply or be useful for other groups or schools [12].

Recent research has shown how machine learning (ML) could be utilised to overcome the limitations of traditional approaches [12]. ML models could be trained using academic data regardless of its size to identify at-risk students earlier compared to the traditional approach, and it can be applied on a large scale. Despite the fact that ML has an advantage over the traditional approach, the most common approach is training an ML model on the original features of a dataset. Unfortunately, this method fails to capture additional insights within the data, such as relationships between instances, overfitting, and high dimensional features, which could be computationally expensive and irrelevant features that could negatively impact the performance of the model. To overcome the limitations of ML models trained on the original features of a dataset, this study proposes to extract topological features and generate node embeddings from the data to train various ML models and evaluate their performance. There is strong evidence showing the positive impact of generating node embeddings from a dataset to improve the performance of ML models in other domains, such as fake review detection [18]. Therefore, the objective of this study is to develop various ML classifiers to assess the extent to which incorporating topological features and node embeddings with the original dataset improves the early detection of at-risk students in a higher education setting.

The organization of the remaining sections in this chapter is as follows: Sect. 2 provides a review of relevant and literature on the identification of at-risk students in higher education settings. It explores key existing approaches and methods, highlighting their limitations. Section 3 introduces the proposed methodology, which involves augmenting traditional ML classifiers with topological features and node embeddings derived from graph-based representations of student data from programming courses. The methodology is accompanied by in-depth discussions on its design and implementation. In Sect. 4, the results and discussions of the experiments are presented, analysing the performance and effectiveness of the proposed approach. Finally, Sect. 5 concludes the chapter by summarizing the key findings, discussing their implications, and suggesting avenues for future research.

2 Literature Review

The importance of early identification of at-risk students in higher education settings cannot be overstated. Early detection facilitates the provision of timely and targeted support, potentially improving student outcomes and reducing drop-out rates. Over the years, numerous studies have focused on predicting student success using ML techniques, leading to the emergence of an expansive body of literature in this field. These works have utilized various types of data, from demographic characteristics to course performance metrics, and have applied an array of ML algorithms to predict student performance. While traditional ML approaches have shown promise, they often fail to account for the complex interactions and relationships between students, which can provide valuable insights into student behaviour and performance. Recently, graph-based methods have been proposed to address this gap, leveraging the rich, interconnected structure of student data to enhance prediction models.

In the following sections, we critically review existing literature in this domain, focusing on traditional ML-based prediction models and emerging graph-based approaches. Through this review, we aim to identify gaps in the current state of research and position our study within this context, highlighting the novelty and significance of our proposed approach.

The article by Waheed et al. [16] highlights the potential benefits of using technology-enhanced learning platforms to mine data on student learning behaviours. This data can be used to optimise the educational environment and enable data-driven decision-making. The authors of the study use a deep artificial neural network (Deep ANN) to predict at-risk students by analysing unique handcrafted features extracted from virtual learning environments' clickstream data. The study shows that this model achieves a classification accuracy of 84% to 93%, outperforming baseline models like logistic regression and Support Vector Machine (SVM). Legacy and assessment-related data are found to have a significant impact on the model's performance. Additionally, students who access the content of previous lectures tend to perform better. The study aims to help educational institutes develop frameworks for pedagogical support and facilitate higher education decision-making processes towards sustainable education.

The article by Marbouti et al. [13] discusses the use of predictive modelling methods to identify at-risk students early and inform instructors and students. The authors compare different predictive methods to identify at-risk students in a course that used standards-based grading, using only in-semester performance data available to instructors. They emphasise the importance of minimising false negative errors while not significantly increasing false positive errors when identifying at-risk students. To increase the accuracy of predictions, a feature selection method is used to reduce the number of variables in each model. The Naive Bayes (NB) Classifier model and an Ensemble model using a sequence of models had the best results among the seven tested modelling methods, emphasising the importance of selecting appropriate models to improve the accuracy of predicting at-risk students in courses that use standards-based grading.

The study by He et al. [8] focuses on predicting students' performance in a specific course using their personal biographical information and sequential behaviour data in a virtual learning environment (VLE). The researchers propose a novel recurrent neural network (RNN)-gated recurrent unit (GRU) joint neural network that can fit both static and sequential data. They compare the performance of three deep neural network algorithms (simple RNN, GRU, and LSTM) in identifying at-risk students and find that simple methods like GRU and simple RNN have better results than the more complex LSTM model. The proposed joint model achieves over 80% prediction accuracy of at-risk students at the end of the semester, and the results show that different models have different peak performance times. This study highlights the importance of considering both static and sequential data to predict students' performance in VLEs.

The study by Albreiki et al. [2] focuses on detecting at-risk students early on has numerous benefits for educational institutions, including improving student retention rates, enrollment management, and institutional effectiveness. This paper proposes a customized rule-based system (RBS) that identifies and visualizes at-risk students using risk flags (RF). The system provides instructors with a dashboard to monitor students' performance and suggests remedial actions based on severity and time flagged. Implementing the system is expected to improve students' achievement, benefit underperforming students, educators, and academic institutions.

The paper by Albreiki et al. [4] proposes a method that uses graph representations of student data to predict students' academic performance and risk of failure. The technique involves converting tabulated data into graphs using distance measures and extracting graph topological features. The resulting data can be used alone or jointly with the initial dataset to improve the predictive power of ML methods. The proposed method was tested on an educational dataset and returned superior results, with the use of graph topological features improving prediction accuracy by 2% in terms of AUC and 3.3% in terms of accuracy. By incorporating graph-embedding features through a graph convolutional network (GCN), at-risk students can be identified with 87.4% accuracy, providing a tool for the early detection of at-risk students, and benefiting universities by improving their prediction performance and reputation.

The article by Hu et al. [9] discusses the importance of an early warning system to predict student performance and identify at-risk students. Previous studies have focused on post-course performance evaluation and neglected the practical value of an early warning system during the course. The authors collected data from an online undergraduate course and used data mining techniques to develop an early warning system. The results showed that time-dependent variables extracted from the learning management system (LMS) are critical factors for online learning, and the classification and regression tree (CART) model, supplemented by AdaBoost, is the best classifier for evaluating learning performance. The study highlights the usefulness of data mining techniques in developing early warning systems to improve student performance in online courses.

The article by Albreiki et al. [3] introduces a model designed to accurately identify and aid students at risk of underperforming. Using explainable ML techniques, the model identifies factors linked to underperformance and suggests tailored remedial strategies based on course learning outcomes. By integrating this with a rule-based risk flagging system, the model improves the precision of performance predictions. The model has proven highly accurate, with an Area Under the ROC Curve of 1.0 when trained with four checkpoints. The implementation of this model can enhance student performance and contribute to institutional effectiveness, potentially improving reputation and rankings.

The study by Chui et al. [6] proposed a reduced training vector-based support vector machine (RTV-SVM) for predicting at-risk and marginal students in university settings. Although the RTV-SVM achieved high accuracy rates in predicting at-risk and marginal students, the study's limitations include limited analysis of other potential features and data sources that could enhance the predictive model.

Adejo and Connolly [1] investigated the use of multiple data sources, different classifiers, and ensemble techniques for predicting student academic performance. While the study demonstrated the effectiveness of using multiple data sources and ensemble techniques in accurately predicting student performance and identifying at-risk students, the generalizability of the findings may be limited due to the relatively small sample size.

Wakelam et al. [17] focused on predicting individual student performance in small university cohorts with limited attributes. The study's limitations include the small sample size and the exclusive reliance on lecture/tutorial attendance, virtual learning environment access, and intermediate assessments as data sources, which may not capture the full range of factors influencing student performance.

Lakkaraju et al. [12] presented a ML framework for identifying at-risk students in high schools. While the framework demonstrated promise in identifying at-risk students and providing appropriate support, the study's limitations include the reliance on data from only two U.S. school districts, which may limit the generalizability of the findings to other educational contexts.

The main weaknesses of the mentioned studies that primarily rely on conventional ML with limited adoption of graph data structures and solely use table-based data include the potential lack of capturing complex relationships and interactions among variables. Conventional ML approaches often struggle to handle interconnected data effectively, which can lead to limited predictive power and insights. By not fully leveraging the rich structural information present in graph data, these studies may miss out on uncovering valuable patterns and correlations that could enhance the accuracy and depth of predictions. Furthermore, the exclusive reliance on table-based data may overlook the holistic nature of student performance, as it may not capture the dynamic and interdependent aspects of student behavior and learning processes. To overcome these weaknesses, here we explore the integration of graph-based representations and methodologies to better capture the complex nature of educational data and improve the prediction and support of at-risk students.

3 Methodology

In this research, we utilized an education dataset as presented by Albreiki et al. [4]. This dataset encompasses 230 entries and 36 characteristics, detailing student demographics and their performance metrics in a programming course conducted by the Department of Computer Science and Software Engineering at the United Arab Emirates University (UAEU). The data spans several semesters, covering the academic years from 2016/2017 through 2020/2021.

Figure 1 provides an inclusive visual representation of the adopted methodology, which is primarily intended to determine the effectiveness of integrating topological features and node embeddings with the original dataset to optimize performance outcomes. The study's main goal is to harness these additional data dimensions to provide a more comprehensive and accurate depiction of the data, consequently enhancing the precision and reliability of the analytical outcomes.

In order to guarantee the appropriateness and relevance of the dataset, data pre-processing is indispensable. This step enhances the quality of data and optimizes it for further analysis. Figure 2 provides a comprehensive representation of the chosen features, along with their respective descriptions. This not only aids in a deeper understanding of the variables under scrutiny, but also sets the stage for subsequent analysis and modeling.

The data pre-processing process involved several distinct stages to ensure the quality and consistency of the data. The stages can be summarized as follows:

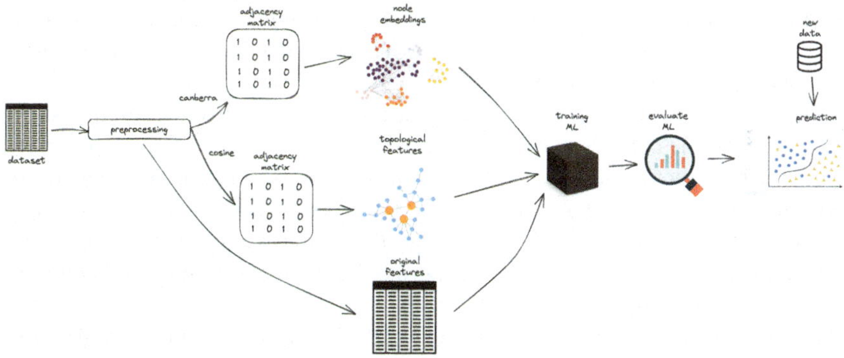

Fig. 1 Proposed methodology: integrating topological features and node embeddings with the original sample dataset

	Features	Description
course related	AgeCourseStart	Student age at course start
	Quiz1Norm	1st quiz score (normalized)
	Quiz2Norm	2nd quiz score (normalized)
	HW1Norm	1st homework score (normalized)
	MTNorm	Midterm score (normalized)
	TGNorm	Total grade score (normalized)
historical features	AgeAdmitted	Student age at university admission
	RegHrs	Registered hours count
	CumGPA	Cumulative GPA
	TermGPA	Term GPA
	HsGPA	High school GPA
	Math	High school math score
	Phys	High school physics score
	Repeated Grade (ITBP119/CSBP119)	Course repeat count
	ITBP119-CSBP119-max	Max course score
	Repeated Grade (ITBP219/CSBP219)	Course repeat count
	ESPN	ESPN score
	CSBP121	CSBP121 score
	MATH105	Math105 score
	CENG205	CENG205 score
	PHYS105	PHYS105 score
	CENG202	CENG202 score
	CIT	CIT student status
	AcademicStanding	Academic standing
	Citizenship	Citizenship status
	Gender	Student gender
	Sponsor	Sponsorship status
	Resident	City residency status

Fig. 2 Selected features and their descriptions for data analysis and modeling

1. **Data Combination**: Course assessment files, student information from the Banner system, and manually collected documents were combined to create a comprehensive dataset.

2. **Data Cleansing**: The gathered data were thoroughly cleansed to remove any unnecessary or irrelevant entries, ensuring that only relevant information remained for further analysis.
3. **Data Standardization**: To address inconsistencies arising from different instructors teaching the courses, the data underwent standardization. This step ensured a consistent structure across all files, enabling seamless integration and analysis.
4. **Missing Data Handling**: Any missing data values were addressed using an imputation method. Missing entries were replaced with the average value of the corresponding coursework components, ensuring that the dataset remained complete for subsequent analysis.
5. **Data Aggregation**: The data were aggregated and merged into a single file, facilitating efficient analysis and processing.
6. **Data Normalization**: Categorical data were transformed into numerical values, and the grades were normalized using min–max normalization. This normalization technique rescaled the grades to a range of [0, 1], allowing for fair comparison and analysis.

Moreover, an additional column was introduced based on specific rules and critical milestones in student performance. This column categorized students into three main groups based on their total grade (TG): Good ($TG \geq 0.7$), At-Risk ($0.6 \leq TG < 0.7$), and Expected to Fail ($TG < 0.6$). The distribution of the target class revealed that approximately 28% of the grades fell under the "At-Risk" category as shown in Fig. 3. It is important to note that only course scores obtained before the midterm (MT) were considered in the analysis, as the objective was to identify at-risk students in advance. This rigorous data pre-processing process ensures the quality, consistency, and relevance of the dataset, setting the foundation for subsequent analysis and identification of at-risk students.

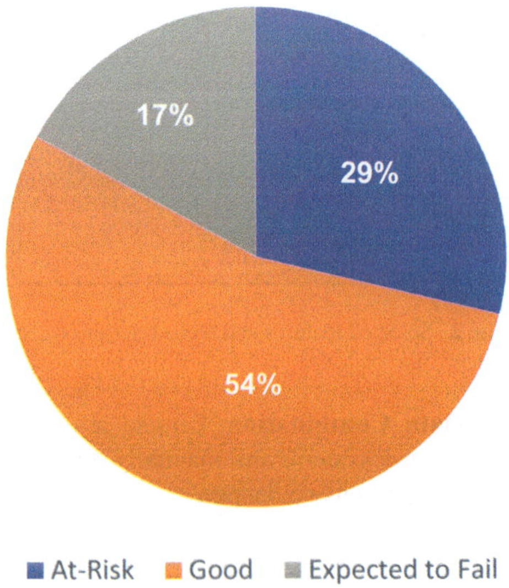

Fig. 3 Distribution of the target classes

3.1 Graph Representation

To develop a comprehensive graph representation of the data, we must first examine the interconnections between students, based on their demographic traits and academic performances. This necessitates computing an adjacency matrix utilizing distance metrics like cosine, Euclidean, and Manhattan. In this section, we delve into the procedures for creating the graph G.

The first step is to isolate relevant features from the dataset, encompassing both demographic details (such as age and gender) and course-related information (for instance, scores from quizzes and homework). Subsequently, we calculate distance metrics for each student pair in the dataset. We've chosen two specific metrics for this purpose: cosine and Canberra.

The cosine distance, which ranges from 0 (perfect similarity) to 1 (complete dissimilarity), measures the cosine of the angle between two vectors of features, as depicted in formula (1). This metric effectively encapsulates the relationships between students, grounded in their demographic and course attributes, via the angle between their respective feature vectors. Conversely, the Canberra distance is essentially a weighted Manhattan distance, where the total of absolute differences between the corresponding elements of two feature vectors is divided by the aggregate of the absolute values of these components. Given its emphasis on non-zero elements, this metric is particularly beneficial when comparing sparse vectors. The Canberra distance formula is expressed in (2).

$$d(u, v) = 1 - \frac{u \cdot v}{\|u\|_2 \|v\|_2} \tag{1}$$

$$d(u, v) = \sum\nolimits_i \frac{|u_i - v_v|}{|u_i| + |v_i|} \tag{2}$$

After determining these distance metrics, we proceed to create an adjacency matrix for the graph. Each element within this matrix signifies the similarity between two students (i.e., the nodes of the graph) as measured by the cosine and Canberra distances. Initially, this matrix gives rise to a weighted graph that encapsulates the subtle differences between student pairs.

However, to facilitate further analysis, we simplify the graph structure by establishing the matrix average as a threshold, thus converting the graph to an unweighted format. This transformation fosters a more efficient extraction of topological features and the generation of node embeddings in subsequent stages.

Finally, using the computed adjacency matrix, we create the graph representation of the dataset. This graph, denoted as G, comprises nodes representing students and edges indicating their interconnected relationships.

Through this methodical approach, we generate a graph representation that captures the complex relationships between students. This acts as a foundation for enriching the dataset via the extraction of topological features and creation of node embeddings. Consequently, this potentially augments the predictive accuracy of ML models.

3.2 Topological Features

In this research, we contemplate a broad range of topological features to encapsulate different attributes of the nodes in the unweighted graph G. The topological features selected for consideration include In-Degree, Out-Degree, Degree, Weighted In-Degree, Closeness Centrality, Harmonic Centrality, Betweenness Centrality, Authority, Hub, Modularity, PageRank, Triangles, and Eigenvector Centrality. For more details about the topological features please refer to [4]. The aim of extracting these topological features is to depict various facets of the graph structure and to highlight the significance of individual nodes. This comprehensive approach, in combination with the original dataset, could potentially enhance the predictive accuracy of ML models.

3.3 Node Embeddings

In the present study, we utilize the Node2Vec algorithm to generate low-dimensional, continuous node embeddings representing our graph's nodes. Node2Vec is an unsupervised method for generating graph embeddings that efficiently capture the local and global structural attributes of the graph. It essentially extends the principles of the Skip-gram model, traditionally used in natural language processing, to network data. By leveraging random walks to navigate each node's vicinity, Node2Vec fine-tunes the embeddings to ensure nodes sharing similar neighborhoods correspondingly share similar embeddings. The purpose of deploying Node2Vec is to condense the complex structural and relational features of the graph into a more manageable format. When these embeddings are combined with topological features and the original dataset, they provide a more comprehensive insight into the network structure. This enhanced representation could potentially augment the predictive accuracy of diverse ML models.

3.4 Feature Selection

To ensure that our predictive model is influenced by only the most informative features, we employ a feature selection process, focusing on the selection of the top 'k' features based on their univariate statistical significance. This process leverages the SelectKBest method, which earmarks the top 'k' features according to their respective scores. We experiment with both the chi-squared (chi2) and ANOVA F-value (f_classif) tests to ascertain which criterion best suits our enhanced dataset. This step assists us in pinpointing the most pertinent topological features and node embeddings. By implementing feature selection, we strive to mitigate the complexity of the dataset and bolster the performance of our ML models, sidestepping the need for dimensionality reduction techniques such as PCA. Consequently, our models are trained on a condensed yet informative feature set, which could potentially enhance prediction accuracy.

3.5 Model Training and Evaluation

During the model selection and training phase, we intend to examine a diverse array of ML algorithms. The list encompasses SVM, Logistic Regression (LR), Random Forest (RF), Gradient Boosting (GB), AdaBoost, XGBoost (XGB), Decision Tree (DT), Naïve Bayes (NB), Multilayer Perceptron (MLP), and K-Nearest Neighbor algorithms (KNN).

In the subsequent model evaluation and comparison phase, our strategy involves evaluating the performance of the trained models using a robust collection of metrics. These include accuracy, balanced accuracy, precision, recall, F1-score, and Cohen's kappa. Accuracy, specifically, represents the ratio of correct predictions made by the model, which is expressed by the following formula:

$$Accuracy = \frac{TP + TN}{TP + TN + FP + FN} \tag{3}$$

Precision, sometimes referred to as the positive predictive value, is a measure of the model's ability to correctly identify positive instances. It is calculated as the ratio of true positives (actual positives correctly identified) to the combined total of true positives and false positives (negative instances incorrectly classified as positive). The formula is as follows:

$$Precision = \frac{TP}{TP + FP} \tag{4}$$

Recall, also known as sensitivity or the true positive rate, is a metric that quantifies the model's capacity to correctly identify the actual positive instances. It does this by determining the proportion of true positives in relation to the sum of true positives and false negatives (actual positives that were incorrectly classified as negatives). The calculation is given by the following formula:

$$Recall = \frac{TP}{TP + FN} \tag{5}$$

The F1-score represents the harmonic mean of precision and recall. It is an effective measure that balances both these components, thereby providing a more comprehensive evaluation of a model's performance. The computation of the F1-score involves the following formula:

$$F1 = 2 \times \frac{Precision \times Recall}{Precision + Recall} \tag{6}$$

Cohen's kappa is a robust statistical measure that gauges the level of agreement between categorical classifications, while adjusting for any potential agreement that might occur by chance. In essence, it provides a more nuanced perspective on model performance, particularly when dealing with imbalanced datasets. The formula for calculating Cohen's kappa is as follows:

$$\kappa = \frac{(P_o - P_e)}{(1 - P_e)} \tag{7}$$

In the formula for Cohen's kappa, the observed accuracy represents the proportion of instances on which the raters are in consensus, while the expected accuracy signifies the proportion of instances where agreement would be expected solely by random chance. In this study, we give particular emphasis to the F1-score as the decisive metric. Given its design to proficiently handle imbalanced datasets, the F1-score provides a balanced measure of the model's performance, striking a harmony between precision and recall, two crucial aspects in model evaluation.

4 Results and Discussions

This research aims to construct various ML classifiers to scrutinize the enhancement in early detection of students at risk in a tertiary education environment when topological features and node embeddings are integrated with the original dataset. The application of graph representations in our study provides a richer and deeper understanding of student interactions compared to conventional ML methodologies. After excluding grade-related features ranging from midterm to final grades, our refined dataset consisted of 28 features. In order to ascertain the benefits of unifying topological features (TF) and node embeddings (ND) with the original dataset (OD), we devised and executed an array of experiments. These included initial trials with just the original features, followed by examinations exclusively with topological features and then solely with node embeddings. Subsequently, we combined the original features with both the topological features and node embeddings separately, and ultimately, amalgamated all three sets of features for a comprehensive analysis.

In our investigation, we tried different distance metrics to extract topological features and generate node embeddings. Notably, the Cosine distance metric proved most effective in extracting topological features, and the Canberra distance metric

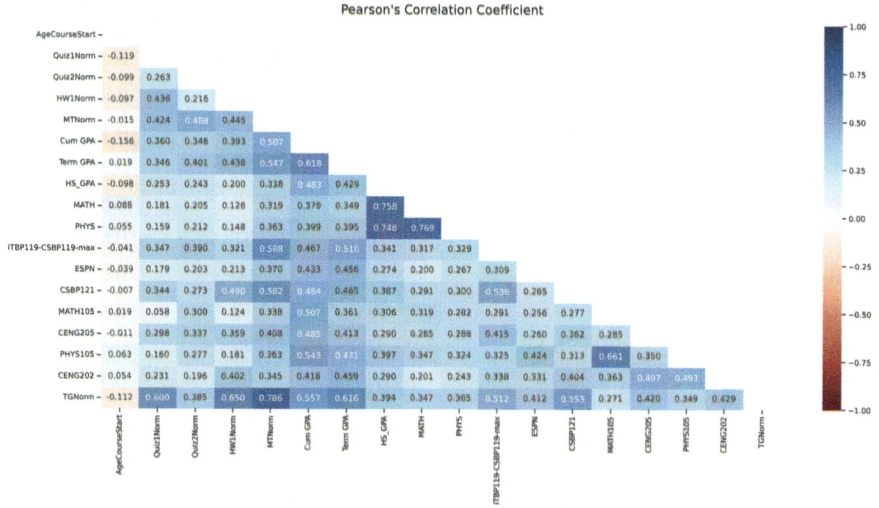

Fig. 4 Correlation heatmap of selected features for graph representation

demonstrated superior results in generating node embeddings. The 80–20 train-test split sampling technique was used in our analysis. For the training phase, nine classifiers were implemented, including SVM, RF, LR, MLP, DT, XGB, GB, KNN, and NB. Of particular note, distinct feature sets were used to create graph representations for topological features and node embeddings to avoid introducing noise that could impede effective graph feature extraction. Certain features with higher correlations, such as Quiz1, Homework 1 (HW1) and Midterm Exam (MT), were found to enhance graph feature capture when constructing topological features, as shown by Pearson's Correlation Coefficient in Fig. 4. Moreover, strong positive correlations were found between midterm and term grade scores (0.786), and high school GPA with math (0.758) and physics (0.748) scores, suggesting high-achieving students consistently perform well across these areas.

In Table 1, we detail the performance evaluation of various models utilizing only the initial dataset features, termed as Original Dataset (OD). Strikingly, the SVM model stood out across all key metrics, delivering the highest accuracy (84.78%), recall (84.78%), F1-score (84.20%), and Cohen's Kappa (0.73), thereby indicating its superiority in accurately predicting and classifying the target classes within the OD context. Interestingly, while the NB model registered the lowest accuracy (63.04%), it surprisingly achieved the peak precision (85.82%) among all evaluated models, suggesting its high reliability when it correctly predicted a class. However, it is subpar accuracy and other metrics that hint towards its probable high rate of misclassifications. These initial results serve as a crucial benchmark, offering valuable insights for comparing the potential improvements upon the integration of topological features and node embeddings with the original dataset in the forthcoming experimental stages.

Table 1 Performance evaluation of ml models using original dataset features only

Model	Accuracy	Precision	Recall	F1-score	Kappa
SVM	84.78	84.98	84.78	84.20	0.73
RF	82.61	82.41	82.61	81.80	0.70
LR	80.43	79.67	80.43	79.42	0.66
MLP	78.26	77.53	78.26	77.57	0.62
DT	78.26	77.89	78.26	77.31	0.62
XGB	78.26	77.53	78.26	77.00	0.62
GB	76.09	74.07	76.09	74.10	0.58
KNN	73.91	76.44	73.91	70.70	0.52
NB	63.04	85.82	63.04	59.99	0.44

Subsequently, we proceeded to conduct experiments solely utilizing the extracted topological features from the original dataset to train the classifiers. Table 2 displays the evaluation results obtained for these topological features. Intriguingly, the MLP model emerged as the top performer among all the models when trained solely on topological features. It achieved the highest accuracy (80.43%), precision (80.39%), F1-score (79.71%), and Cohen's kappa (0.66). This suggests that the MLP model possesses a particular proficiency in leveraging the structural information embedded within the topological features. In contrast to the previous findings where the SVM model displayed the best performance, it appears to be less effective when trained exclusively on topological features. The accuracy, precision, F1-score, and kappa values for the SVM model noticeably dropped to moderate levels. Notably, none of the models in this experiment surpassed the highest accuracy achieved by the SVM model when trained on the original dataset. This implies that while topological features offer valuable insights, they may not be as potent as the original features in predicting at-risk students when utilized in isolation. These outcomes will serve as a significant reference point as we evaluate the performance of models trained on combinations of different feature types.

In Fig. 5, we present the Pearson's correlation coefficients for the extracted topological features. Notably, certain features, including Degree, ClosenessCentrality, HarmonicCentrality, Authority, Hub, and EigenvectorCentrality, demonstrate a strong positive correlation with coefficients close to 1. This implies that these features may capture similar aspects of the data, suggesting the potential presence of redundancy. Consequently, the consideration of dimensionality reduction techniques or feature selection methods may be warranted to eliminate this redundancy and enhance the model's performance.

Remarkably, the Modularity feature displays a notably low correlation with other features as well as the target variable. This suggests that it may have limited predictive power in its current state. These insights can guide the feature selection process, enabling a focus on the most informative features for accurately predicting student grades. However, it is crucial to note that correlation does not establish causation.

Table 2 Evaluation results of machine learning models trained on topological features only

Model	Accuracy	Precision	Recall	F1-score	Kappa
MLP	80.43	80.39	80.43	79.71	0.66
KNN	73.91	76.64	73.91	72.53	0.53
SVM	73.91	72.95	73.91	72.03	0.53
LR	71.74	70.65	71.74	69.69	0.49
XGB	69.57	69.06	69.57	67.30	0.44
NB	67.39	65.49	67.39	65.60	0.42
RF	63.04	62.23	63.04	61.62	0.34
GB	60.87	58.70	60.87	57.63	0.27
DT	58.70	55.96	58.70	56.90	0.29

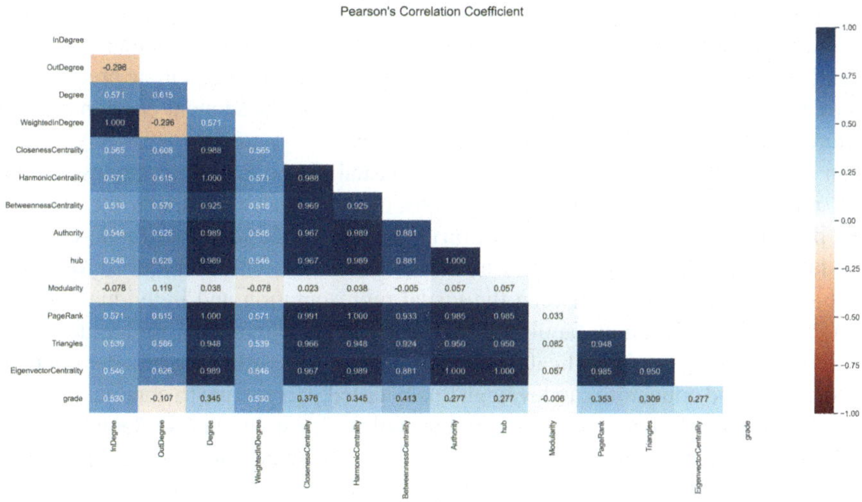

Fig. 5 Correlation analysis of extracted topological features using Pearson's coefficients

Thus, further analysis is warranted to gain a deeper understanding of the relationships between these features and the target variable. This correlation analysis provides valuable insights for feature selection, emphasizing the importance of selecting non-redundant and highly informative features to enhance the predictive capabilities of the model.

In Table 3, we present the performance evaluation results of various machine-learning models trained solely on node embeddings. Remarkably, the NB model exhibited substantial improvement when trained on node embeddings, achieving the highest accuracy (89.13%), precision (89.43%), F1-score (89.19%), and Cohen's kappa (0.82) among all evaluated models. This exceptional performance suggests that the NB model effectively leverages the information captured within the node embeddings for accurate predictions.

Interestingly, in contrast to the previous experiment where the MLP model excelled with topological features, its performance decreased when using node embeddings, as indicated by a decline in all evaluation metrics. However, it is noteworthy that the NB model trained on node embeddings surpassed the performance of the best model SVM trained on the original dataset. This implies that the node embeddings encapsulate crucial information that may not be present in the original features or topological features alone. The remarkable performance boost can be attributed to the effective random walks performed by the Node2Vec algorithm, as evidenced by the t-SNE visualization depicted in Fig. 6a. This visualization provides a clear understanding of the high-dimensional node embeddings in a two-dimensional space. Although there is some overlap among data points from different classes, which is common in real-world data, the discernible separation between students from different classes suggests that the node embeddings capture meaningful information. This information proves valuable for the ML models, particularly the NB model, in effectively distinguishing between classes and making accurate predictions. These results offer valuable insights that will guide the subsequent combination of features for further model training and evaluation.

In Fig. 6b, we present an enriched t-SNE visualization that incorporates student IDs, allowing for a more detailed analysis of the data. To explore the interactive plot, please visit https://www.zahiriddin.com/projects/stu_perf/node-embeddings.html. Each data point in the scatter plot is labeled with a unique student ID, establishing a direct link between the visual representation and individual student data. This interactive view facilitates a granular understanding of the grade distribution by enabling the identification of specific students within the broader context. Researchers and educators can leverage this detailed perspective to conduct targeted analyses and develop intervention strategies tailored to individual students' needs.

Table 4 presents a comparison of evaluation results when integrating topological features (TF) and node embeddings (ND) with the original dataset features (OD).

Table 3 Evaluation results of ML models trained on node embeddings

Model	Accuracy	Precision	Recall	F1-score	Kappa
NB	89.13	89.43	89.13	89.19	0.82
LR	82.61	83.05	82.61	82.78	0.71
SVM	82.61	82.92	82.61	82.67	0.71
RF	78.26	80.02	78.26	77.02	0.61
KNN	76.09	75.44	76.09	75.49	0.59
XGB	76.09	74.36	76.09	74.65	0.58
GB	76.09	74.86	76.09	74.57	0.57
MLP	73.91	73.19	73.91	73.41	0.55
DT	65.22	67.59	65.22	66.00	0.44

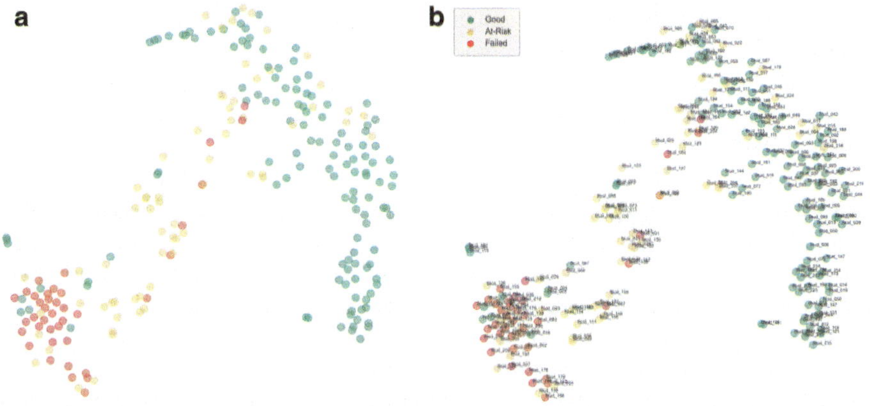

Fig. 6 Visualization of node embeddings using t-SNE and Node2Vec algorithm with Canberra distance metric: understanding student grade distribution. **a** Scatter plot of the t-SNE embeddings. Each point represents a student, and the colour of the point indicates the student's grade. **b** Scatter plot of the t-SNE embeddings with student IDs. Each point is labelled with the corresponding student's ID, providing a more detailed view of the student distribution

The purpose of this analysis was to examine the impact of combining these feature sets on model performance.

Among the models tested, the MLP model demonstrated the most significant improvement when utilizing all three feature sets (OD, TF, and ND). It achieved an impressive accuracy increase of 15.22% compared to using the original dataset features alone. This suggests that the MLP model effectively leverages the additional information provided by the topological features and node embeddings to enhance its prediction accuracy.

Table 4 Performance comparison of models with different feature combinations

Model	Accuracy						
	OD	+TF	Δ_{ACC}	+ND	Δ_{ACC}	+TF +ND	Δ_{ACC}
DT	78.26	78.26	0.00	69.57	−8.69	71.74	-6.52
GB	76.09	78.26	2.17	82.61	6.52	80.43	4.34
KNN	73.91	78.26	4.35	80.43	6.52	84.78	10.87
LR	80.43	86.96	6.53	86.96	6.53	86.96	6.53
MLP	78.26	86.96	8.69	80.43	2.17	93.48	15.22
NB	63.04	76.09	13.05	91.30	28.26	82.61	19.57
RF	82.61	82.61	0.00	82.61	0.00	82.61	0.00
SVM	84.78	86.96	2.18	82.61	-2.17	84.78	0.00

OD: Original Dataset; TP: Topological Features; ND: Node Embeddings; Δ_{ACC}: change in accuracy

The NB model also exhibited notable improvements when incorporating node embeddings, with a substantial accuracy increase of 28.26%. However, its performance slightly declined when adding topological features to the node embeddings. This suggests that the NB model might be more sensitive to the inclusion of certain types of features.

In contrast, the DT and SVM models did not show any improvement or experienced a slight decrease in accuracy when incorporating the additional features. Similarly, the RF model demonstrated no change in accuracy across all feature combinations. These results indicate that these models might not benefit significantly from the added complexity of the topological features and node embeddings.

Overall, the findings highlight the model-dependent nature of the impact of integrating topological features and node embeddings. The MLP and NB models demonstrated the most substantial performance improvements, emphasizing the importance of selecting appropriate models for specific feature combinations.

Figure 7 displays the impact of integrating different feature sets on the accuracy and F1-score of the models. The inclusion of topological features and node embeddings led to notable improvements, resulting in performance gains of up to 10%. Particularly, the integration of node embeddings demonstrated the most significant change, highlighting their effectiveness in capturing intricate network relationships and interactions among students. As a result, the classifiers exhibited enhanced predictive capabilities, leading to higher accuracy and F1-scores.

The summary table, displayed in Table 5, offers a comprehensive overview of the experimental results, showcasing the impact of different feature combinations

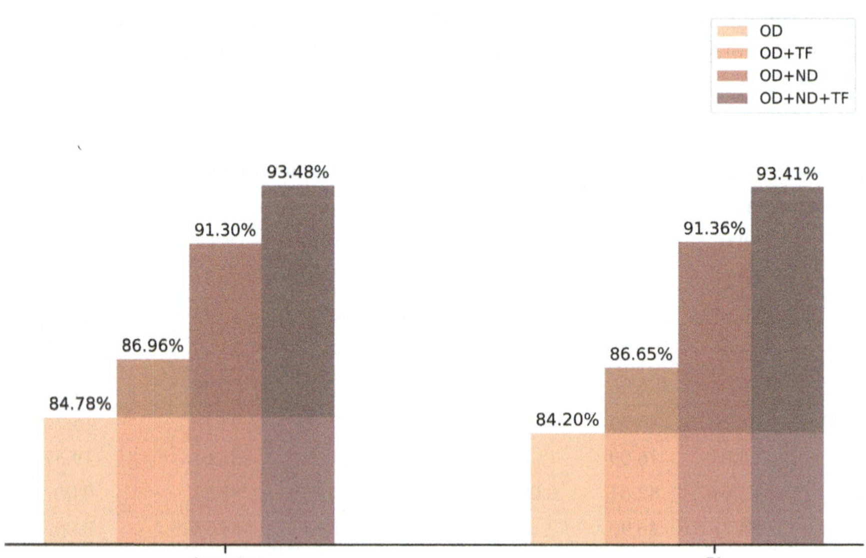

Fig. 7 Evaluation comparison of incorporating topological features and node embeddings on original dataset features: bar chart analysis

on various models. Notably, the MLP model, utilizing a combination of Original Dataset (OD), Topological features (TP), and Node Embeddings (ND), demonstrated the highest performance across all metrics. With an accuracy of 93.48%, precision of 93.46%, F1-score of 93.41%, and kappa of 0.89, the MLP model proved its proficiency in handling the complexities and patterns within this combined feature set, resulting in superior predictive performance.

The NB model showcased exceptional performance when incorporating the Original Dataset (OD) with Node Embeddings (ND), achieving an accuracy of 91.30%. Furthermore, even when using Node Embeddings (ND) alone, the NB model achieved an accuracy of 89.13%. This indicates the NB model's capability to effectively leverage node embeddings, which capture essential information that significantly enhances prediction accuracy.

The performance of the SVM model using the Original Dataset (OD) alone is noteworthy. Despite not incorporating any additional features, the SVM model achieved an accuracy of 84.78%, which is competitive when compared to the performance of other models utilizing combined features.

The LR model exhibited strong performance when combining the Original Dataset (OD) with Topological features (TP), achieving an accuracy of 86.96%. However, its performance did not surpass that of the MLP and NB models with additional features. This suggests that LR effectively utilizes topological features, but the inclusion of node embeddings could potentially further enhance its performance.

Interestingly, the MLP model's performance with only Topological features (TP) was significantly lower compared to its performance with combined features. This highlights the importance of incorporating multiple feature sets to maximize the model's predictive capabilities.

This underscores the significance of integrating node embeddings with topological features to enhance the model's performance. These findings emphasize the potential of combining traditional data features with topological features and node embeddings to improve the performance of ML models. However, it is important to note that the effectiveness of these combined features heavily relies on the choice of the ML model. While models such as MLP and NB demonstrated substantial improvements when incorporating these features, other models like SVM and LR exhibited varying

Table 5 Summary of the evaluation results on different feature sets

Feature	Model	Accuracy	Precision	Recall	F1-score	Kappa
OD + TP + ND	MLP	93.48	93.46	93.48	93.41	0.89
OD + ND	NB	91.30	91.98	91.30	91.36	0.86
ND	NB	89.13	89.43	89.13	89.19	0.82
OD + TP	LR	86.96	87.10	86.96	86.65	0.77
OD	SVM	84.78	84.98	84.78	84.20	0.73
TP	MLP	80.43	80.39	80.43	79.71	0.66

OD: Original Dataset; TP: Topological Features; ND: Node Embeddings degrees of enhancement. This highlights the importance of thoughtful consideration and experimentation when selecting the suitable model and feature combination for specific prediction tasks.

The study highlights the significant potential of incorporating topological features and node embeddings alongside traditional dataset features to enhance the prediction accuracy of ML models. The experiments demonstrate that the inclusion of these additional features can lead to substantial improvements in model performance. However, the effectiveness of the combined features relies heavily on the choice of the ML model.

Among the models tested, the MLP model stood out as highly effective in leveraging the combination of all three feature sets (Original Dataset, Topological features, and Node Embeddings), achieving the highest performance across all metrics. Similarly, the NB model demonstrated exceptional performance with the incorporation of node embeddings, indicating that these features capture crucial information that significantly enhances prediction accuracy. Conversely, the SVM and LR models exhibited varying degrees of enhancement when using the combined features. Notably, the SVM model performed competitively using only the original dataset features, highlighting its robustness in this prediction task without the need for additional features.

These results emphasize the importance of thoughtful model selection and feature combination in early prediction of at-risk students. They also underscore the need for further research to investigate the interactions between different types of features and ML models, and how they can be optimally combined to improve prediction accuracy in various contexts. Such research efforts have the potential to lead to more sophisticated and accurate prediction models, thereby enhancing our ability to identify at-risk students in a higher education setting and provide them with effective support.

5 Conclusion

This study aimed to assess the potential of integrating topological features and node embeddings to enhance the prediction accuracy of ML models in identifying at-risk students in a higher education setting. The results demonstrated significant improvements in predictive performance when incorporating these additional features. The MLP model showed remarkable performance with the combined feature set, while the NB model exhibited strong performance with the inclusion of node embeddings. However, the effectiveness of the combined features varied across different models, highlighting the importance of careful model selection and feature combination.

The findings of this study have important implications for early identification and support of at-risk students. By leveraging the rich information captured by topological features and node embeddings, institutions can improve their ability to predict student outcomes and intervene proactively to provide timely support.

These insights contribute to the development of effective support systems in higher education, ultimately improving student success and well-being.

Further research is needed to explore additional graph-based methods and incorporate diverse datasets to enhance the predictive capabilities of the models. The inclusion of real-time engagement metrics, socio-economic factors, and personal characteristics could provide valuable insights for refining the models and achieving even higher prediction accuracy. This study serves as a foundation for future investigations in this field and opens avenues for practical applications in student support and retention programs.

The integration of topological features and node embeddings with traditional dataset features holds great promise in improving the prediction accuracy of ML models for identifying at-risk students. This research contributes to the growing body of knowledge in the field of educational data analytics and paves the way for data-driven interventions that can positively impact student outcomes in higher education.

References

1. Adejo OW, Connolly T (2018) Predicting student academic performance using multi-model heterogeneous ensemble approach. J Appl Res High Educ 10(1):61–75. https://doi.org/10.1108/JARHE-09-2017-0113
2. Albreiki B, Habuza T, Shuqfa Z, Serhani MA, Zaki N, Harous S (2021) Customized rule-based model to identify at-risk students and propose rational remedial actions. Big Data Cogn Comput 5(4):71. https://doi.org/10.3390/bdcc5040071
3. Albreiki B, Habuza T, Zaki N (2022) Framework for automatically suggesting remedial actions to help students at risk based on explainable ML and rule-based models. Int J Educ Technol High Educ 19(1):49. https://doi.org/10.1186/s41239-022-00354-6
4. Albreiki B, Habuza T, Zaki N (2023) Extracting topological features to identify at-risk students using machine learning and graph convolutional network models. Int J Educ Technol High Educ 20(1):23. https://doi.org/10.1186/s41239-023-00389-3
5. Azmat G, Iriberri N (2010) The importance of relative performance feedback information: evidence from a natural experiment using high school students. J Public Econ 94(7–8):435–452. https://doi.org/10.1016/j.jpubeco.2010.04.001
6. Chui KT, Fung DCL, Lytras MD, Lam TM (2020) Predicting at-risk university students in a virtual learning environment via a machine learning algorithm. Comput Human Behav 107(December 2017):105584. https://doi.org/10.1016/j.chb.2018.06.032
7. Fritz J (2011) Classroom walls that talk: Using online course activity data of successful students to raise self-awareness of underperforming peers. Internet High Educ 14(2):89–97. https://doi.org/10.1016/j.iheduc.2010.07.007
8. He Y, Chen R, Li X, Hao C, Liu S, Zhang G, Jiang B (2020) Online at-risk student identification using RNN-GRU joint neural networks. Information 11(10):474. https://doi.org/10.3390/info11100474
9. Hu Y-H, Lo C-L, Shih S-P (2014) Developing early warning systems to predict students' online learning performance. Comput Hum Behav 36:469–478. https://doi.org/10.1016/j.chb.2014.04.002
10. Iqbal Z, Qadir J, Mian AN, Kamiran F (2017) Machine learning based student grade prediction: a case study, pp 1–22. http://arxiv.org/abs/1708.08744

11. KH S, Ravikumar KM (2019) Student risk identification learning model using machine learning approach. Int J Electr Comput Eng (IJECE) 9(5):3872. https://doi.org/10.11591/ijece.v9i5.pp3872-3879
12. Lakkaraju H, Aguiar E, Shan C, Miller D, Bhanpuri N, Ghani R, Addison KL (2015) A machine learning framework to identify students at risk of adverse academic outcomes. In: Proceedings of the 21th ACM SIGKDD international conference on knowledge discovery and data mining, pp 1909–1918. https://doi.org/10.1145/2783258.2788620
13. Marbouti F, Diefes-Dux HA, Madhavan K (2016) Models for early prediction of at-risk students in a course using standards-based grading. Comput Educ 103:1–15. https://doi.org/10.1016/j.compedu.2016.09.005
14. Sekeroglu B, Abiyev R, Ilhan A, Arslan M, Idoko JB (2021) Systematic literature review on machine learning and student performance prediction: critical gaps and possible remedies. Appl Sci (Switzerland) 11(22):10907. https://doi.org/10.3390/app112210907
15. Tjandra E, Kusumawardani SS, Ferdiana R (2022) Student performance prediction in higher education: a comprehensive review. AIP Conf Proc 2470(July):050005. https://doi.org/10.1063/5.0080187
16. Waheed H, Hassan SU, Aljohani NR, Hardman J, Alelyani S, Nawaz R (2020) Predicting academic performance of students from VLE big data using deep learning models. Comput Human Behav 104:106189. https://doi.org/10.1016/j.chb.2019.106189
17. Wakelam E, Jefferies A, Davey N, Sun Y (2020) The potential for student performance prediction in small cohorts with minimal available attributes. Br J Edu Technol 51(2):347–370. https://doi.org/10.1111/bjet.12836
18. Zaki N, Krishnan A, Turaev S, Rustamov Z, Rustamov J, Almusalami A, Ayyad F, Regasa T, Iriho BB (2023) Node embedding approach for accurate detection of fake reviews: a graph-based machine learning approach with explainable AI. Research Square. https://doi.org/10.21203/rs.3.rs-2841712/v1

Machine Learning in Assessment Processes

Applying Topic Modeling to Understand Assessment Practices of U.S. College Instructors in Response to the COVID-19 Pandemic

Teresa M. Ober, Xiangyu Xu, Madelynn Denner, Maxwell R. Hong, and Ying Cheng

Abstract We sought to understand the U.S. college instructors' perceptions of the impact of COVID-19 on their instructional and assessment practices of college instructors. Between February and July 2021, 145 faculty teaching at over 80 different U.S. institutions ($Mean_{age} = 41.29$ years, $SD_{age} = 10.95$; $\%female = 69.7$) completed an online survey. Given the novelty of the circumstances, and the lack of existing measures to adequately study the context at the time, the survey consisted mainly of constructed response questions. To analyze such a volume of written responses, we conducted topic modeling using latent Dirichlet allocation (LDA), a machine learning approach Blei et al. J Mach Learn Res 3:993–1022 2003 [8]. This method is appropriate for identifying key themes within a set of responses. Most instructors reported that during Spring 2020 at least one course they taught shifted from an in-person to an online format (88.7%). Topic models based on instructors' responses to open-ended questions provided additional insights about how instructors prioritized content, adjusted grading policies, prepared students to complete exams, navigated the challenges of administering online assessments, and addressed concerns of academic integrity. Even in non-emergency teaching situations, recommendations for best practices for online and remotely delivered exams in higher education contexts are influenced by an understanding of assessment-related challenges during COVID-19. The findings from the present study not only provide an application of a machine learning approach such as topic modeling, but also contribute to a growing understanding of how COVID-19 affected assessment practices in higher education.

Keywords Online and remote assessment · Faculty perspectives · Assessment design · COVID-19 · Topic modeling

T. M. Ober (✉)
Educational Testing Service, Princeton, NJ, USA
e-mail: tober@nd.edu

X. Xu · M. Denner · M. R. Hong · Y. Cheng
University of Notre Dame, Notre Dame, IN, USA

M. S. Khine (ed.), *Machine Learning in Educational Sciences*,
https://doi.org/10.1007/978-981-99-9379-6_11

1 Introduction

During the outbreak of the COVID-19 pandemic in the spring of 2020, many instructors in higher education throughout the U.S. faced difficult decisions about how to adapt to online and remote teaching. One critical aspect of instruction that has received far less attention than other factors is the changes around classroom-based assessment of student learning. From a pedagogical standpoint, classroom assessment provides crucial information about students' current levels of mastery in a subject area, helping the instructor adapt to meet the needs of learners on an individual or collective basis [54]. Alongside many other instructional practices, the COVID-19 pandemic disrupted assessment of student learning. A shift from in-person to a fully online and remote instructional format occurred in many institutions of higher education soon after the World Health Organization announced the pandemic status in mid-March 2020 [55]. Yet, as numerous higher education institutions resumed in-person courses during the subsequent 2020–2021 academic year [5], hesitancy to return to in-person instruction lingered due to concerns about the transmissibility of the virus. Even after the 2019–2020 academic year, many college courses continued to be taught in a fully remote and online mode as the pandemic continued. For instance, there was a reported 93% increase in the number of students taking only distance education courses in Fall 2020 compared to the Fall 2019 semester [35]. While many institutions shifted towards offering more in-person courses during the Spring 2021 semester, some estimates suggest that at least 40% of colleges and universities were primarily offering courses online [5]. As such, there is a continued need to understand how the swift transition to an online and remote instructional format influences teaching and assessment in higher education.

1.1 Assessment Amidst Emergency Remote Teaching During the Pandemic

Well before the pandemic, there has been a trend toward offering more courses in a fully online format in higher education institutions across the U.S. [32]. Approximately 70% of higher education institutions had indicated that online learning was critical to the institution's long-term strategic plan [4]. Some estimates indicated that about 37% of all college students in the U.S. were enrolled in at least one online course and about 20% were completing all their courses online at the start of the 2019-2020 academic year [41]. Though these estimates suggest an inevitable trend towards online and remote instruction, the emergency remote teaching circumstances created by the pandemic complicated many aspects of teaching in higher education.

1.1.1 Role of Situated Cognition in Online Learning Assessment

One such pedagogical consideration affected by the pandemic is the administration of online classroom assessments [12, 29]. According to *situative theories of cognition*, the primary purpose of educational assessments is to provide evidence of the transfer of learning (i.e., from a learning environment) to a novel environment (i.e., the transfer environment) [23, 34]. Unlike traditional classroom-based instruction, teaching, learning, and assessment in an online and remote modality may lack some of the critical benefits of learning within a situated context. Given this lack of a place-based setting, it is therefore critical to provide authentic and relevant tasks for assessment in a practical context for the application. In an online assessment context, instructors must ensure that students can overcome the constraints of an online platform to complete assessment activities that provide opportunities for skill application.

Aside from the lack of a place-based setting, when administering online assessments, instructors may also need to recognize that their immediate presence is effectively absent. According to the *community of inquiry framework* [19], presence is a critical factor that supports students' educational experience and can be divided into three distinct dimensions: social, cognitive, and teaching presence. Social presence generally refers to a sense of belonging in a learning environment. Cognitive presence refers to the ability to reflectively think about the learning process. Teaching presence emerges in the form of both learning activities and support that ensures learners progress towards knowledge of skill mastery. Since online assessment is inherently remote (i.e., not situated in a place-based context), instructors must overcome this lack of presence in assessing student learning for it to serve as an effective tool for pedagogical purposes.

1.1.2 Challenges Administering Assessments Online

Given the situated nature of learning, administering online assessments of learning is challenging for college instructors for several reasons. Instructors may have fewer opportunities to communicate and interact directly with students, so there is often less potential for instructors to gauge how students are progressing [25, 38, 57]. Innovations in Internet-enabled technologies within the past several decades have improved the way in which instructors and students can communicate, which has drastically enhanced the quality of learning in asynchronous, remote, and online environments [36].

Though improvements have been made, even in non-emergency circumstances there remain opportunities for improving online assessment of student learning. Assessment of student learning in fully remote and online contexts requires a rethinking of conventional approaches to the format and method of administration of assessments, particularly as it changes the *technical* and *collaborative demands* of assessment [15]. For example, both instructors and students must have basic technical proficiency to use an Internet-enabled assessment platform or arrangement or must be able to communicate (whether in-person or online) with others to overcome obstacles to technical proficiency. Such obstacles may include students' unfamiliarity with the assessment format, instructors' own unfamiliarity with setting up and creating parameters for online assessments, and a lack of immediacy for students would have had in asking for support (e.g., using technology, learning material) when needed.

Aside from the technical proficiency needed to navigate an online assessment platform, instructors and students must both work collaboratively and develop a sense of trust to complete assessments on Internet-enabled platforms [10, 12]. Threats to this type of collaboration include actions that involve breaches in academic integrity by students, instructors' suspicions of students' breaches in academic integrity, the ineffectiveness of honor codes [50], students' discomfort in being surveilled [18, 30], and students' concerns being incorrectly flagged as cheating via online proctoring [6, 11].

On top of these existing issues in remote and online classroom assessment, the onset of the COVID-19 pandemic also depleted instructors of another essential resource: time. Given the abruptness of the shift to remote and online instruction, instructors lacked the critical time to adequately adapt instruction for teaching online [1, 13]. For these reasons, remote and online assessment via Internet-enabled devices is complicated in ways that can threaten the validity and interpretation of assessment outcomes and thus deserves further investigation [39].

2 Research Aims

We wanted to understand the factors that instructors considered when assessing student learning during the COVID-19 pandemic. In undertaking this investigation, we had two general aims. *First*, we wanted to know more about the perceptions of the changes to assessment practices. *Second*, we wanted to develop a clarity regarding some of the main challenges instructors faced while administering assessments online. We overall wanted to understand the challenges and methods of adaptation in various aspects of assessment, including the instructors' decisions concerning the prioritization of content covered, adjustment to the course grading scheme/policy, preparation for exams,[1] difficulties administering assessments online, and the handling of academic integrity concerns.

To achieve these aims, we administered an online survey to a sample of college instructors teaching across a variety of disciplines at higher education institutions throughout the U.S. Given that context of COVID-19 brought about many unprecedented changes in teaching and learning in higher education, we wanted to capture as much unconstrained information from the instructors as possible. As such, the

[1] Aligning with the terminology used by Authors [42], in the context of this investigation, terms such as "tests" or "exams" are distinguished from the broader category of "assessments." While "test" or "exam" refers to more conventionally structured assessment formats, "assessment" is a more encompassing term which includes these in addition to essays or capstone projects, for example.

survey consisted of many open-ended questions. We used topic modeling to identify themes in the responses to many of these questions. To aid in the interpretation of these topics, we also conducted sentiment analysis of the high frequency words for each topic.

Without understanding what changes are occurring to assessment practices because of the pandemic, it is impossible to examine what challenges or concerns such changes bring, let alone gauge the extent to which we can trust the validity and fairness of assessments. Such an understanding can help determine how the COVID-19 epidemic is affecting the core processes of teaching and learning, but it can also help determine how to administer assessments in online and remote higher education settings.

3 Methods

The sections below describe the methods used in the present study. Additional information about the methods and results can be found in the Supplemental Materials available online.

3.1 Participants

Demographics. Participants ($N = 145$; *Mean age* = 41.29 years, *SD age* = 10.95; *%female* = 69.7) resided in the U.S. and identified as current instructors of college/university courses. Responses were provided by instructors affiliated with over 86 different U.S. institutions. Key demographic variables for the sample of participants are reported in Table 1. Most of the sample (86.30%) fell between the ages of 25 and 54. The sample included faculty who identified as White/European American (86.2%), Asian/Asian American (11.0%), Hispanic/Latinx (4.8%), Black/African American (0.7%), Middle Eastern or North African (0.7%), Native Hawaiian or other Pacific Islander (0.7%), or a member of multiple groups (4.1%).

Educational and Instructional Rank. Table 2 provides a summary of respondents' educational experience and rank. Most of the sample (80%) had earned a doctoral or professional degree, while all but one of the remaining respondents had earned a master's degree as their highest level of education (19.3%). Many participants (57.30%) had between 2 and 19 years of teaching experience, with a mode of between 5 and 9 years of experience (19.3%). Just over three-quarters (78.6%) of respondents worked full-time, with most of the remainder working part-time (19.3%). The instructors were about equally distributed across instructional ranks, ranging from adjunct/graduate student instructor of record (16.6%) to tenured full professor (21.4%).

Recent teaching experience. Table 3 provides a summary of the respondents' most recent teaching experience. Almost all respondents had taught across multiple

Table 1 Respondent demographic information

Variable	Count	Percent (%)
Gender		
Female	101	69.7
Male	44	30.3
Age (years)		
24 or younger	1	0.7
25–29	18	12.4
30–34	25	17.2
35–39	30	20.7
40–44	24	16.6
45–49	14	9.7
50–54	14	9.7
55–59	8	5.5
60–64	6	4.1
65–69	3	2.1
70 or older	2	1.4
Race/ethnicity		
American Indian or Alaska Native	1	0.7
Asian/Asian American	16	11
Black or African American	1	0.7
Hispanic/Latinx	7	4.8
Middle Eastern or North African	1	0.7
Native Hawaiian and Other Pacific Islander	1	0.7
White or European American	125	86.2
Other	0	0
Multiple	6	4.1

semesters between Fall 2019 and Spring 2021 (97.2%). Most respondents indicated they taught courses within a four-year undergraduate degree-granting program (86.9%), while about a third indicated teaching graduate-level courses (31.0%). While most respondents taught within a social science discipline (62.8%), other disciplines were represented within the sample, including education (15.9%), math and statistics (11.0%), humanities (8.3%), biological and medical sciences (7.6%), physical sciences (6.2%), engineering (5.5%), health professions (3.4%), business (2.8%), communication and journalism (1.4%), and the visual and performing arts (1.4%). About one-quarter of respondents (24.8%) indicated that they taught across multiple disciplines.

Table 2 Respondent educational attainment and professional teaching experience

Variable	Count	Percent (%)
Highest degree attainment		
Bachelor's degree (B.A., B.S., etc.)	1	0.7
Master's degree	28	19.3
Doctoral or professional degree	116	80
Years of teaching experience		
1 year	4	2.8
2–4	19	13.1
5–9	28	19.3
10–14	22	15.2
15–19	14	9.7
20–29	14	9.7
30–39	11	7.6
4 0–49	1	0.7
Missing	32	22.1
Full/Part-time teaching position		
Full-time	114	78.6
Part-time	28	19.3
Other	3	2.1
Instructional rank		
Adjunct/Graduate Student Instructor of Record	24	16.6
Lecturer/Instructor/Teaching Professor (non-tenure track)	34	23.4
Assistant professor (tenure track)	25	17.2
Associate professor (tenure track)	30	20.7
Full professor (tenure track)	31	21.4
Other	1	0.7

3.2 Data Collection Procedure

Institutional Review Board approval was granted prior to all data collection. Eligible participants were automatically screened based on meeting the following criteria between Spring 2020 and Spring 2021: (a) residency within the U.S., (b) hold teaching responsibilities as part of their job, and (c) currently working in a college/university or other adult education institution.

Participants were recruited in two phases. Data were collected between February 27 and July 12, 2021, via Prolific[2] ($n_1 = 18$) and direct contact with various academic departments and teaching-focused listservs ($n_2 = 127$). We initially attempted to recruit via Prolific but soon realized we would need to conduct more direct outreach

[2] https://www.prolific.co/.

Table 3 Semester, program, and disciplines recently taught (Fall 2019–Spring 2021)

Variable	Count	Percent (%)
Semester(s) taught		
Fall 2019	123	84.8
Spring 2020	128	88.3
Fall 2020	138	95.2
Spring 2021	131	90.3
Multiple semesters	141	97.2
Degree program(s) of students		
Non-degree seeking program	0	0
Technical/community college (AA/etc.) or another 2-year program degree	10	6.9
Undergraduate degree or another 4-year program degree (BA/BS/etc.)	126	86.9
Graduate or Professional level degree (MA/MS/MBA/etc.)	45	31.0
Doctoral degree (Ph.D./JD/etc.)	45	31.0
Multiple	57	39.3
Discipline(s) of courses taught		
Biological and biomedical sciences	11	7.6
Business	4	2.8
Communication and Journalism	2	1.4
Education	23	15.9
Engineering	8	5.5
Health professions and related programs	5	3.4
Humanities	12	8.3
Math and statistics	16	11.0
Physical sciences	9	6.2
Social sciences	91	62.8
Visual and performing arts	2	1.4
Multiple disciplines	36	24.8

to meet the target recruit goals of approximately 150 respondents. To expand recruitment of respondents beyond our direct network of contacts, who are mostly within social science disciplines, we also contacted department administrators at institutions affiliated with the Big 10 Conference.[3] Between 20 and 25 department administers were contacted from each of institution across a variety of disciplines (24% Biological and biomedical sciences; 7% Business; 7% Communication, Journalism, and related programs; 20% Engineering; 8% Math and statistics; 7% Physical sciences; 17% Social sciences; 10% Visual and performing arts).

All participants were given the option to receive a $10 gift card to offset their time in completing the survey. Most participants completed the survey in under 20 min

[3] https://en.wikipedia.org/wiki/big_ten_conference.

to complete the survey, with a median response time of 12.45 minutes. Aside from questions asking about demographic information and teaching experience, the survey consisted of Likert-type, multiple choice, and open-ended questions created by the researchers based on items from a previous survey administered to students [42]. We administered open-ended questions to help us confirm whether the questions we posed using other question formats were sufficiently comprehensive or otherwise to learn new information not already captured by responses to the other types of questions. Some questions were conditionally administered based on whether teachers indicated having taught during a particular semester. The survey, de-identified data, and analysis script and output associated with this manuscript are available in an online repository [44][4].

3.3 Analytic Procedure

Tabular summaries of responses to Likert-type and multiple-choice questions were summarized for descriptive purposes. For constructed response items, topic modeling was used to assess text responses to open-ended questions using packages in the *R* statistical environment [57]. We chose a quantitative data-driven approach to analyzing the open-ended responses as opposed to a more qualitative approach because we were interested in gathering insight about emergent themes. Before conducting topic modeling analyses, the text data were pre-processed according to the following steps: (1) misspellings were corrected, (2) words were stemmed to form a common root (e.g., "made" → "make"); (3) general stop-words (e.g., "the", "and", etc.) and words from question stem (i.e., question-specific) stop-words (e.g., "exam") were removed; (4) numbers and punctuations were removed. Misspellings were corrected and word stemming was conducted using the *hunspell* package in *R* [45]. The text corpus was created using the *tm* package [17]. Topic modeling was conducted using the latent Dirichlet allocation (LDA) approach [8] with the *topicmodels* package in *R* [24]. This procedure takes into account both the most often used terms and the context of the responses. As such, some terms may be present in multiple topics given that topic modeling is able to model words that are polysemic (i.e., terms that possess multiple meanings) or homonymic (i.e., terms that possess multiple unrelated meanings) [33]. In addition, responses may not be mutually exclusive to one topic or another as responses, particularly lengthier responses, can touch on multiple themes.

We conducted topic modeling using the latent Dirichlet allocation (LDA) approach [8]. This machine learning enabled method is appropriate for identifying topics (i.e., key themes) within a set of responses (i.e., documents). Each response is viewed as a combination of topics with the presence of substantive words indicative of a certain topic or topics. LDA first assumes a multinomial distribution given a certain number (k) of topics in the documents, then another multinomial distribution of words in

[4] https://osf.io/5hrpw/?view_only=b56920b04f58476abca009a870b94177.

each topic. The parameters of these two multinomial distributions are considered as following two Dirichlet distributions. A document (i.e., a respondent's answer to a specific question) was deemed uninformative if it did not provide a valid response to the question (e.g., "I don't think so," "yes," "no," "N/A," etc.) and was removed prior to analysis. Cross-validation (fivefold) was conducted assuming all possible solutions consisting of 2–9 topics ($2 < k < 9$). We used sample perplexity to evaluate the solutions, which is a measure of how well a topic model predicts an out of sample text. Sample perplexity estimates were calculated for each number (k) of topics in the LDA models and were used for selecting the number of topics to determine the final model. A smaller sample perplexity estimate suggests a more optimal solution. Note that the Dirichlet distribution is a distribution over multinomial parameter vectors, and thus the vectors of positive values that sum to one. In this context, a topic beta value refers to a parameter of one of the multinomial topic distributions over a fixed set of vocabulary terms. Larger beta values suggest that a term has a stronger association with a particular topic.

Using the words most strongly associated with the latent topic, we also conducted sentiment analysis to aid in the interpretation of each topic. We used either the top 25% (or the top 150 words in instances where document term matrices were relatively small) of words associated with each topic. Sentiment analysis was performed using the Bing lexicon and the NRC lexicon from the *textdata* package in R. The Bing lexicon consists of over 6000 words from the English language identified as reflecting either a positive or negative sentiment. The NRC lexicon includes over 14,000 words from the English language that are categorized according to eight basic emotions (anger, fear, anticipation, trust, surprise, sadness, joy, and disgust).

4 Results

The sections that follow provide a summary of the information gathered from survey responses, organized around the aims described previously. We also conducted sentiment analysis based on key words associated with each construct. In general, there did not appear to be noticeable differences in the extent to which positive or negative sentiments expressed in each topic. Additionally, most responses appeared to include words associated with sentiments including "anticipation" and "trust," likely reflecting both the uncertainty of the circumstances and the trust needed considering the shift to remote online teaching, learning, and assessment. Additional analyses describing the context of instructors as well as the graphs illustrating the results of the sentiment analysis can be found in Figs. (9, 10, 11 and 12) in Supplemental Materials. Additional findings from the survey not reported here are also available [43].

4.1 Perceived Changes to Assessment Practices

In the present section, we focus on the perspectives of instructors during and after the transition period with respect to assessment, in particular. This section relies solely on constructed response data from answers to open-ended questions about the effects of the pandemic on respondents' assessment practices.

4.1.1 Administration of Exams

We asked instructors several questions about how the pandemic affected their exam practices. Instructors appeared divided in their concern about students' access to online course content. Regarding other aspects of administering exams, there also appeared to be a lack of uniformity of opinions. Nearly half of respondents (49%) either agreed or strongly agreed that the pandemic had not affected their decision to administer a cumulative course exam, while another one-third of respondents (33%) either disagreed or strongly disagreed, and the remainder (18%) indicated they had no opinion on this issue (see Fig. 1).

We were interested in further understanding how the preparation and administration of exams changed during the pandemic. Respondents were also asked to provide additional information in response to the prompt, "*Please provide any additional information about how the preparation and administration of exams have been affected during the pandemic.*" To find underlying themes in the responses, we again used topic modeling. Based on smaller perplexity values and the content of the responses, a two-topic solution was deemed most appropriate for the responses

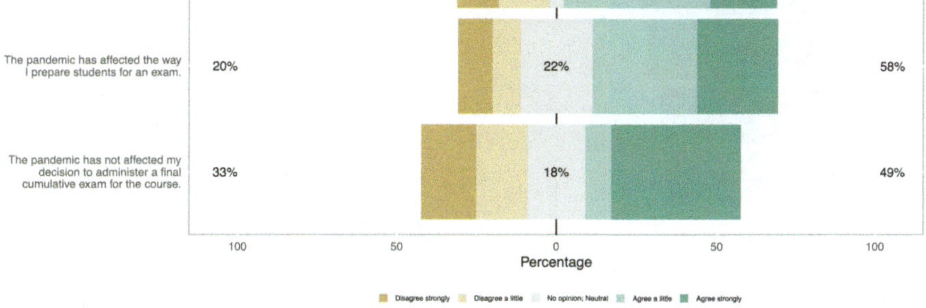

Fig. 1 Responses to questions about the effects of the pandemic on assessment practices

to this question. Figure 2 shows the 10 most frequent terms associated with both topics with associated beta values. As mentioned previously, larger beta values indicate a term which has a stronger association with a topic. Inspecting both the most commonly used terms for each topic and the responses most reflective of those topics, two general themes related to classroom assessment appeared to emerge. One topic appeared to include responses that emphasized creating open and accessible exams to support learning (topic 1). Alternatively, responses also appeared to emphasize the design of assessments which included varied items or test formats, such as multiple item types, multiple exam forms, or being administered over a longer time span (e.g., several days, as opposed to hours) (topic 2). Excerpts of responses most reflective of the two topics are shown in Table 4.

While these two themes highlight the different approaches and views on the purpose of classroom assessment, there were several recurring themes. In particular, many instructors mentioned concerns over cheating and modifying question types on exams to require greater application (as opposed to memorization). Three general factors may contribute to the likelihood of student cheating [40]. These include the lack of perceived consequences of cognitive offloading (i.e., low value assigned to content mastery), high motivation to cheat (e.g., high difficulty to complete lengthy

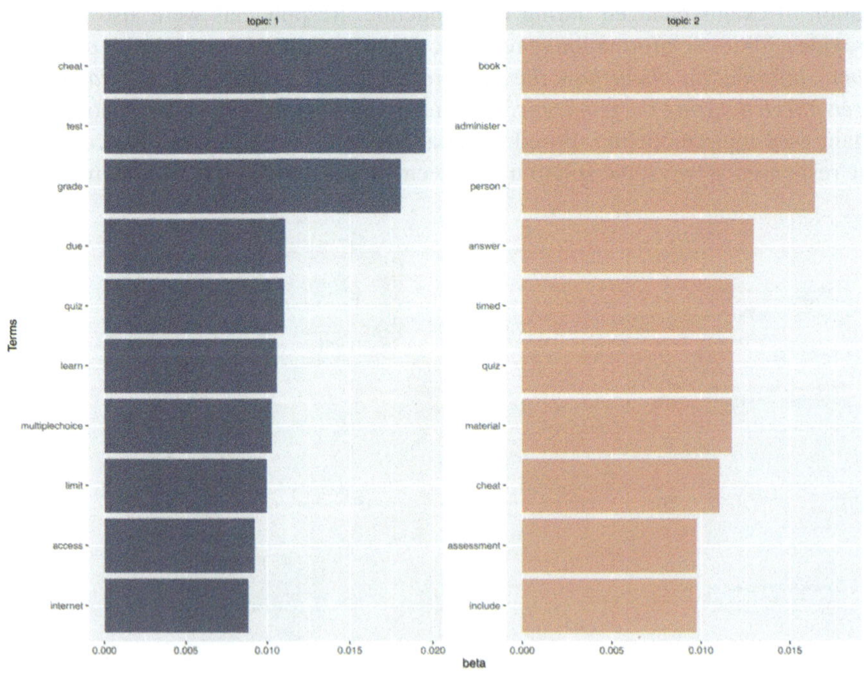

Fig. 2 Ten most frequent terms used to describe two topics in response to the question, "*Please provide any additional information about how the preparation and administration of exams have been affected during the pandemic*"

Table 4 Themes and quoted excerpts of unedited responses to the question, "*Please provide any additional information about how the preparation and administration of exams have been affected during the pandemic*"

Open and accessible exams to support learning (Topic 1)	Varied item/test formats and flexible administration (Topic 2)
"I emphasize the developmental nature of my exams; they are both an assessment and a learning opportunity (they can be taken twice; different exams, same content, highest score kept). Students are encouraged to use the first exam as a learning opportunity and budget time to complete a second exam. I stress my availability for question discussion if there are concepts that remain unclear ..." *"I used to use timed, multiple-choice, closed-book exams during a regular class lecture meeting. With the pandemic and shift from weekly in person lecture meetings to online asynchronous lectures delivered through our LMS, I first shifted to keep a timed multiplechoice exam through our LMS but giving students a 48-h window during which they could complete the open-book exam in fall 2020. I further adapted in spring 2021 to use a take-home short essay, untimed, open-book format. While this created a significant increase in my workload from a grading perspective, students responded favorably to the change that was more accessible and inclusive of their diverse needs and I saw deeper levels of learning."* *"... My students were very stressed out and the online proctoring system increased that stress, so I decided not to use it. That means it was unreasonable to expect that my students were not looking up information, so I changed all the exam questions to application of concepts instead of definitions or factual information. I spent a lot of time thinking about how to help my students learn and ways to use the exams to reinforce their learning. ..."*	*"I have two exam - midterm and final, each covering one half of the semester. They are uploaded and submitted online. The students had 4–7 days to complete them during the fall, and 7–14 days during the spring following students' feedback. [..] My exams have multiple-choice, short answer, and some analysis/application sections. ..."* *"There were vast changes in preparation and administration of exams:\n\n* *- Quantity of content assessed in exams was reduced.\n* *- The amount of time provided to students was reduced.\n* *- The difficulty and complexity of the questions was increased (higher on Bloom's scale).\n* *- The exams were administered over a period of days, in which students could choose their timeslot for the exam."* *"The hybrid and online courses I taught during this time frame included weekly quizzes (open-book, timed, and LockDown Browser with Webcam) and exams (closed-book, timed, and Lockdown Browser with Webcam). A final exam was given but it was optional and could be used to replace an earlier grade. Quizzes and exams were generally open for 2–3 days. I use randomization on my quizzes/exams and draw from pools of questions organized by content area. ..."*

exam within allotted timespan), and capability to cheat (i.e., low difficulty finding information from unauthorized sources). Drafting assessment questions in a manner that requires test-takers to applying their knowledge or skills in a way that doesn't readily allow them to find answers from unauthorized sources may both reduce the potential to cheat and given the emphasis on application of knowledge may also promote value of content mastery. Developing questions in this manner has the potential to improve the rigor of online and remotely administered assessments even beyond the pandemic circumstances.

4.1.2 Exam Preparation

Aside from changes to content coverage and grading, we were interested in knowing more about whether instructors changed how they prepared students for exams. We found that most instructors either agreed or strongly agreed that the pandemic had affected the way they prepared students for an exam (58%; see Fig. 1). Respondents were asked, "*If you have changed the way you are preparing students to take an exam, what has changed?*" The ten most frequent terms associated with the two topics are shown in Fig. 3, while Table 5 shows responses most closely related to these two topics. The results of the topic modeling analysis revealed two topics: one that emphasized providing study guides and facilitating the application of fundamental knowledge and making assessment goals clear (topic 1), with a second topic that expressed the importance of teaching effective study strategies and treating exam preparation as learning opportunities (topic 2). These two topics likely reflect not only changes to the format of assessments and administration of exams but also different pedagogical approaches. While the first topic reflects an approach that views them as more summative (i.e., as a tool for decision-making), the second topic reflects an approach to the purpose of assessment as one which is formative (i.e., to guide instruction for student learning).

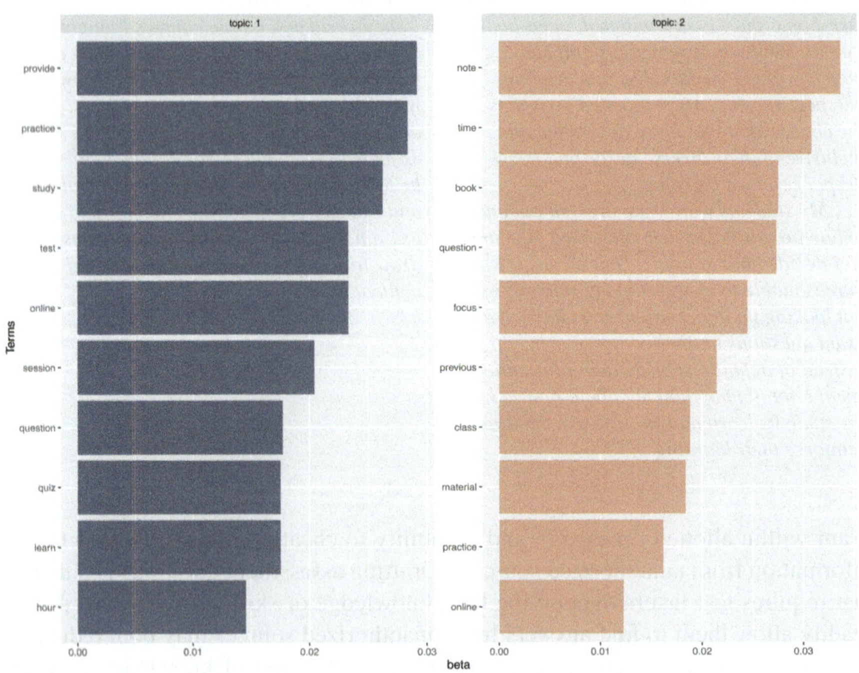

Fig. 3 Ten most frequent terms used to describe two topics in response to the question, "*If you have changed the way you are preparing students to take an exam, what has changed?*"

Table 5 Themes and quoted excerpts of unedited responses to the question, "*If you have changed the way you are preparing students to take an exam, what has changed?*"

Make exam goals explicit (Topic 1)	Exams as learning opportunities (Topic 2)
"I have made my study guides more detailed and I now provide video reviews and host additional office hours for students with questions as they study" *"The pandemic pushed me to rethink the format of exams. [...] By sharing with students the reasons why I now have essay take home exams and giving students a choice among a larger bank of questions to answer, I found that students performed better, demonstrated deeper learning, and some even shared that they enjoyed taking the exams. I think they appreciated the flexibility and less pressure to prepare. They could focus their attention on completing the exam, not on hours of preparation."* *"They always has [sic] problem set or practice HW and formative assignments (like essays) that gave them explicit and implicit goals that re precisely what was assessed in the summative assignments"*	*"The major change is that exams are open-book and open-note. So I spend a bit more time encouraging students to take detailed/organized notes that will be useful for them as they are searching for information. \n\n However, the main emphasis in my courses remains comprehending the material and developing the ability to apply that information to unique problems."* *"Spent more time with freshman on how to apply concepts from the class to real-world examples. Spent time during lecture explicitly discussing how to study for exams. Previously, would integrate examples and application into lecture when students were confused, but less feedback during a virtual lecture meant less examples, and students paid less attention so were more confused and surprised by the format of the short-answer, application-heavy exams."* *"Previously, I have relied on their self-ownership of doing the practice they need and focused on student questions and more challenging material. Now, I am more closely monitoring individual progress as I do not see them in person as often which allowed me to do so previously."*

Particularly in relation to the second topic, several instructors mentioned they had spent class time explicitly teaching test-taking skills, particularly in the online format. During the pandemic, many students likely benefited from direct instruction in using learning management systems for completing activities, particularly students with limited prior experience [2, 3]. Yet even in recent pre-pandemic circumstances, there is considerable variability in students' familiarity and ability to navigate online environments [20], in addition to variability in basic organizational and test-taking skills [53]. To ensure the validity of online and remotely administered assessments, instructors may need to provide additional support for all students to comfortably navigate online test taking.

4.2 Challenges Administering Assessments Online

We were also interested in knowing more about some of the greatest challenges instructors faced while attempting to administer assessments during the pandemic. We asked instructors, "*What is the most challenging aspect of administering an assessment online?*" In general, responses reflected either the effort needed to prepare and give adequate feedback (topic 1) or ensuring engagement and application of knowledge (topic 2). Figure 4 shows the top ten key terms associated with each topic. Table 6 shows excerpts of responses associated with each theme. Notably, not all responses related to the second topic were negative, with some highlighting the relative ease of giving feedback with hyperlinks to relevant course materials.

Several responses related to both topics, particularly topic 2, noted the increased time and effort needed to develop and administer assessments, as well as provide adequate feedback to students. Both themes touched on the difficulty of balancing workload and expectations, with one respondent noting, "*I'm doing the best I can, but it's hard knowing that [...] I'm either letting students cheat, invading their privacy, or not assessing true authentic learning.*" Other aspects that emerged in some of the unique responses addressed concerns over technical issues noting that "*[i]t is also more time consuming to field technical issues during a long assessment window rather*

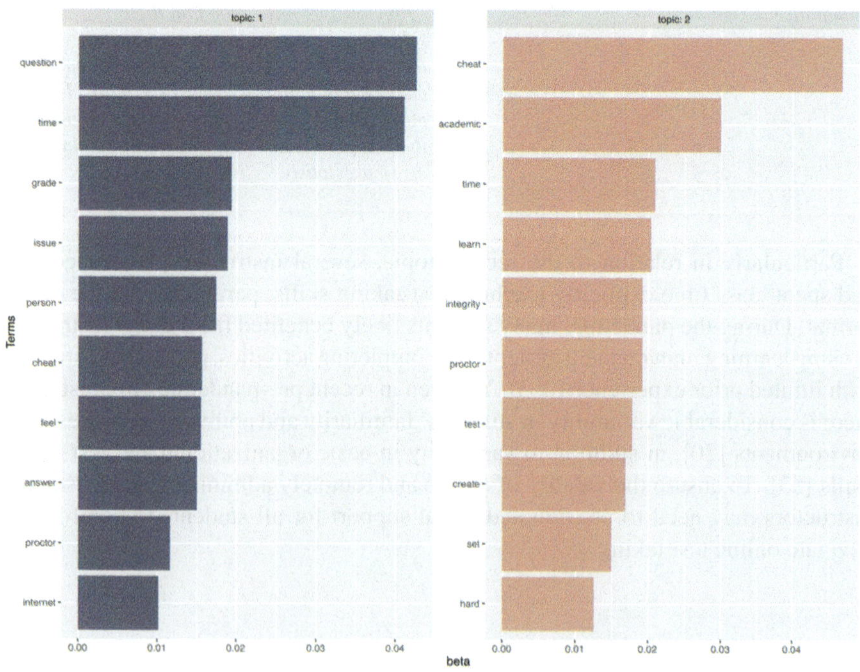

Fig. 4 Ten most frequent terms used to describe two topics in response to the question, "*What is the most challenging aspect of administering an assessment online?*"

Table 6 Themes and quoted excerpts of unedited responses to the question, "*What is the most challenging aspect of administering an assessment online?*"

Effort preparing and giving feedback (Topic 1)	Ensuring engagement and application of knowledge (Topic 2)
"The shear [sic] number of multiple-choice questions needed to ensure that students are not cheating. Creating and revising all the questions so that the answers can't be easily found on the internet. Also, it is more difficult to assess the reliability of the assessment when students are receiving different questions. I wish I didn't have to use webcam monitor too. There can also be legitimate issues, like the internet going out, that don't occur with in-person exams." *"Concerns about cheating. Extra effort is required to mitigate such possibilities, for example structuring or framing questions in a different way so as to minimize the risk. It is also more time consuming to field technical issues during a long assessment window rather than to proctor an in person exam."* *"I find giving meaningful feedback the most difficult. It takes a tremendous amount of time to type up quality feedback, and there is no guarantee the students will read it. I much prefer the discussion after class or in office hours, because at least in person I can read their reactions and we can go through followup questions quickly."*	*"I'm somewhat concerned that students may be utilizing methods of cheating (essentially, mostly by working together or obtaining materials in some illicit manner) on the quizzes that I now utilize instead of mid-term exams. Setting up assessments way ahead of when they are administered (essential in my online teaching) also is challenging in how the forward planning can squash spontaneous melding of course material with current events,* etc.*"* *"Getting students to engage it seriously. I give students a week and two attempts because, for some students, it's hard. And still, some students do one attempt, thirty minutes before it's due. I have added time management assignments and studying assignments at the beginning of my semesters to help students understand that success is both mastery of the material and demonstrating it in the way required by this class (which requires studying and time management), but I still struggle against the "online exam = easy exam" assumption."* *"It is hard to administer proctored exams or quizzes to make sure students grasp the basic knowledge and skills. Without that students lose the motivator to master the basics before attempting to apply the knowledge at a higher cognitive level."*

than to proctor an in-person exam." While administering assessments in an online format seems that it should reduce time in terms of question scoring and grading, several comments provided by instructors note the lack of time efficiency for various reasons. Such challenges may highlight opportunities for professional development and training for instructors not otherwise familiar with designing online assessments or who otherwise have difficulty managing the needs of such assessments such as providing immediate support to students.

4.2.1 Handling Academic Integrity

We were interested in learning more about the approaches faculty used to mitigate or address issues related to academic integrity. Respondents were asked, "*How are you handling concerns over academic integrity?*" Topic modeling revealed two key themes, one which emphasized redesigning the test and including honor statement (topic 1) and another which concentrated on proctoring exams and identifying cheating (topic 2). Figure 5 shows the top ten terms associated with these two topics. Table 7 provides examples of responses indicative of these two topics. Some respondents noted the challenge in handling all lapses of academic integrity, with some even commenting that the experience of administering exams remotely online changed their perspective on the purpose of assessments, with one respondent noting:

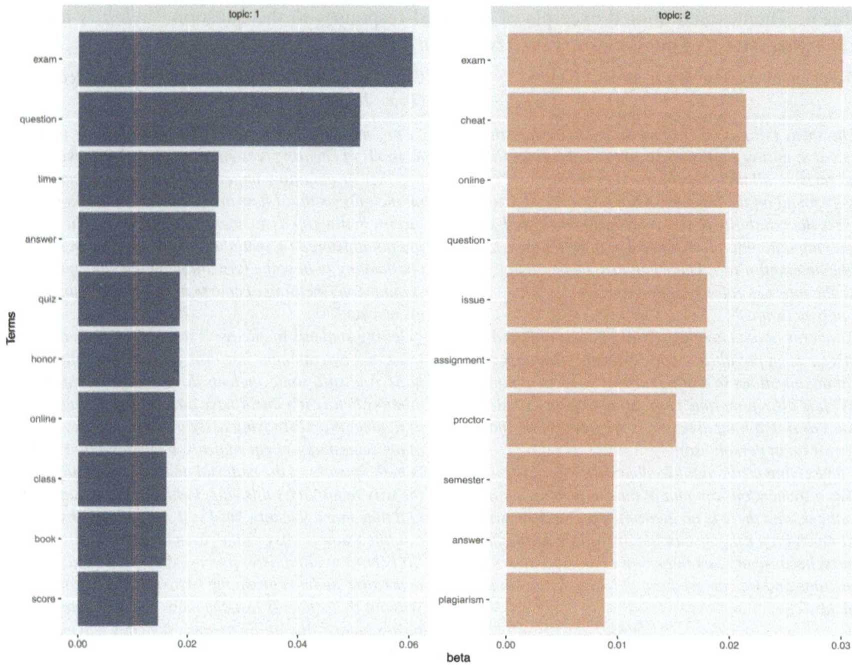

Fig. 5 Ten most frequent terms used to describe two topics in response to the question, "*How are you handling concerns over academic integrity?*"

> … I think this changed how I think about outside information. Our students live in a world where information is available to them and maybe the way [is] to give them skills to use that information successfully. I think the idea that they would not use the information that is so easily available is sort of unreasonable.

About three-quarters of instructors (78%) indicated they did not use a virtual proctor to administer online assessments while the remaining (22%) indicated they had. Those that had indicated they used a virtual proctor were asked to provide more details in response to the question, "*If you use a virtual proctor, please describe your experience using the virtual proctor.*" Some respondents indicated that they proctored the exams themselves through a video conference application such as Zoom. Others described their experiences using various virtual proctoring software, with several comments emphasizing students' apprehension about being recorded, problems incorrectly flagging cheating behaviors, technical difficulties, and lack of technical support. While using methods designed to detect cheating such as virtual proctors and others described within the first topic may reduce the student's perceived ability to cheat, doing so doesn't necessarily change student's orientations towards the value of learning the content nor their motivation to cheat. By contrast, methods described in the second topic such as implementing an honor code or expanding authorized resources for test-taking via an open-book format may be more likely to

Table 7 Themes and quoted excerpts of unedited responses to the question, "*How are you handling concerns over academic integrity?*"

Redesign test and include honor statement (Topic 1)	Proctor exams and identify cheating (Topic 2)
"… Students have to check a box saying that they read and understood and agree to the honor statement before taking an exam online. I also assumed that the temptation to look up questions would be very great for some students, so I just said that everyone could use their books and notes (but I tried to develop questions that could not be easily looked up online in a search engine). For short–answer questions, I asked for examples, not just definitions. Those are easy to look up online and see if students copied them from particular definitions. …" *"I reiterated the importance of trust and the honor code. I also structured the exam to try to decrease the temptation: limited the time students were able to spend on a higher number of multiplechoice questions, so they did not have time to look up answers and still complete them all; over 50% of each exam was an essay; questions were worded in a way to make them difficult to google answers to."* *"I give weekly quizzes that have a mix of multiple answer, multiplechoice, short answer, and fill-in-the-blank. Averages on these quizzes are around 70% with partial credit and there are about 6 questions per quiz. These are graded by TAs following a well-defined rubric…"*	*"Web proctoring and lock down. Exam design where possible (random questions, choices of 1 out of 4 essays). Include assignments and assessment in class that involve oral presentation and participation. Initial module on academic honesty, including how it can hurt them or affect them to have a culture of cheating."* *"I require that students have their cameras on during completion of the exam remotely since the majority of the students are taking it in class. I write all of the questions myself so that students can't easily Google the question to find it in an online repository. I try to create a test that doesn't leave a lot of time for searching online or exchanging emails with fellow students to get answers."* *"I used a warning that all exams would be compared to all the other exams and evidence of collaboration will result in a 0 for all parties"*

place greater value on learning and ideally reduce motivation for cheating. As such, when developing online and remotely administered assessments, virtual proctors may be best viewed as a last resort.

5 Discussion

While many higher education institutions in the U.S. have resumed in-person classes, there is still much to be understood about the experiences of faculty in administering assessments in an online and remote context during the emergency teaching circumstances surrounding the COVID-19 pandemic. Research on and recommendations for best practices in administering assessments in an online and remote setting is still emerging, yet situative theories of learning help to draw attention to some of the potential opportunities for improvement. Notably, within the community of inquiry framework [19], such distributed environments may challenge the benefits of social, cognitive, and teaching presence that students otherwise readily experience when participating in-person assessments.

Using LDA topic modeling, we sought to learn more about the experiences of college-level instructors during the semesters surrounding the outbreak of the COVID-19 pandemic, particularly with respect to the challenges and adaptations for assessment during online and remote teaching during the 2019–2020 and 2020–2021 academic years. Given that the context and circumstances were so novel, there were limited measures that could adequately capture the experiences of instructors. As such, many of the survey items were constructed as response type questions. Such a method as LDA topic modeling is appropriate for analyzing such a volume of text data at scale.

Developing such an understanding of the circumstances of the COVID-19 pandemic and its impact on teaching, learning, and assessment in higher educational contexts is imperative for multiple reasons. First, it provides insight into the ways in which student learning and academic progress in college-level courses were not adequately assessed. Second, it may provide information about the difficulties college instructors faced as well as the adaptations they made to address students' instructional needs despite the barriers they faced in gathering information about student learning. Thirdly, this understanding can provide insight into how to better design and administer assessments in the future remote and online learning contexts.

Under normal circumstances, post-secondary teaching professionals' jobs are already highly demanding such that their time is limited, and expectations are abundant [58], with online and remote education also posing an additional set of work demands [22]. By the end of March 2020, it was estimated that over 1,000 colleges and universities in the U.S. had closed their campuses due to the spread of COVID-19, impacting at least 14 million students across the country [26].

Recognizing the unique circumstances surrounding the pandemic, our first aim was to gather insights on how the pandemic changed classroom assessment from the perspective of instructors. We gathered information about the instructors' decisions concerning how they prioritized content, adjusted the course grading scheme/policy, prepared for exams, dealt with difficulties administering assessments online, and handled concerns about academic integrity. Instructors were about evenly divided in their belief that students may not have access to assignments administered online. Even so, instructors mostly agreed that they had changed some aspects of their grading policy, grading scheme for the course, and how they prepare students for an exam. These findings are consistent with those of a survey administered in June 2020 by the National Institute for Learning Outcomes Assessment, which found that 97% of college faculty surveyed had made some change to their grading [31]. We found that many of these changes persisted even as in-person instruction resumed. Instructors provided additional details about the ways in which they prioritized content, adjusted their grading, prepared students for exams, navigated the difficulties administering assessments online, and handled academic integrity. While similar themes appeared to emerge among response sets in terms of the challenges instructors faced given the circumstances (e.g., limited time to adapt, sense of feeling overburdened, difficulty engaging students, etc.), there was a variety in the ways instructors responded to the circumstances. For example, when asked about how they prioritized content, some indicated they reduced content coverage while others emphasized that they instead chose to focus on greater opportunities for practice and skill development. When asked about changes to their grading practices, some instructors noted adjusting penalties for assignments submitted late, most often by being more lenient. Others implemented grading practices that emphasized participation, such as including more low-stakes assignments or evidence of attending synchronous online sessions.

The second aim of the present investigation was to better understand the challenges instructors faced during and their responses to the pandemic circumstances, particularly in adapting assessment for online and remote administration. While we are presently beginning to understand the negative influence of the pandemic on students' academic progress, relatively little is understood about the specific instructional challenges and adaptations instructors made during this period. Despite the prevalence of the change in instructional modality among instructors' courses, there were relatively minimal changes in the assessment format between Fall 2019 (the semester immediately preceding the pandemic) and Spring 2021, though instructors appeared slightly less inclined to assign multiple projects to students. Such projects may either require material resources or collaborative arrangements that were not feasible then, given the circumstances. Changes to the assessment administration were more obvious. During the pandemic-affected portion of Spring 2020, instructors appeared more likely to assign untimed, open-book assessments that were administered outside of scheduled class time and either proctored online or even more likely, not proctored at all. Such preferences appeared to persist even into the 2020–2021 academic year, as in-person instruction resumed at many U.S. institutions.

We wanted to further deepen an understanding of how instructors adapted their practice considering an online assessment administration and thus posed several open-ended questions. The themes identified in our analysis resonate with those of previous investigations. Unlike previous studies exploring instructors' approaches to assessing student learning online, our method of analysis allows us to systematically examine distinct themes. Instructors were asked to share if their efforts to prepare students ahead of exams had changed, with some noting an emphasis on promoting organizational skills, particularly in terms of notetaking, and others concentrating on ways students could apply fundamental knowledge and skills. When asked about challenges faced administering assessments online, some instructors expressed concern over issues of academic integrity, noting that many of their efforts to adapt to an online format were rooted in apprehension that students could or would cheat. Other instructors noted some of the affordances of the online modality, with several highlighting how providing feedback based on assessment performance had fundamentally changed. We also asked instructors how they were handling concerns over academic integrity. Some instructors responded indicating they had derived methods for detecting cheating. Others acknowledged that online exam administration made cheating a likely inevitability and thus either attempted to change students' attitudes around cheating (e.g., signing an honor code) or modified assessments to be more open-book with questions focusing more on application than strictly knowledge retention.

5.1 Implications and Recommendations for Practice

Assessment in an online environment is obviously challenging, particularly during a global pandemic. Even so, the circumstances surrounding the COVID-19 pandemic may have accelerated a movement towards online learning and assessment [52]. Understanding the challenges instructors faced and their approaches to circumvent such difficulties during this period could help to address issues in online assessment in the future. Our findings highlight different approaches to assessment that occur within an online and remote learning context. Some instructors became more stringent, particularly on issues related to timing or proctoring exams. Others accepted that their ability and role in monitoring good test-taking practices had fundamentally changed. Professional development opportunities that focus on an online modality, especially regarding assessment development, delivery, and feedback, could help better prepare instructors for the future [9, 14, 16, 21, 48].

Supporting college instructors in developing a range of online and remote teaching skills may be most beneficial. For example, in the present analysis, we found that instructors had differing views about how to approach issues of academic integrity. In response to the questions about handling academic integrity, some instructors write about methods of detecting cheating (e.g., correlating test-takers' answers) while others describe ways to mitigate cheating (e.g., more conceptual, and open-ended exam questions). Depending on the context and purpose of the assessment, both types of approaches could be useful for instructors interested in gauging and making decisions based on the evidence of student learning. Rather than focusing only on the purpose of assessment, assessment design principles that focus on the situative nature of learning and emphasize engaging learners in the process may be most effective in remote and online learning environments.

5.2 Limitations

Despite the potential significance of these findings, there are several limitations stemming from the sample and methods which may limit the generalizability of our findings. First, our sample is obviously limited to a small subset of U.S. faculty and instructors who experienced the transition to online and remote instruction amidst the COVID-19 pandemic. As such, the perspectives reported are not necessarily representative of all college instructors' experiences in the U.S. or across the world. For instance, most of the respondents indicated they taught courses in social science disciplines. Our efforts highlight some of the challenges faced in recruiting specific populations, particularly those whose work obligations over the past year have been strained by many of the issues described in the present study [51]. Further work should attempt to recruit faculty not only from a more geographically diverse region but also from a variety of disciplines.

To capture information outside the scope of the forced-choice questions that were administered, we conducted text mining to analyze responses from open-ended questions on the survey. Though we believe the responses and subsequent analyses provide insights into the major dilemmas college instructors faced during this time, there are limitations in generalizing beyond the present questions posed and sample. For one, we note that all the solutions produced a two-topic solution. This likely emerged because of the limited sample size—both in terms of participants ($\leq$145 responses) and length of the responses (i.e., number of terms). Thus, with a larger sample, a greater number of distinct themes are likely to have emerged. In the process of analyzing the text, we also noticed that the qualities of the question influenced the richness and variety of responses provided. Questions that are somewhat suggestive of a certain number of topics (e.g., "preparation" and "administration" in "*Please provide any additional information about how the preparation and administration of exams have been affected during the pandemic.*") are less likely to elicit a range of responses compared to questions that are specific enough for participants to draw from personal experience yet not suggestive of a particular set of themes (e.g., "*What is the most challenging aspect of administering an assessment online?*"). Even so, the responses provided overall, and the themes gleaned from them provide useful potentially actionable information about the unique challenges experienced and the various ways instructors adapted to the circumstances.

5.3 Conclusion

Understanding what changes in assessment and the experiences behind such changes from the perspective of college faculty may provide information about the specific difficulties instructors faced collecting information on student learning as well as the ways they were able to use resources or otherwise adapt to the circumstances. Such knowledge may shed light on issues of fairness and test validity of classroom test administration amidst emergency teaching circumstances that necessitate a transition to online and remote instruction. The goal of the project is to inform recommendations for best practices for online and remotely administered assessments.

Supplemental Materials

Understanding the Context of Teaching and Assessment

Considering research documenting stress experienced by students due to the pandemic-related shift to online learning [37], we wanted to understand the extent to which instructors were similarly impacted. We wanted to understand how instructors in the sample were personally affected by the many possible stressors created by the pandemic (see Table 8). Most indicated that their physical or mental health had suffered because of the pandemic, eve nn if they were not ill (71.3%), and that they knew someone who had lost a loved one because of complications from COVID-19 (69.9%). However, most indicated that they had *not* tested positive (91.9%), had *not* personally lost a loved one as a result of COVID-19 complications (86.8%), and had *not* been quarantined (69.5%).

Respondents were asked about changes in the instructional modality of the courses they were teaching during the Spring 2020 semester (see Table 9). Most indicated that at least one of their courses had transitioned to an online instructional modality (70.3%), while only 2.1% indicated their course continued to meet in person for at least part of the time, and another 6.9% indicated their courses were already taught online. We found that instructors were almost equally divided when asked, "*If I administer assignments online, I am concerned some students will not be able to access the content.*" While 47.0% disagreed or strongly disagreed with this statement, 49.0% agreed or strongly agreed.

Table 8 Effects of pandemic on assessment practices

	Yes		No	
Item	Count	Percent (%)	Count	Percent (%)
Though I may or may not have been sick, my physical or mental health has suffered as a result of this pandemic	97	71.3	39	28.7
Someone I know has lost a loved one due during the pandemic due to complications of COVID-19	95	69.9	41	30.1
I was quarantined during the semester	36	30.5	82	69.5
I have lost a loved one during the pandemic due to complications of COVID-19	18	13.2	118	86.8
I have either tested positive or was symptomatic	11	8.1	125	91.9

Note The last item ("I was quarantined during the semester.") was only administered during the second wave of data collection. Thus, the sample size is smaller than the other items shown in the table above

Table 9 Responses to question: "During the Spring 2020 semester, did the format or modality change for the course(s) you were teaching?"

Level	Count	Percent (%)
No, my course(s) continued to meet in person at least part of the time	3	2.6
No, my course(s) that semester were already taught online	10	8.7
Yes, at least one of my course(s) transitioned to having more online meetings	102	88.7
Note Percentages are based on the number of faculty teaching during the Spring 2020 semester ($N = 115$)		

Effects of the Pandemic on Instructional Practices

We were interested in knowing more about how the pandemic affected general instructional practices, particularly with respect to content coverage and approaches to grading. Recognizing changes in the scope and depth of content coverage provides a context for understanding the decisions that led to adjustments in assessment practices. Given that grading policies are often treated as a component of classroom assessment practices, understanding changes in these aspects of teaching would provide greater insight about the challenges instructors responded to in adapting their assessment practices.

Prioritizing content. The scope and depth of content coverage are likely to be reflected on assessments administered in a course. Given the changes in the instructional modality, we expected that faculty needed to prioritize content, and we were interested in learning more about how they approached this issue. In particular, faculty were asked, "*If the pandemic has affected the coverage of topics in your class, how are you prioritizing the topics that your students must master?*" The responses appeared to reflect either the tendency to retain the amount of content covered during the semester (topic 1) or to reevaluate the course expectations and change the content accordingly (topic 2). Figure 6 shows the ten most frequent terms used per topic and Table 10 shows example responses with respect to both topics. Several associated responses indicated that instructors did not really need to prioritize content, but instead had already used a pedagogical approach that emphasized skill acquisition over strictly content learning.

Grading. Most instructors either agreed or strongly agreed that the pandemic caused them to change some aspects of their grading policy in response to the transition to teaching online (82%) or had affected the grading scheme for their course grades (67%). In response to the question, "*Since the pandemic began, have you changed the way you grade assignments?*", 62.8% of instructors indicated they had changed the way they graded assignments while the remainder (37.2%) indicated they had not. Those who responded affirmatively were asked to provide more detail in a follow-up open-ended question. Respondents were asked, "*If you have changed the way you grade assignments, please explain:* (1) *how they changed?*; (2) *whether*

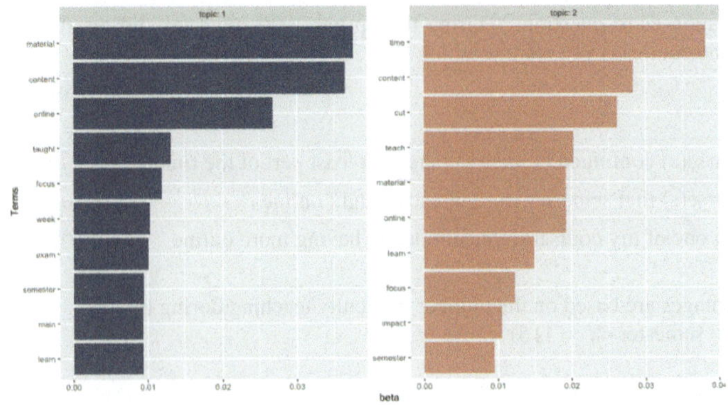

(a)

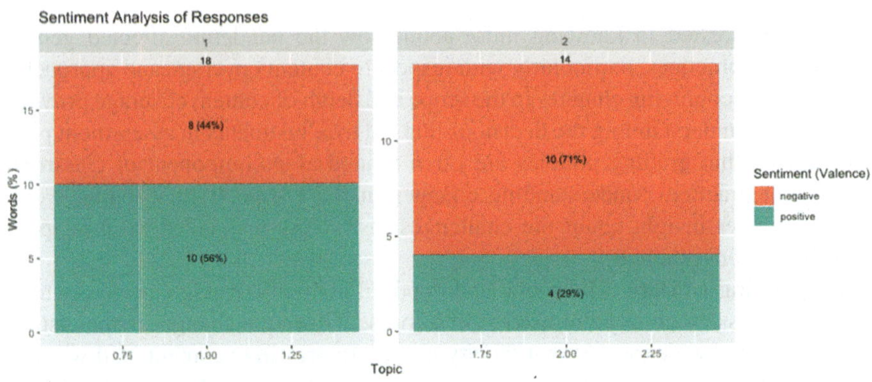

(b)

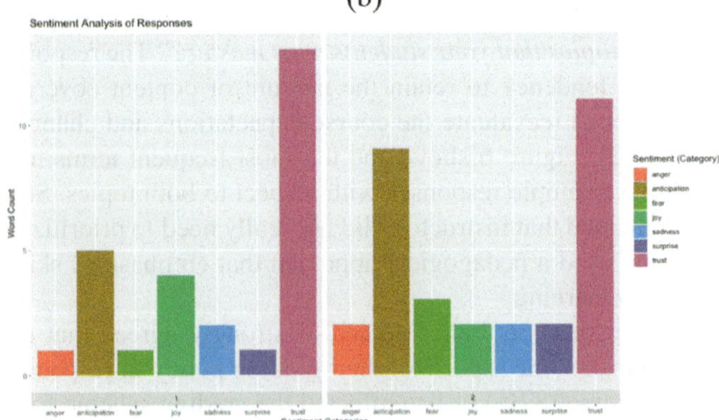

(c)

Fig. 6 **a** Ten most frequent terms used to describe two topics, **b** Bing sentiments, and **c** NRC sentiments reflected in those top frequently used words for each topic in response to the question, "*If the pandemic has affected the coverage of topics in your class, how are you prioritizing the topics that your students must master?*"

Table 10 Themes and quoted excerpts of unedited responses to the question, "*If the pandemic has affected the coverage of topics in your class, how are you prioritizing the topics that your students must master?*"

Retained content coverage (Topic 1)	Evaluated course expectations and changed content (Topic 2)
"has not really changed what I cover - just trying to be more flexible in how students get material (e.g., offer readings, discussions in class, discussions online, videos) - more of a "choose your own adventure" approach for students." *"... the question phrasing here about "prioritizing" implies that you are assuming I covered fewer topics, and that is not the case; very little changed in my syllabus (or the 26 90-min small-group discussion sections that constitute the class time each semester)."* *"National events have more generally affected the emphasis that I place on certain topics (e.g., public health, poverty, race and racism) but the pandemic did not change my coverage of material."*	*"I looked more closely at the course descriptions of my classes as well as my learning objectives. I made sure I emphasized the topics that were covered by those areas [..]. I also prioritized areas that are part of my expertise [...], sharing more real-life examples and stories from my research/clinical/ professional experiences."* *"For my intro-level classes I looked at the disciplinary norms and recommendations [...], then made sure my coverage reflected the key themes, and then selected two units/chapters per theme and let the rest go. For my upper-level courses, I am covering the same amount of content as usual by getting rid of exams, so I regained some class time* via *that method.."* *"Since I teach undergraduate courses, I focus on a smaller amount of topics and go in-depth to gain student interest and understanding in the subject matter (rather than present a shallow broad and all-encompassing course). I typically ask at the beginning of a course for student ambitions, career goals, and interests in the topics, and then pick the topics that will attract the majority of the class. I also prioritize topics for which I have good video clips and engaging examples that will work well in an online format."*

this change will be continued when classes return to regular classroom instruction?" Instructors generally emphasized two different aspects of their changes to grading assignments, as indicated by the results of the topic modeling analysis (see Fig. 7 and Table 11). Several responses described either including or relaxing penalties for submitting assignments late (topic 1) while other responses described changes to grading practices to emphasize or encourage participation (topic 2). Some responses also highlight the different approaches in course sections of different sizes.

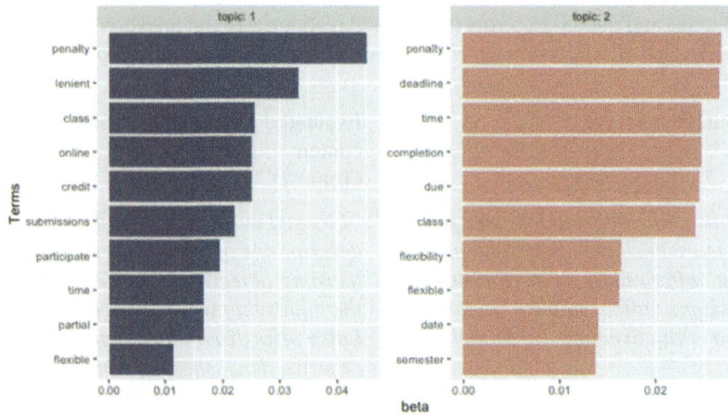

(a)

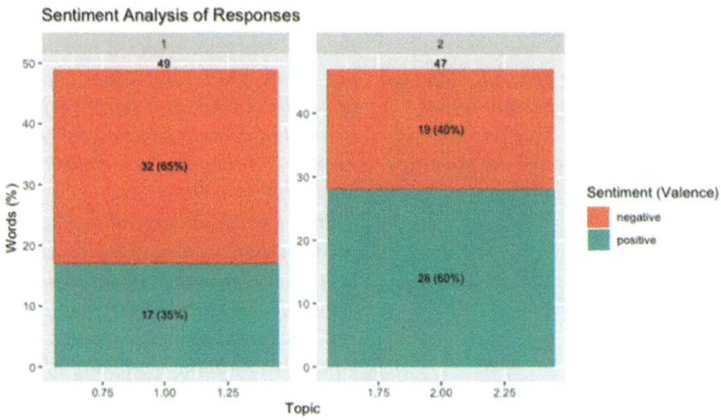

(b)

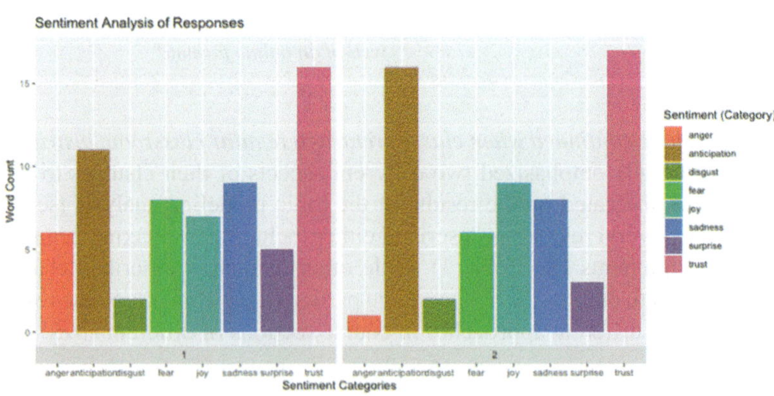

(c)

Fig. 7 **a** Ten most frequent terms used to describe two topics, **b** Bing sentiments, and **c** NRC sentiments reflected in those top frequently used words for each topic in response to the question, "*If you have changed the way you grade assignments, please explain… how they changed?*"

Table 11 Themes and quoted excerpts of unedited responses to the question, "*If you have changed the way you grade assignments, please explain... how they changed?*"

Late assignment penalties (Topic 1)	Emphasis on participation (Topic 2)
"Automatic late penalty deductions through Canvas to help with administration in large courses. - All but eliminated late penalties for smaller enrollments. ..." *"... I didn't accept any late assignments when hard copies of assignments were turned in, but I did allow late submissions with a penalty when I switched to the online submission format. I also began to use online multiple choice quizzes and allowed students to attempt these a second time after feedback and hints were provided. ..."* *"I have a late assignment penalty in the syllabus, but I have effectively stopped enforcing it, and I have invited students explicitly to request extensions for major assignments. ..."*	*"... I also added more smaller stakes assignments and built in bonus points so that students could miss an assignments without a negative impact on their final course grade. For example, the final grade in a course was out of 1,000 possible points but all of the assignments added up to 1,050 points. Lastly, I also allowed students to revise and resubmit papers to earn back partial points they missed on their first attempt."* *"I shifted fully into standards based learning with students having the opportunity to revise everything until they reached an acceptable standard (approximately B level); this had previously been used in some classes but not all. ..."* *"I significantly changed the weighting of assignments vs. exams. Weekly assignments counted much more towards the final grade and exams were weighted less. Assignments were primarily graded based on completion. I also offered more extra credit opportunities than past semesters and was extremely flexible with extensions."*

Pre-pandemic Professional Development

Recognizing that some instructors may have been better prepared to adapt to teaching online having had prior experience, the respondents were asked, "*Had you received any professional development or another form of preparation to teach online courses prior to the pandemic?*" The responses were about evenly divided, with about 52.2% indicating they had, and the remaining 47.8% indicating they had not.

We wanted to know more about instructors' professional development experiences. In follow-up, respondents who indicated they had received professional development were asked, "*If you have received professional development, please explain:* (1) *what skills it addressed?*; (2) *whether or not it was helpful for your teaching?*" Those who indicated they had received some form of professional development were asked to elaborate on the types of skills that were addressed. As described previously, LDA topic modeling was used to uncover themes in the responses to this open-ended question. Figure 8 shows the 10 most frequent terms (i.e., unigrams) associated with both topics. Large beta values associated with each term suggest a stronger association with one of the themes. Table 12 shows responses most closely associated with the two topics. This analysis revealed two distinct themes, one which emphasized

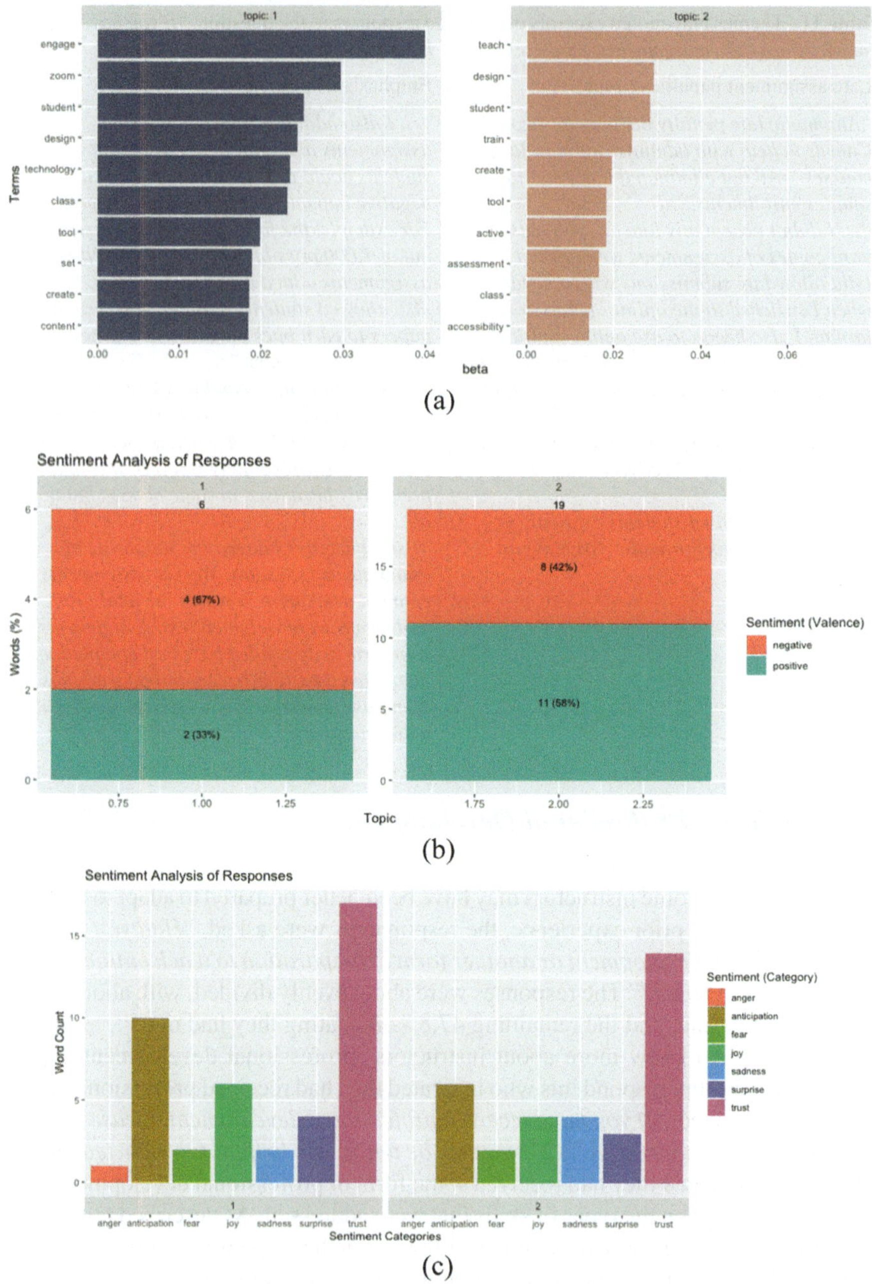

Fig. 8 **a** Ten most frequent terms used to describe two topics, **b** Bing sentiments, and **c** NRC sentiments reflected in those top frequently used words for each topic in response to the question, "*If you have received professional development, please explain… what skills it addressed?*"

Table 12 Themes and quoted excerpts of unedited responses to the question, "*If you have received professional development, please explain... what skills it addressed?*"

Technology skills (Topic 1)	Pedagogical skills (Topic 2)
"Our university offered a 5 week (mostly asynchronous) course in Summer 2020 re: creating and teaching online courses. e.g., how to set up course website, structure synchronous vs. asynchronous time, assessments." *"Our Digital Education department offers multiple, optional workshops throughout the academic year (and summer) to make faculty aware of and train them in the use of various online instructional tools–especially the campus learning management system (LMS). Within these workshops, topics covered have included lesson design, assessment creation and administration, grading, peer collaboration, student engagement, and others.."* *"using learning management systems, developing online content and assessments, using new platforms such as gradescope and Campuswire, using university online proctoring service."*	*"As a graduate student I completed a certificate program and coursework that was in part focused on pedagogy and technology. In addition to exploring specific educational technology tools (e.g., e-portfolios) we also spent a lot of time discussing teaching and learning methods, dynamics, challenges,* etc.." *"Our Teaching Center did workshops on Zoom, online assignments and grading in BB,. I also attended the Remote Conference (I think ASU sponsored it), which was free and all about teaching online - assessments, exams, creating a sense of belongingness, communication,* etc.." *"how to teach online??? honestly most of it was incredibly inadequate and I mostly muddled through online teaching through trial-and-error and applied what I knew worked well in f2f classes but adapted for online setting."*

skills for technology tool use (topic 1) and another which highlighted pedagogical skills such as engaging students in an online format (topic 2).

The instructors were also asked whether they found the professional development experience helpful ("*If you have received professional development, please explain... whether or not it was helpful for your teaching?*"). Most indicated they found the professional development experience helpful. One instructor noted, "*[t]he professional development was extremely helpful, especially for a new teacher. [...]*" Others found the experiences less helpful given that such prior professional development sessions had not addressed teaching in a fully online format. For example, one respondent wrote, "*I did not find this course to be very helpful because most of us were expected to teach our 2020–2021 courses in hybrid mode - i.e., in the classroom but with students zooming in. The information we learned re: teaching a fully online course was really not applicable to hybrid teaching* (Figs. 9, 10, 11 and 12)."

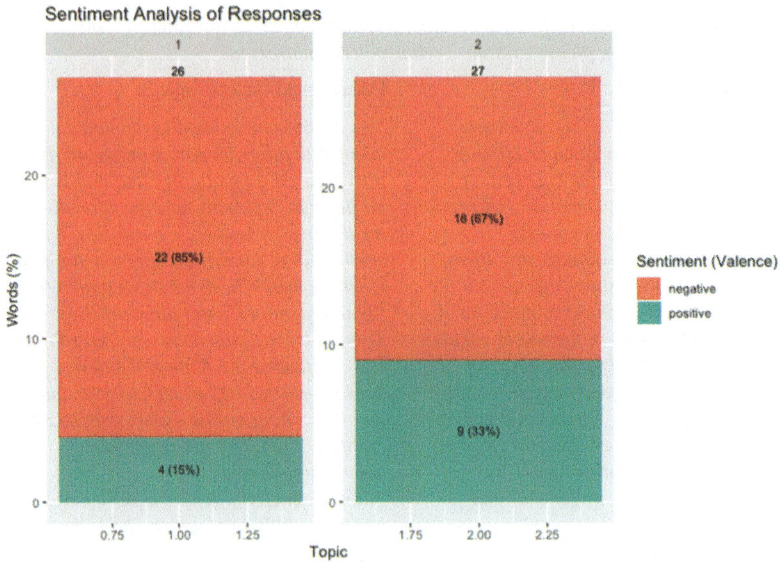

(a)

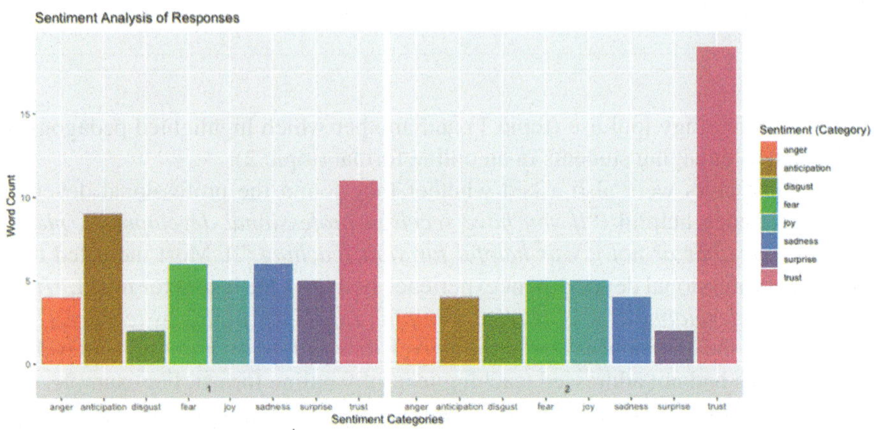

(b)

Fig. 9 Results of **a** Bing and **b** NRC sentiments analysis based on the top frequently used words for each topic using the in response to the question, "*Please provide any additional information about how the preparation and administration of exams have been affected during the pandemic*"

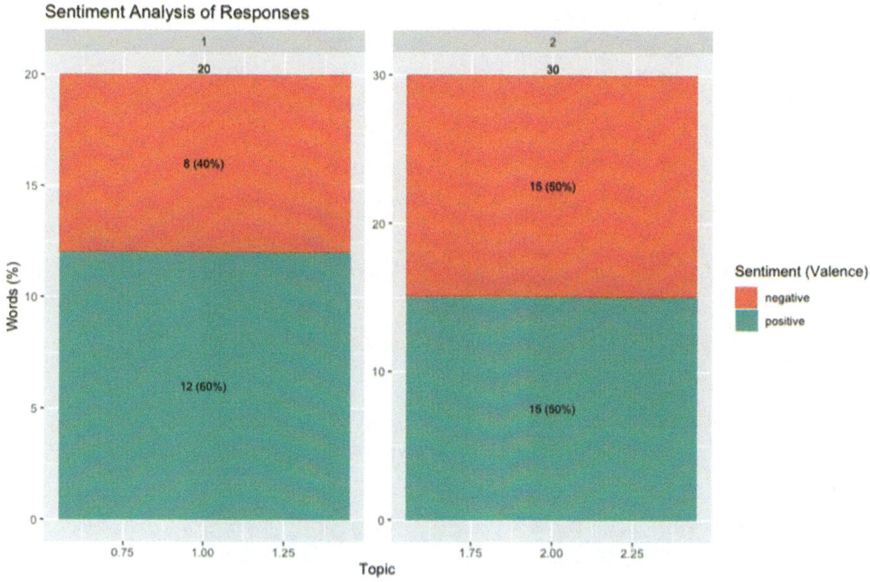

(a)

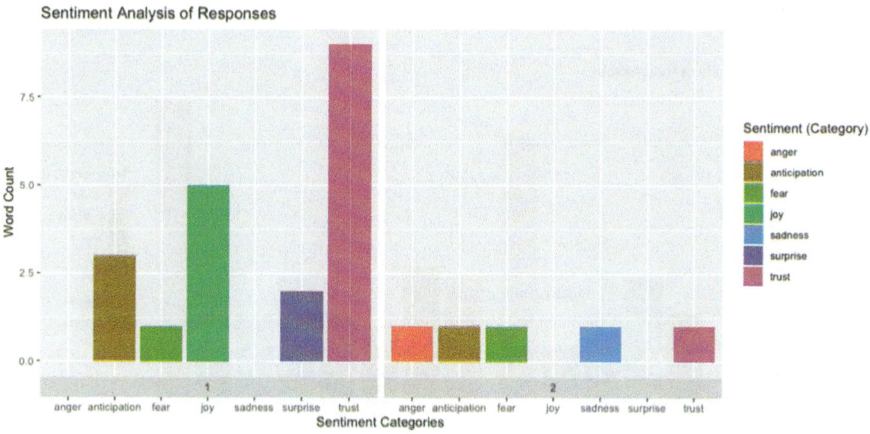

(b)

Fig. 10 S results of **a** Bing and **b** NRC sentiments analysis based on the top frequently used words for each topic using the in response to the question, "*If you have changed the way you are preparing students to take an exam, what has changed?*"

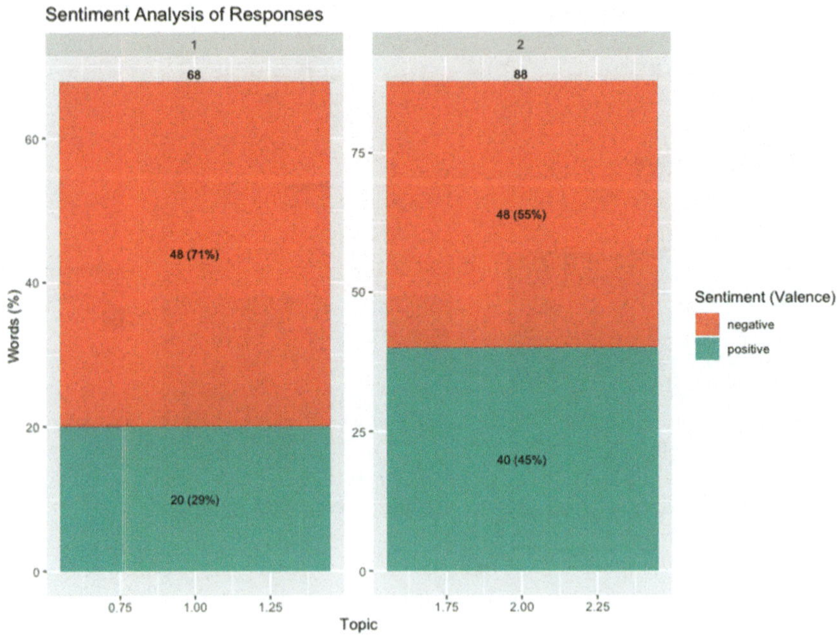

(a)

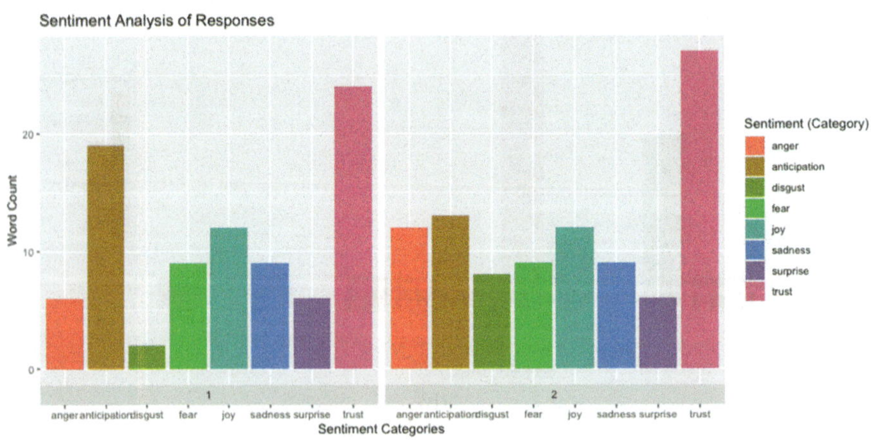

(b)

Fig. 11 Results of **a** Bing and **b** NRC sentiments analysis based on the top frequently used words for each topic using the in response to the question, "*What is the most challenging aspect of administering an assessment online?*"

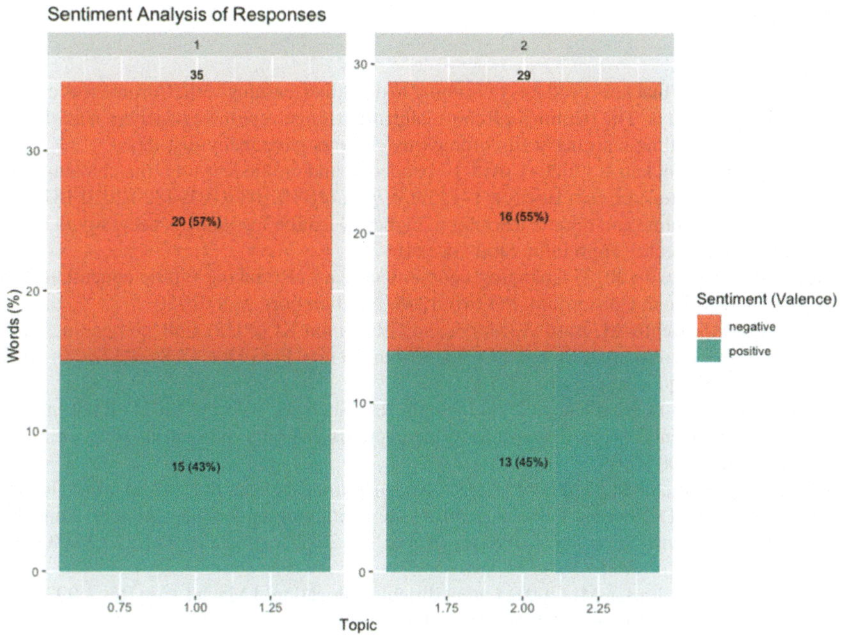

(a)

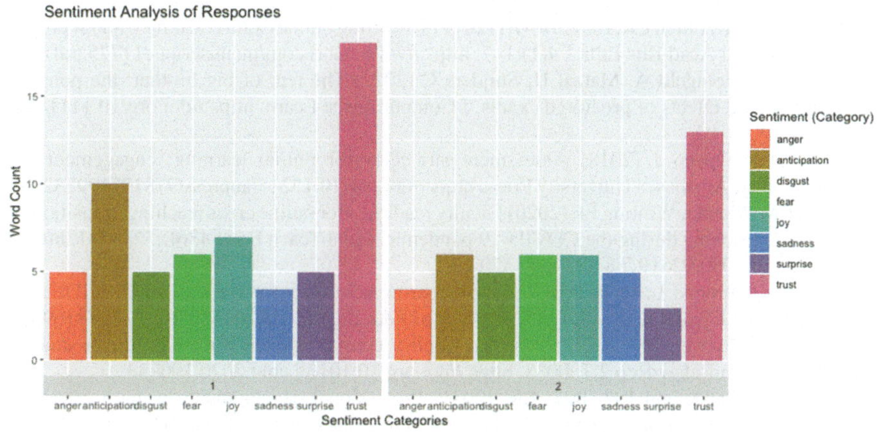

(b)

Fig. 12 Results of **a** Bing and **b** NRC sentiments analysis based on the top frequently used words for each topic using the in response to the question, "*How are you handling concerns over academic integrity?*"

References

1. Adams C (2020) Teachers need lots of training to do online learning well. Coronavirus closures gave many just days. The Hechinger Report. https://hechingerreport.org/teachers-need-lots-of-training-to-do-online-learning-well-coronavirus-closures-gave-many-just-days/
2. Adedoyin OB, Soykan E (2020) Covid-19 pandemic and online learning: the challenges and opportunities. Interact Learn Environ 1–13.https://doi.org/10.1080/10494820.2020.1813180
3. Ali W (2020) Online and remote learning in higher education institutes: a necessity in light of COVID-19 pandemic. High Educ Stud 10(3):16–25
4. Allen IE, Seaman J (2013) Changing course: ten years of tracking online education in the United States. Sloan Consortium. PO Box 1238, Newburyport, MA 01950
5. Andersen MS, Bento AI, Basu A, Marsicano CR, Simon KI (2022) College openings in the United States increase mobility and COVID-19 incidence. PloS One 17(8):e0272820. https://doi.org/10.1371/journal.pone.0272820
6. Balash DG, Kim D, Shaibekova D, Fainchtein RA, Sherr M, Aviv AJ (2021) Examining the examiners: students' privacy and security perceptions of online proctoring services. https://arxiv.org/abs/2106.05917
7. Bird KA, Castleman BL, Lohner G (2020) Negative impacts from the shift to online learning during the COVID-19 crisis: evidence from a statewide community college system. Annenberg Institute at Brown University, EdWorkingPaper, pp 20–299. https://doi.org/10.26300/gx68-rq13
8. Blei DM, Ng AY, Jordan MI (2003) Latent dirichlet allocation. J Mach Learn Res 3:993–1022. https://www.jmlr.org/papers/volume3/blei03a/blei03a.pdf
9. Bragg L, Walsh C, Heyeres M (2021) Successful design and delivery of online professional development for teachers: a systematic review of the literature. Comput Educ 104158.https://doi.org/10.1016/j.compedu.2021.104158
10. Cluskey Jr GR, Ehlen CR, Raiborn MH (2011) Thwarting online exam cheating without proctor supervision. J Acad Bus Ethics 4(1):1–7. https://www.aabri.com/manuscripts/11775.pdf
11. Conijn R, Kleingeld A, Matzat U, Snijders C (2022) The fear of big brother: the potential negative side-effects of proctored exams. J Comput Assist Learn. https://doi.org/10.1111/jcal.12651
12. Conrad D, Openo J (2018) Assessment strategies for online learning: engagement and authenticity. Athabasca University Press. https://doi.org/10.15215/aupress/9781771992329.01
13. Cutri RM, Mena J, Whiting EF (2020) Faculty readiness for online crisis teaching: transitioning to online teaching during the COVID-19 pandemic. Eur J Teach Educ 43(4):523–541. https://doi.org/10.1080/02619768.2020.1815702
14. Darling-Hammond L, Hyler ME (2020) Preparing educators for the time of COVID… and beyond. Eur J Teach Educ 43(4):457–465. https://doi.org/10.1080/02619768.2020.1816961
15. Dendir S, Maxwell RS (2020) Cheating in online courses: evidence from online proctoring. Comput Hum Behav Rep 2:100033. https://doi.org/10.1016/j.chbr.2020.100033
16. Eddy PL, Macdonald RH, Baer EM (2021) Professional development during a crisis and beyond: lessons learned during COVID. New Directions Commun Coll 2021(195):199–212. https://doi.org/10.1002/cc.20477
17. Feinerer I (2020) Introduction to the tm package text mining in R. CRAN R Project. https://cran.r-project.org/web/packages/tm/vignettes/tm.pdf
18. Flaherty C (2020) Big proctor. Inside Higher Ed. https://www.insidehighered.com/news/2020/05/11/online-proctoring-surging-during-covid-19
19. Garrison DR, Anderson T, Archer W (2010) The first decade of the community of inquiry framework: a retrospective. Internet High Educ 13(1–2):5–9. https://doi.org/10.1016/j.iheduc.2009.10.003
20. Geng S, Law KM, Niu B (2019) Investigating self-directed learning and technology readiness in blending learning environment. Int J Educ Technol High Educ 16(1):1–22. https://doi.org/10.1186/s41239-019-0147-0

21. Gottenborg E, Yu A, Naderi R, Keniston A, McBeth L, Morrison K, Burden M (2020) The personal and professional impact of COVID-19 on faculty and staff at an academic institution and a blueprint for the future: a multi-methods study. https://doi.org/10.21203/rs.3.rs-106319/v1
22. Gous IG, Roberts JJ (2015) About time: a metacognitive view of time and workload created by technological advancements in an ODL environment. Distance Educ 36(2):263–281. https://doi.org/10.1080/01587919.2015.1056334
23. Greeno JG, Moore JL, Smith DR (1993) Transfer of situated learning. In: Detterman DK, Sternberg RJ (eds) Transfer on trial: intelligence, cognition, and instruction. Ablex Publishing, pp 99–167
24. Grün B, Hornig K (2021) Topicmodels: an R package for fitting topic models. CRAN R Project. https://cran.r-project.org/web/packages/topicmodels/vignettes/topicmodels.pdf
25. Guangul FM, Suhail AH, Khalit MI, Khidhir BA (2020) Challenges of remote assessment in higher education in the context of COVID-19: a case study of Middle East College. Educ Assess Eval Account 32(4):519–535. https://doi.org/10.1007/s11092-020-09340-w
26. Hess A (2020) How coronavirus dramatically changed college for over 14 million students. https://www.cnbc.com/2020/03/26/how-coronavirus-changed-college-for-over-14-million-students.html
27. Hickey DT, Chartrand GT, Andrews CD (2020) Expansive framing as a pragmatic theory for instructional design. Educ Tech Res Dev 68(2):751–782. https://doi.org/10.1007/s11423-020-09759-4
28. Hickey D, Harris T (2021) Reimagining online grading, assessment, and testing using situated cognition. Distance Educ 42(2):290–309. https://doi.org/10.1080/01587919.2021.1911627
29. Hickey DT, Ingram-Goble AA, Jameson EM (2009) Designing assessments and assessing designs in virtual educational environments. J Sci Educ Technol 18(2):187–208. https://doi.org/10.1007/s10956-008-9143-1
30. Hylton K, Levy Y, Dringus LP (2016) Utilizing webcam-based proctoring to deter misconduct in online exams. Comput Educ 92:53–63. https://doi.org/10.1016/j.compedu.2015.10.002
31. Jankowski NA (2020) Assessment during a crisis: responding to a global pandemic. Urbana, IL: University of Illinois and Indiana University, National Institute for Learning Outcomes Assessment. https://files.eric.ed.gov/fulltext/ED608798.pdf
32. Kentnor HE (2015) Distance education and the evolution of online learning in the United States. Curric Teach Dialogue 17(1):21–34. https://digitalcommons.du.edu/cgi/viewcontent.cgi?article=1026&context=law_facpub
33. Kjell ONE, Kjell K, Garcia D, Sikström S (2019) Semantic measures: using natural language processing to measure, differentiate, and describe psychological constructs. Psychol Methods 24(1):92–115. https://doi.org/10.1037/met0000191
34. Lave J, Wenger E (1991) Situated learning: legitimate peripheral participation. Cambridge University Press. https://doi.org/10.1007/s10649-014-9587-6
35. Lederman D (2021) Detailing last fall's online enrollment surge. Inside Higher Ed. https://www.insidehighered.com/news/2021/09/16/new-data-offer-sense-how-covid-expanded-online-learning. Accessed 16 Sept 2021
36. Lee K (2021) Openness and innovation in online higher education: a historical review of the two discourses. Open Learn J Open Distance e-Learn 36(2):112–132
37. Lemay DJ, Bazelais P, Doleck T (2021) Transition to online learning during the COVID-19 pandemic. Comput Hum Behav Rep 4:100130. https://doi.org/10.1016/j.chbr.2021.100130
38. Lopez KM, Solano DM (2021) Ethics of cheating: effects of the COVID-19 pandemic on academic honesty. In: Schelble SM, Elkins KM (eds) International ethics in chemistry: developing common values across cultures, vol 1401, pp 63–77. American Chemical Society
39. Messick S (1995) Validity of psychological assessment: validation of inferences from persons' responses and performances as scientific inquiry into score meaning. Am Psychol 50(9):741–749. https://doi.org/10.1037/0003-066X.50.9.741
40. Munoz A, Mackay J (2019) An online testing design choice typology towards cheating threat minimisation. J Univ Teach Learn Pract 16(3):5. https://doi.org/10.53761/1.16.3.5

41. National Center for Education Statistics [NCES], U.S. Department of Education (2020) Digest of Education Statistics, 2020. Table 311.15: Number and percentage of students enrolled in degree-granting postsecondary institutions, by distance education participation, location of student, level of enrollment, and control and level of institution: Fall 2018 and fall 2019. https://nces.ed.gov/programs/digest/d20/tables/dt20_311.15.asp
42. Ober, T. M. & Cheng, Y. (2023). Adapting to online and remote learning: Examining the educational assessment experiences of U.S. college students Amidst COVID-19. *IEEE Internet Computing*. https://doi.org/10.1109/MIC.2023.3334025
43. Ober, T. M., Xu, X., Kane, E., Hong, M. R., & Cheng, Y. (2023). Classroom-based assessment practices of U.S. college faculty in response to emergency teaching circumstances during the COVID-19 pandemic. *Assessment Update*. https://doi.org/10.1002/au.30349
44. Ober, T., & Cheng, A. (2022, December 16). Educational assessment practices of college instructors during COVID-19. https://osf.io/5hrpw/https://osf.io/5hrpw/?view_only=b56920b04f58476abca009a870b94177
45. Ooms J (2020) Hunspell: high-performance stemmer, tokenizer, and spell checker. CRAN R Project. https://cran.r-project.org/web/packages/hunspell/vignettes/intro.html
46. Openo J (2020) Education's response to the COVID-19 pandemic reveals online education's three enduring challenges. Can J Learn Technol 46(2):1–12. https://doi.org/10.21432/cjlt27981
47. Rodríguez-Planas N (2021) COVID-19 and college academic performance: a longitudinal analysis. IZA Discussion Paper No. 14113. SSRN. https://ssrn.com/abstract=3789380
48. Sadler TD, Friedrichsen P, Zangori L, Ke L (2020) Technology-supported professional development for collaborative design of COVID-19 instructional materials. J Technol Teach Educ 28(2):171–177. https://www.learntechlib.org/primary/p/216087/
49. St-Onge C, Ouellet K, Lakhal S, Dubé T, Marceau M (2022) COVID-19 as the tipping point for integrating e-assessment in higher education practices. Br J Edu Technol 53(2):349–366. https://doi.org/10.1111/bjet.13169
50. Supiano B (2020) Students cheat. How much does it matter? The Chronicle of Higher Education. https://www.chronicle.com/article/students-cheat-how-much-does-it-matter
51. Taylor DG, Frechette M (2022) The impact of workload, productivity, and social support on burnout among marketing faculty during the COVID-19 pandemic. J Mark Educ 02734753221074284. https://doi.org/10.1177/02734753221074284
52. Tesar M (2020) Towards a post-Covid-19 'new normality?': Physical and social distancing, the move to online and higher education. Policy Futures Educ 18(5): 556–559. https://doi.org/10.1177/1478210320935671
53. Venezia A, Jaeger L (2013) Transitions from high school to college. Future Child 117–136. https://www.jstor.org/stable/23409491
54. Wiliam D (2006) Formative assessment: getting the focus right. Educ Assess 11(3–4):283–289. https://doi.org/10.1080/10627197.2006.9652993
55. World Health Organization [WHO] (202) Rolling updates on coronavirus disease (COVID-19). Updated 31 July 2020. https://www.who.int/emergencies/diseases/novel-coronavirus-2019/events-as-they-happen
56. Wuthisatian R (2020) Student exam performance in different proctored environments: evidence from an online economics course. Int Rev Econ Educ 35:100196. https://doi.org/10.1016/j.iree.2020.100196
57. Zhang Z (2018) Text mining for social and behavioral research using R: a case study on teaching evaluation. https://books.psychstat.org/textmining.
58. Ziker J (2014) The long, lonely job of homo academicus. Blue Rev 31:03–14. https://www.boisestate.edu/bluereview/faculty-time-allocation/

Applying Machine Learning to Augment the Design and Assessment of Immersive Learning Experience

Chih-Pu Dai

Abstract The use of machine learning has seen a remarkable rise in education research with extraordinary potential to enhance immersive learning experience. An immersive learning experience, in which learners participate in simulated virtual environments, can promote deep learning as learners actively explore and construct knowledge within the learning environments. Despite the growing interest and increasing applications, the ways in which machine learning can be used to augment the design and assessment of immersive learning experience remain an open area of exploration. Machine learning can be used to provide adaptive and personalized learning, increase interactivity and engagement, and track learning activities in immersive learning environments. In this chapter, the author describes the current state of research on machine learning in immersive learning environments, including adaptive and personalized learning, natural language processing and conversational artificial intelligence, and data and learning analytics. The author also outlines the potential future directions for the applications of machine learning in designing and assessing immersive learning experiences to inform educational sciences. This chapter serves as a useful reference for educational researchers, practitioners, and policy makers seeking to make informed decisions on the design and assessment of immersive learning experiences.

Keywords Machine learning · Natural language processing · Immersive learning environments · Artificial intelligence · Personalized learning

1 Introduction

Immersive learning experience afforded by learning technologies—for example, virtual reality, augmented reality, mixed reality, and simulation games—has gained attention in education research due to its affordances to offer a learning experience

C.-P. Dai (✉)
University of Hawaiʻi at Mānoa, Honolulu, HI 96822, USA
e-mail: cdai@hawaii.edu

M. S. Khine (ed.), *Machine Learning in Educational Sciences*,
https://doi.org/10.1007/978-981-99-9379-6_12

that is not limited or constrained by time and space. Education researchers and practitioners have also been putting efforts into exploring how to provide more personally meaningful and adaptive immersive learning experiences [12, 41]. However, providing such learning experience for individual learners is not an easy task because each learner is unique in their diverse and dynamic learning backgrounds, learning states, and learning needs. Further, as learners actively interact with the immersive learning environments, they vary in terms of their learning trajectories and interactions leading to diverse performance outcomes; hence it is difficult to adapt to individual needs with machines or computer systems. For example, despite decades of extensive efforts, a previous meta-analysis found no significant difference between adaptive mechanism and non-adaptive mechanism in the context of educational games [41]. There is a critical need to improve these technologies in sound ways so that adaptive and personalized learning can be realized. Machine learning—a component of artificial intelligence, has experienced significant growth in recent years and emerged as a potential solution for designing and assessing personally meaningful and adaptive immersive learning experiences.

Jordan and Mitchell [30] proposed that machine learning deals with two major questions: "How can one construct computer systems that automatically improve through experience? and What are the fundamental statistical-computational-information-theoretic laws that govern all learning systems, including computers, humans, and organizations?" (p. 255). Based on the suggestion of Jordan and Mitchell [30], when designing and assessing immersive learning experience, machine learning is used as a holistic approach, involving the computers, learners, and learning settings, to learn automatically when learners have interacted in the systems and adapt to learners' interactions driven by learners' input data and algorithms. As such, designing and assessing immersive learning experience is an interdisciplinary effort that requires expertise and input from disciplines such as education, psychology, cognitive science, computer science, and data science. Machine learning should also be used purposefully to address problems. In particular, "a learning problem can be defined as the problem of improving some measure of performance when executing some tasks, through some type of training experience" (Jordan and Mitchell [30], p. 255). In designing and assessing immersive learning experience, specifically defined learning outcomes is of critical importance while the process-oriented interactions were collected and recorded as evidence and baseline conditions. The process-oriented interactions can be demonstrated in the forms of natural language and embodied movements. For example, learners can use text input and verbal chat to interact with the immersive learning environments. They can also perform actions to improve knowledge or skills.

In the following sections of this chapter, I introduce the current state of research on the integration of machine learning for the purpose of designing and assessing immersive learning environments. In particular, I explore how natural language processing, conversational artificial intelligence, multimodal learning analytics, and data and learning analytics have been used to enhance learning in immersive learning environments. The chapter will conclude with potential future directions for the applications of machine learning for designing and assessing immersive learning experiences, aiming to provide applicable insights for the field of educational sciences.

2 Machine Learning Approaches in Educational Sciences Research

The applications of machine learning can be identified in many different industries, such as health, business, military, and education. According to Thormundsson [65], the market of artificial intelligence is projected to grow to 126 billion U.S. dollars by 2025, machine learning is part of this tremendous market. In education, machine learning can be useful as it can be applied to provide personalized and adaptive learning as well as analyze large-scale datasets to inform educational practices [9].

2.1 What is Machine Learning?

Machine learning has grown in the past few years. There are several techniques or solutions in machine learning that can be used to tackle various problem domains, such as clustering problems, predictive problems, detecting problems, optimization problems, or association rules. With these solutions, machine learning can be applied in education to provide adaptive learning, personalized learning, and assessment. There are different applications of machine learning in education research, sometimes coupled with natural language processing. One characteristic of immersive learning is that the interactions are multimodal. Therefore, machine learning has been applied in multiple ways to enhance the learning experiences.

Machine learning can be classified broadly as supervised learning, unsupervised learning, and reinforcement learning [30]. Jordan and Mitchell [30] also pointed out that there are variant approaches in between this broad classification. For example, semi-supervised learning has been used to predict learning performance [42]. Semi-supervised learning is useful when a portion of data is labeled while a large portion of data is unlabeled [42].

Supervised learning is when a machine learns to accurately identify labeled data with machine recognizable vector scores [37]. For example, if given a house, the machine will identify it as a house. To achieve this, a large amount of data will be collected and labeled with its category (e.g., house as a house), this dataset is a training dataset. During the training stage, machines will learn the patterns of the dataset. During the testing stage, when provided with an unseen dataset, the machine will use the training experience for problem-solving. That is, if given a house, the machine is hopefully to identify it as a house with a high accuracy rate; if given a mouse, it will hopefully identify it as not a house. In addition to these pattern identification problems, supervised learning can also be used to address predictions or decision-making problems.

Unsupervised learning, on the other hand, refers to the type of machine learning that does not use labeled data, instead, raw data are provided, and the machine finds patterns or relations of the data on its own. Generally speaking, unsupervised learning is used for exploratory purposes. For example, Amershi and Conati [2] collected log data from a learning environment; without labeling, they used unsupervised learning to cluster students' behaviors in the learning environments for the purpose of user model building to enhance the learning experience.

Reinforcement learning, which shares similar concepts to behaviorism in human learning, is when machines learn from the processes of rewards and punishments by an agent based on the actions taken. In other words, machines are engaged in pattern-finding processes with a policy and determined values. When the machines complete desired actions, they will get a reward, if the actions are undesired, they will be punished. For example, Bassen et al. [5] used reinforcement learning to automatically give relevant assignments to the students in an online learning system. Based on the traced students' performance and their interactions with the course materials in the system, reinforcement learning agents update students' learning states and assign personalized tasks for the students.

2.2 *Application of Machine Learning in Educational Sciences*

While machine learning for immersive learning experience has begun to grow, machine learning has been utilized in various educational research contexts. Examining its applications in broad educational sciences contexts offers valuable insights and implications for the use of machine learning in immersive learning environments, particularly as the specific designs, development, and applications of machine learning in immersive learning are still in the early stages of innovation. By leveraging the experiences and successes of other educational contexts, educators can more effectively integrate machine learning into their immersive learning environments and drive advancements in the field as a whole.

To understand the types, contexts, disciplines, and targeted populations that machine learning has been used in educational sciences, Luan and Tsai [44] conducted a review. They first mapped out *different categories* that machine learning has been applied in education. Prediction dominates the included studies, with approximately 63% of the included studies aimed to provide prediction, followed by diagnosis or profiling (23%). 45% of the included studies is in online settings, 25% in STEM and another 25% belongs to multiple domains. Another dominance of machine learning applications is that it has been mostly applied in higher education settings for university students (60%). The findings of the study underscored the limited presence of machine learning applications tailored specifically for K-12 students. Moreover, it emphasized the importance of careful deliberation when integrating machine learning into K-12 education, particularly in terms of addressing ethical concerns and ensuring developmental appropriateness. The study revealed that deploying machine learning in K-12 settings requires extensive efforts and considerations to navigate these crucial aspects effectively.

To further explore the applications, Shah et al. [53] summarized the different ways machine learning can be applied in educational sciences. In general, prediction, clustering, and semi-supervised learning are the three most frequently used applications. Specifically, Shah et al. [53] proposed "Academic Performance Predictions" (p. 6), "Data Mining to Find Out Hidden Patterns" (p. 7), "Improving Student Results on Basis of Past Experience" (p. 8), and "Semi-supervised Learning in Education" (p. 8). Generally speaking, predication leverages historic data or past students' performance to generate future insights. Specifically, predicting *"at risk" students* in the learning environments is one of the prominent areas that can genuinely help learners and improve learning. Mining the data to understand the patterns of students' learning is also a valuable way machine learning can be used to provide data-driven insights. For example, topic modeling is an unsupervised machine learning approach with a natural language processing technique to mine text-based data for educational use. In health education and promotion, topic modeling has been used with social media data [66]. Gencoglu et al. [21] used topic modeling to analyze students' open-ended responses to teachers' teaching behaviors. By comparing machine learning-based topic models and human ratings, the authors finalized eight topics (i.e., *Topic 1*, Clear explanation, *Topic 2*, Student-centered supportive learning climate, *Topic 3*, Lesson Variety, *Topic 4*, Likable characteristics of the teacher, *Topic 5*, Evoking interest, *Topic 6*, Monitoring understanding, *Topic 7*, Inclusiveness and equity, and *Topic 8*, Lesson objectives and formative assessment).

3 The Design and Assessment of Immersive Learning Experience

3.1 Immersive Learning experience—How Learning Occurs?

Learning is a complex process, and it can be understood and studied from different learning perspectives, paradigms, and theories, such as behaviorism, cognitivism, and constructivism. Moreover, given its complex nature, learning is also being studied with its connections with cognition, development, technology, motivation, and neuroscience [19]. As learning spaces and technologies evolved, connectivism [22, 61] and learning principles related to *metaverse* [27] have also emerged. Connectivism emphasized that learning is not an individualistic activity, instead, it connects nodes and knowledge sources [22]. In essence, knowledge development and management occur within *people* and *contexts* [61]. Learning and knowledge activities in massive open online courses (MOOCs) are usually explained with connectivism [22]. Similar to *connectivism*, in the context of artificial intelligence and metaverse, learning activities are "shared," "persistent," and "de-centralized" (Hwang and Chien [27], p. 1). In the metaverse, learners interact with multiusers (shared). Further, learning, working, and living are sustained activities (persistent), and unauthorized modifications to personal property and logs are not possible (de-centralized) [27]. These characteristics of metaverse call for new and updated learning theories and paradigms to explain the learning phenomena [27].

While contemporary learning theories and paradigms are still in need of innovations, to explain the learning phenomena in immersive learning environments, *situativity theory* is often used as a perspective [20]. Situativity theory suggests that "knowledge and thinking (cognition; i.e., situated cognition), as well as learning (i.e., situated learning), are situated in experience" (Durning and Artino [20], p. 188). Situativity theory is an extension of the theories of Vygotsky, Bandura, and Dewey; it contains multiple theoretical perspectives, including situated cognition, situated learning, distributed cognition, and embodied cognition [20]. These theoretical perspectives emphasize different context interactions and learners' experiences that are well-suited for explaining learning in immersive learning environments.

Another theoretical underpinning for the immersive learning experience is experiential learning [35]. According to Kolb, learning occurs through a four-stage cycle of concrete experience, reflective observation, abstract conceptualization, and active experimentation. This cycle emphasizes the importance of both concrete experiences and reflection in the learning process [35]. It involves engaging in concrete experiences, reflecting on those experiences, and applying the insights gained to new situations [35]. Experiential learning is often used to facilitate deep learning and promote learning transfer (including knowledge and skills). Experiential learning can take many forms, including internships, apprenticeships, simulations, and field trips. The goal of experiential learning is to create opportunities for learners to engage in authentic, real-world experiences that are relevant to their learning goals. By reflecting on these experiences and applying the insights gained to new situations, learners can deepen their understanding and develop practical skills that are transferable to different contexts.

Building upon situativity theory and experiential learning, immersive learning experience can be guided by different learning principles and theories depending on the desired achievements and outcomes. To elaborate, learning principles and theories should be intrinsically integrated and strongly aligned when considering the design and assessment mechanism in an immersive learning environment in conjunction with machine learning. In the subsequent sections, I will delve into the intricacies of the design and assessment of immersive learning environments with essential learning theories and principles to effectively serve educational purposes.

3.2 *Augmenting the Design of Immersive Learning Experience with Machine Learning*

"Immersive learning allows learners to freely explore, experience, interact with objects and characters, and try out new ideas and solutions in the virtual environment through experiential learning approach" (Ip et al. [28], p. 505). To maximize the affordances of immersive learning environments, the design of an immersive learning experience can be augmented by machine learning as machine learning makes dynamic and adaptive interactions and personalization possible. To this end, learning designers can potentially assist learners in achieving and optimizing learning outcomes and experiences in technology-enhanced learning environments. These types of learning experiences have been theorized and promoted for decades. With the advancement of technology, the affordances of immersive learning environments elevated by machine learning can be naturally linked to sociocultural theory [68]. Grounded in Vygotsky's [68] sociocultural theory, the notion of the zone of proximal development provided the foundations for adaptive and personalized learning.

Different machine learning techniques have been applied to drive and augment the design of adaptive and personalized learning. Adaptive learning is thus arguably a key focus of machine-learning-driven immersive learning experience. Vaughan et al. [67] discussed the principles of adaptive systems, defining an adaptive system as "a set of interacting entities that together are able to respond to changes" (p. 3). Liu et al. [41] suggested that adaptive learning in simulation games considers *what to adapt* and *how to adapt*. *What to adapt* involves adaptation to learners and the instructional approaches; whereas *how to adapt* entails a competency-based approach using threshold and decision algorithms as well as a preference-based approach using classification machine learning techniques to tailor to students' interests. Similarly, Zahabi and Abdul Razak's [74] review of adaptive virtual reality-based training suggested various variables that can be adaptive to the learners. Adaptivity can be implemented at different timing—before the training experience and during the training experience [74]. In addition to adapting the content at different timing, providing adaptive feedback is another important variable in machine-learning-driven immersive learning experience design [74]. With a similar emphasis as Zahabi and Abdul Razak's [74] on adaptive feedback, [67] also pointed out that a key mechanism in an adaptive system is feedback loops. Citing Stenudd [64], the machine learning feedback loop in the adaptive system utilized four categories—prediction, recognition, detection, and optimization. In these loops, the inputs are determined by the outputs from previous loops. Through constant update and response mechanisms driven by machine algorithms in the contexts, the functional adaptation to learners' behaviors and parameter adjustments is realized.

Recently, innovative machine learning approaches have been possible in addition to the traditional a priori competency-based approach, in which a competency model has been determined so that learners' data and performance will be examined accordingly. The recent machine learning models can be built with learners' in-situ generated data for more real-time and authentic designs (see Fig. 1 for an example).

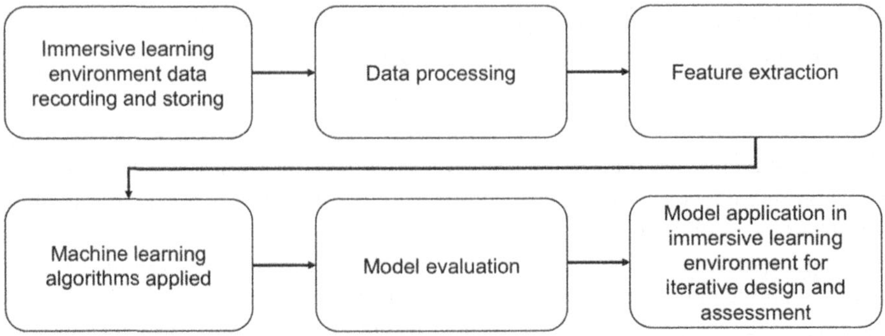

Fig. 1 Machine learning approach for design and assessment in immersive learning environment

To enhance students' learning experiences with machine learning, using a sound procedure is vital. Machine learning approaches follow a systematic procedure to be realized and implemented for the purpose of enhancing teaching and learning. As mentioned previously, machine learning is based on using large amounts of data to train or allow machines to learn, therefore, in real-world application of machine learning, the procedures have a focus on data wrangling and processing to solve different machine learning problems. For example, Wu et al. [73] used a machine learning approach to classify text for social-media-based online discussion. In the context of an immersive learning environment, Rogers et al. [52] proposed novel techniques integrating machine learning and virtual reality for surgical training. Both propositions from Wu et al. [73], p. 68 and Rogers et al. [52], p. 1251 suggested useful procedures that apply innovative machine learning approaches for human learning enhancement.

To be more specific, the first step is to gather data as learners engage in immersive learning and training experiences. These data can be automatically logged into the computer and can be very messy and complex. Next, data are to be processed and cleaned for feature extraction. Feature extraction can reduce the number of information used for machine learning and improve the relevance of the machine learning outcomes [25]. Afterward, machine learning algorithms can be selected based on the purposes, problems, expected outcomes, and accuracy rate of the algorithms. Rogers et al. [52] provided a few examples—K-nearest neighbor Gradient Boosting Logistic Regression Support Vector Machines Discriminant Analysis. In this process, the machine learning models and algorithms are trained and tested using the provided data, and the learning outcomes are analyzed. Appropriate machine learning algorithms can be applied to learning contexts such as providing personalized, adaptive, and real-time feedback, classifying learners' interaction for personalized learning, and assessing learning ([11, 52]).

There has been a growing interest in these applications among researchers in educational sciences. With a similar approach, Abouelenein and Nagy Elmaadaway [1] used ANN to design neuro-computerized virtual learning environments for math preservice teachers to develop computational thinking. They followed a three-stage design process. The first stage involved exploratory factor analysis with input data and framework development. The second stage focused on process-oriented learning model building with preservice teachers to reinforce the development. The third stage aimed to generate adaptive neuro networks based on an input–output function that preservice teachers can use to fit into their learning progress.

Other innovations focused on using machine learning to guide learners or adapt to learners' profiles. For instance, Lee et al. [38] created a director agent to support the learning experience in a narrative-centered immersive educational game. Supervised learning was used to create a director agent that can make adaptive decisions responding to learners' in-game actions. Dai et al. [13] used the Gaussian Mixture Model (GMM), an unsupervised machine learning model for soft clustering, to classify learners' interaction with learning supports in an immersive simulation game (called *E-Rebuild*, [31]). The machine learning technique coupled with natural language processing can be integrated into learning systems to inform the designs of immersive learning experience for data-driven design decisions. In assessing the accuracy of classification of different supervised machine learning algorithms (i.e., Naive Bayes, *k*-nearest neighbors, and support vector machines), Asbee et al. [3] maintained that adaptive assessment can be best implemented with Naive Bayes in the context of virtual reality. In this context, support vector machines can also perform well in classifying learners, but not when there is missing data.

In Asbee et al. [3] setting, tactile feedback was provided when learners were engaged in virtual reality. This highlighted another growing area of study, that is, machine learning-facilitated multimodal analytics. This area of study is unique in immersive learning environments and the findings can offer innovative insights into the design of immersive learning experience. For example, as the study of hand gestures has been a focus in this area, Bahceci et al. [4] used supervised machine learning to understand and classify learners' hand gestures to improve the immersive learning experience. In a comprehensive review and case study, Philippe et al. [48] highlighted the applications of multimodal interactions in virtual reality. They presented examples of practice training in the pharmaceutical industry and surgical field with hands-on activities in virtual reality. Despite emerging studies, the area is in its nascent stage and has been growing.

3.3 Natural Language Processing and Conversational Artificial Intelligence

Immersive learning experience can be enhanced with the integration of virtual agents using natural language processing (NLP). Virtual agents can be represented as chatbots, virtual humans, conversational agents, pedagogical agents, or virtual beings in the literature (e.g., [12, 33, 62, 63]. Virtual agents can provide interactive experience through dialogic learning [24] as well as coaching and guiding the learners [49]. NLP in virtual agents makes natural interactions possible [49].

There are several different machine learning approaches behind the virtual agents that drive natural language interactions with the learners—Bayesian network [29], logistic regression and long short-term memory (LSTM) [46], latent semantic analysis, and large language models (e.g., [6, 12]. As one of the examples, the Bayesian network allows researchers and learning designers in educational sciences to explore ways to implement natural language interactions with learners in computer-based systems. Bayesian network is a predictive machine learning approach modeling a set of variables that represent learners' characteristics such as motivation, knowledge, and other attributes. As the model learns from learners' data and actions, the model updates its predictive mechanisms for the purpose of recommending adaptive and personalized content that fit into learners' zone of proximal development [68]. In a more complex systems design for the purpose of training for early career teachers' classroom management skills, Delamarre et al. [18] adopted MASCARET framework [50], a type of multi-agent system that utilized Unified Modeling Language (UML) with semantic approach and different modeling architectures, to build a 3D immersive simulation-based learning environment.

Designing tutoring experience in intelligent tutoring systems that immerse learners in conversational knowledge exchange, Graesser et al. [23] adopted expectation-and misconception-tailored (EMT) dialog structure to foster students' learning and scientific reasoning. Using semantic-pattern-completion algorithms, Graesser et al. [23] crafted conversational agents that engage in conversational patterns with expected answers, misconceptions, or alternative answers to constantly negotiate meaning and facilitate reasoning. *Natural language understanding* is another way to realize human–machine conversations. Reviewing the literature on chatbots with conversational artificial intelligence, Wollny et al. [72] indicated that chatbots used large-scale data crawled from the Internet to enable text- and/or voice-based interactions with learners. Their review found that chatbots in education have been predominantly used for skill improvement (32%), followed by efficiency of education (25%) and motivation (13%). They also revealed that chatbots in education assume the pedagogical role of learning the most (49%), followed by assisting (20%) and mentoring (15%). In simulation-based learning settings, Dai and Ke [12]'s review revealed that conversational agents can assume different roles with different AI technologies, for example, providing guidance, acting as teachers, peers, or teachable agents.

There are four distinct types of conversational artificial intelligence to date, ranked by their level of interactivity. These include scripted AI, rule-based AI, module-based AI, and natural language processing/machine learning AI [13]. Scripted AI is the least interactive, while natural language processing/machine learning AI is the most interactive. Despite the perception that natural language processing or conversational artificial intelligence is a recent development, more than fifty years ago, ELIZA was designed to converse in natural language in a manner resembling psychotherapists [69]. In some previous systems, human learners/users have reported perplexity, inappropriateness, irrelevance, and bias as a result of interactions with virtual agents [63]. In recent times, the development of large language models with big data and billions of parameters has been prolific in creating improved natural language understanding.

Large language models have been integrated into different contexts for applications. For example, [63] used a large language model to design virtual partners in an online conferencing tool in public health settings. In educational sciences, large language models have become a transformative tool in conversational artificial intelligence for learning. Large language models' capabilities to carry out and maintain conversations with learners are ideal for educational purposes [12, 14].

For the training of teaching in virtual reality-supported simulation-based learning [15, 34], a large language model (e.g., Generative Predictive Transformers 2, GPT-2) has been used to design and develop virtual humans in *OpenSim* [6, 16, 34]. Ke et al. [34] created *Evelyn*, virtual student agent with conversational artificial intelligence, to assist preservice teachers in practicing ambitious science teaching enactment. Generative artificial intelligence affords real-time authentic dialogs between human learners and virtual agents, whereas virtual reality environments (i.e., OpenSim) can afford immersive and authentic sense of presence with suitable scenario designs for in-situ practices [13, 15, 17]. According to [63], Ke et al. [34], and [11], one way to apply large language models to a local context is by training them on local datasets and integrating them into targeted computer systems. Figure 2 depicts a generic architecture for using large language models in education within a local context. This approach can enable more efficient and accurate natural language processing which could lead to improved educational outcomes.

The designs of the teacher learning system with large language models demonstrated one alternative way to create dialog systems grounded in theories of situativity [20] and experiential learning [35] that help learners with artificial intelligence. By localizing and contextualizing the large language models in educational sciences [12], the system is equipped with capabilities to carry on domain-generic conversations as well as engage in domain-specific interactions. A growing number of studies have suggested that the conversational agents alone provide practice and experiential learning opportunities for inquiry-based learning while scaffolding or

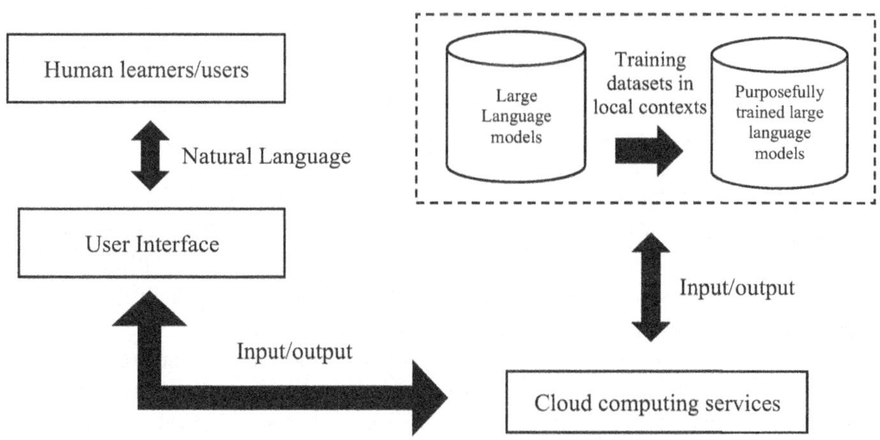

Fig. 2 Conversational artificial intelligence architecture with large language models application

learning support integrated into such learning settings is the most beneficial; in other words, combining conversational agents with additional learning support enhances the effectiveness of the learning experience [75].

The virtual student agent system with artificial intelligence *and* integrated support has been found to be effective [11]. In an experimental design study, [11] examined the effects of learning support on preservice teachers' teaching knowledge. The study also investigated its impacts on preservice teachers' teaching self-efficacy when interacting with virtual artificial intelligence student agents. The results revealed that in comparison to the no-treatment control group, preservice teachers in the learning support group with agents and the agent-only group performed significantly better on teaching knowledge and skills. Although no significant results were found between groups on teaching self-efficacy, the learning support group's teaching self-efficacy improved significantly pre- to post-intervention.

3.4 Assessment for Immersive Learning Experience

Assessment is a critical component of any form of learning. Scholars have been discussing the integral roles of assessment in promoting and encouraging learning [54, 58]. Assessment serves multiple roles for learning [54, 58]. It can be used to determine the learning outcomes (e.g., knowledge, skills, and other traits and capabilities), provide feedback for learners to continuously improve, and maintain and sustain motivation to learn. Essentially, the relations between feedback and assessment have been studied extensively [26, 57]. Hattie and Timperley [26] maintained that feedback received by teachers or learners is the result of formative assessment. The authors suggested that assessment is devised to understand the gap between the current learning state and the determined learning goals at three levels: "about tasks, about the processes or strategies to understand the tasks, and about the regulation, engagement, and confidence to become more committed to learn" (Hattie and Timperley [26], p. 101). Shute [57] further elucidated that feedback and assessment are mutually reinforcing for diagnosis and contribute to learning and performance. In line with a focus on learning and personalization, Shepard [56] emphasized the need to pursue equitable assessment practices that are coupled with ambitious teaching practices. The integration of ambitious teaching practices and equitable assessment urges teachers and designers to gain a thorough and deep understanding of each student, including their learning, emotional, social, and cultural backgrounds as well as the development of disciplinary practice. That is, situated in sociocultural theory, considerations for assessment in immersive learning experience include equitable approach, disciplinary practices, and shared goals.

Assessing immersive learning outcomes with machine learning and artificial intelligence can not only inform learning designers what learners do in the immersive learning environments, but learners themselves can also benefit [55] from unobtrusive assessments designed and developed in the immersive learning environments. There are many different ways to design and blend sound assessment with learning.

In computer-based learning environments, stealth assessment is one way to achieve such goals [59]. "Stealth assessment refers to ECD-based assessments that are woven directly and invisibly into the fabric of the learning environment" (Shute and Ke [60], pp. 52–53). Based on the notion of stealth assessment, Ke and Shute [32] presented assessment design heuristics in immersive games suggesting that machine learning approaches, including, for example, data mining and learning analytics are promising and the integration of assessment and learning task design should be carefully planned in the early stages of the immersive learning environment design and development with iterative testing and refinement. An important driving mechanism behind stealth assessment is evidence-centered design (ECD) [45]. For the stealth assessment to be pedagogically sound and personally meaningful in immersive learning environments, competency and protocols are important, this is called the competency model in stealth assessment [45, 59]. Competency models include the knowledge, skills, and other capabilities and aspects of learning to be assessed. Similar to the assessment in other contexts, stealth assessment in immersive learning environments based on competency models ensures that the shared goals of the learning experience can be achieved [55].

Aligning with the framework of ECD [45], other applications of machine learning in assessment for students' learning have used threshold values, baseline, and indices to perform assessment tasks in immersive learning environments. For instance, in the context of medical education, Winkler-Schwartz et al. [71] used novel processes with machine learning algorithms to generate and extract practice-based evidence and metrics (e.g., with four categories in the operation of surgery: movement, force, bleeding, or tissue) for assessing surgical task in a high-fidelity simulation. The results of the machine learning classification suggested that among four machine learning algorithms (i.e., k-nearest neighbor, Naive Bayes, discriminant analysis, support vector machine), k-nearest neighbor can use the least number of performance metrics (i.e., 6) to classify performances from different expertise groups. Nevertheless, Boulet and Durning [8] have advocated for more psychometric studies to ensure the validity of assessments done by machine learning algorithms.

Understanding the mechanisms of machine learning driving assessments is crucial. The ability to provide equal, just, and inclusive assessments in immersive learning environments largely depends on the machine learning models and the data used to devise these assessments. The efforts to provide deep and meaningful learning for *all*, gave rise to the consideration that the models and data used should include underrepresented groups of learners to capture their interests and promote in-depth and rigorous learning via formative feedback and assessment [12, 59].

More recently, using unsupervised machine learning for data-driven assessment is growing in the context of digital simulation-based learning environments. Using topic modeling, Littenberg-Tobias et al. [40] found that the structural topic model algorithms [51] can properly recognize the natural language text responses by participants that indicate equity practices and mindsets across four simulation modules (i.e., Jeremy's Journal, Coach Wright, Roster Justice, and Layers, see Littenberg-Tobias et al. [40], p. 5 for details).

Aligning with the recent development of machine learning with data-driven assessment, one prominent area of machine learning for immersive learning experience assessment is the application of multimodal learning analytics that afford innovative and unobtrusive assessment and formative feedback [7, 47]. Ouhaichi et al. [47] provided a comprehensive review on the ways multimodal learning analytics can be applied in immersive learning technologies such as virtual reality and mixed reality. Their findings revealed that virtual reality is one key aspect of multimodal learning analytics [47]. Lorenzo et al. [43] used multiple modalities and sources of data to create a comprehensive assessment framework including body movement, voice, eye movement, attention, and empathy. Lee-Cultura et al. [39] used multimodal data capturing young children's interaction with a simulation game (i.e., Marvy learns: Motion-Based Learning Technologies) to inform learning design. They used principal component analysis (PCA) and exploratory factor analysis as fundamental machine learning analyses to identify the relationships between human coded analysis and multimodal learning analytics on young children's problem-solving. They emphasized that the results can be used for providing feedback on learners' cognition and affection. Similarly, using PCA, Kroeze et al. [36] developed an automated assessment tool to identify students' weaknesses in concept maps during inquiry-based learning in virtual immersive laboratories (i.e., with functions that allow students to manipulate variables in science learning). The assessment tool was applied to provide students with feedback on their weaknesses in science conceptualization.

These studies have focused on multiple dimensions of learning and these aspects involve the study of learners' movement, affection, learning products, and learning states. Indeed, the application of multimodal learning analytics to provide formative feedback and in-situ assessment for immersive learning experience requires interdisciplinary expertise such as learning sciences, affective computing, and human–computer interaction [10].

The importance of evaluating the effectiveness of teaching practices when preservice teachers are engaged in immersive teacher learning cannot be overemphasized. While traditional methods of evaluation, such as observation and self-reporting, have their limitations, recent advancements in natural language processing and machine learning offer a promising alternative. In a recent project [6, 11, 34], researchers developed an algorithm that uses natural language processing and deep learning to assess the teaching practices of preservice teachers. By analyzing the language used by these teachers, the algorithm is able to detect and classify the type of teaching practice being employed. To develop the algorithm, the researchers collected natural language data from preservice teachers and used human judgment to identify the specific teaching practices being used, such as lecturing, Socratic questioning, or ambitious teaching practices (e.g., the orchestration of students' ideas or resources). The data was then used to train the algorithm, resulting in an impressive level of accuracy. This innovative approach has several merits and the potential to improve the way teaching practices are supported in immersive teacher learning. By providing preservice teachers with targeted feedback and support for their teaching practices can better prepare them for their careers as educators and contribute to improved educational outcomes for their students. Sharing similar notions of helping teachers with their professional practices, in simulation games, Westera et al. [70] pursued an automatic essay scoring mechanism to reduce teachers' workload in the context of online training. They proposed that "a multilayer perceptron network with two hidden neurons within the hidden layer" (p. 220) was the most appropriate for the assessment purpose after performing the cross-validation with human scoring and excluding biases.

4 Discussion and Future Directions

In this chapter, I have delineated the design and assessment of immersive learning environments augmented by machine learning. Immersive learning environments offer learners valuable opportunities for in-situ practices and knowledge construction. When learners are engaged in these learning environments, they generate a vast amount of data that can be leveraged for machine learning. Drawing on constructivist theories, immersive learning environments supported by machine learning are well-suited for adaptive and personalized learning. However, there is a need for more applicable theories to explain the increasingly complex learning with these learning environments and technologies.

I highlighted several key aspects of immersive learning experiences with machine learning. First, for the design of immersive learning experiences, different machine learning techniques can be integrated. For example, topic modeling (e.g., Gencoglu et al. [21]), GMM (e.g., Dai et al. [13, 15, 17]), and ANN (e.g., Abouelenein and Nagy Elmaadaway [1]). Crucially, a multi-stage approach is essential for producing pedagogically sound designs and validated results. The pipeline of modern machine learning applications is consistent and rigorous in the literature within various contexts (see Fig. 1). Authentic data collection and storage, pre-processing, training, and evaluation are essential steps to ensure the quality of the results.

Second, the design of immersive learning environments can also be enhanced by using conversational agents that are driven by artificial intelligence. Large language models are excellent tools to facilitate human–computer interactions for deep learning. The integration of conversational agents in immersive learning environments also observed creative applications such as in public health settings [63] and teacher education ([11, 34], Training the large language models with localized authentic learning data can ensure the educational values of such learning technologies (see Fig. 2).

Third, in addition to augmenting the designs of immersive learning environments, assessment is another salient aspect of machine learning applications in immersive learning environments. Integrating unobtrusive assessments in immersive learning environments helps to provide feedback and understand the learning outcomes. Machine learning is an ideal tool to accomplish these goals with a data-driven approach that tailors to individual differences, considering *all* learners. That said, to create equal, inclusive, and just assessment in immersive learning environments, the diversity and inclusiveness of the training data is critical. There were different approaches to assessment in immersive learning environments. ECD was one of the pioneering approaches. Novel processes with machine learning algorithms generate and extract practice-based evidence and metrics but not necessarily with an a priori competency model [71]. Finally, considering the characteristics of immersive learning environments, multimodal learning analytics are prominent future directions. Multimodal learning analytics use tracking devices to provide information about learners' actions and behaviors, and therefore, automatic feedback can be provided to individual students based on their performance.

The future of machine learning-integrated immersive learning environments is promising. While existing learning principles and theories from a constructivist perspective can explain machine learning integration in immersive learning well, as the learning landscape grows increasingly complex (e.g., in metaverse), it is necessary to develop novel learning principles and theories to explain learning in such intricate settings. Moreover, the design and assessment of multimodal learning in immersive learning environments are emerging areas that require further development and exploration. Future research is needed to advance these areas. Finally, machine learning-integrated immersive learning environments should be designed with learners' datasets that are inclusive, diverse, and just.

5 Conclusion

By introducing modern approaches of machine learning integration into the design and assessment for immersive learning environments, this chapter provides useful heuristics and applications for learning designers, researchers, and policy and decision makers to consider when using these technologies. The integration of modern machine learning approaches, along with conversational artificial intelligence, into the design and assessment of immersive learning environments holds great promise for educational sciences. By harnessing the power of data-driven techniques, immersive learning experiences can be enhanced in ways that foster more engaging, personalized, and inclusive learning experiences for *all* learners to be successful.

References

1. Abouelenein YAM, Nagy Elmaadaway MA (2023) Impact of teaching a neuro-computerized course through VLE to develop computational thinking among mathematics pre-service teachers. J Educ Comput Res Advance online publication. https://doi.org/10.1177/07356331231165099
2. Amershi S, Conati C (2009) Combining unsupervised and supervised classification to build user models for exploratory learning environments. J Educ Data Mining 1(1):18–71. https://doi.org/10.5281/zenodo.3554659

3. Asbee J, Kelly K, McMahan T, Parsons TD (2023) Machine learning classification analysis for an adaptive virtual reality Stroop task. Virtual Reality 1–17. https://doi.org/10.1007/s10055-022-00744-1
4. Bahceci O, Pena-Rios A, Buckingham G, Conway A (2022) Supervised machine learning hand gesture classification in VR for immersive training. In: 2022 IEEE conference on virtual reality and 3D user interfaces abstracts and workshops (VRW). IEEE, pp 748–749. https://doi.org/10.1109/VRW55335.2022.00225
5. Bassen J, Balaji B, Schaarschmidt M, Thille C, Painter J, Zimmaro D, Mitchell JC et al (2020) Reinforcement learning for the adaptive scheduling of educational activities. In: Proceedings of the 2020 CHI conference on human factors in computing systems, pp 1–12. https://doi.org/10.1145/3313831.3376518
6. Bhowmik S, Barrett A, Ke F, Yuan X, Southerland S, Dai CP, West L, Dai Z (2022) Simulating students: an AI chatbot for teacher training. In: Chinn C, Tan E, Chan C, Kali Y (eds) Proceedings of the 16th international conference of the learning sciences (ICLS). International Society of the Learning Sciences, Hiroshima, Japan, pp 1972–1973. https://repository.isls.org/bitstream/1/8669/1/ICLS2022_1972-1973.pdf
7. Blikstein P (2013) Multimodal learning analytics. In: Proceedings of the third international conference on learning analytics and knowledge, pp 102–106. https://doi.org/10.1145/2460296.2460316
8. Boulet JR, Durning SJ (2019) What we measure… and what we should measure in medical education. Med Educ 53(1):86–94. https://doi.org/10.1111/medu.13652
9. Chen G, Yang J, Hauff C, Houben GJ (2018) LearningQ: a large-scale dataset for educational question generation. In: Proceedings of the international AAAI conference on web and social media (vol 12, No 1), pp 481–490. https://doi.org/10.1609/icwsm.v12i1.14987
10. Cukurova M, Giannakos M, Martinez-Maldonado R (2020) The promise and challenges of multimodal learning analytics. Br J Edu Technol 51(5):1441–1449. https://doi.org/10.1111/bjet.13015
11. Da C-P (2023) Enhancing learning achievements and self-efficacy for preservice teachers using model-based support in simulation-based learning with artificial intelligence-powered virtual agents. Doctoral dissertation. Florida State University.
12. Dai C-P, Ke F (2022) Educational applications of artificial intelligence in simulation-based learning: a systematic mapping review. Comput Educ Artif Intell 3:100087. https://doi.org/10.1016/j.caeai.2022.100087
13. Dai C-P, Ke F, Pan Y, Liu Y (2023) Exploring students' learning support use in digital game-based math learning: a mixed-methods approach using machine learning and multi-cases study. Comput Educ 194:104698. https://doi.org/10.1016/j.compedu.2022.104698
14. Dai C-P, Ke F, Zhang N, Barrett A, West L, Bhowmik S, Southerland SA, Yuan X (2024) Designing conversational agents to support student teacher learning in virtual reality simulation: a case study. In: Proceedings of the ACM CHI conference on human factors in computing systems (ACM CHI '24), Honolulu, HI. https://doi.org/10.1145/3613905.3637145
15. Dai CP, Ke F, Dai Z, Pachman M (2023) Improving teaching practices via virtual reality-supported simulation-based learning: scenario design and the duration of implementation. Br J Educ Technol. Advance online publication. https://doi.org/10.1111/bjet.13296
16. Dai CP, Ke F, Dai Z, West L, Bhowmik S, Yuan X (2021) Designing artificial intelligence (AI) in virtual humans for simulation-based training with graduate teaching assistants. In: de Vries E, Hod Y, Ahn J (eds) Proceedings of the 15th international conference of the learning sciences—ICLS 2021. International Society of the Learning Sciences, Bochum, Germany, pp 1101–1102. https://repository.isls.org/bitstream/1/7418/1/1101-1102.pdf
17. Dai CP, Ke F, Pan Y, Moon J, Liu Z (2023) A meta-analysis on the effects of using artificial intelligence-powered virtual agents in simulation-based learning. Paper Session presented at the 2023 AERA Annual Meeting. Chicago, IL
18. Delamarre A, Shernoff E, Buche C, Frazier S, Gabbard J, Lisetti C (2021) The interactive virtual training for teachers (IVT-T) to practice classroom behavior management. Int J Hum Comput Stud 152:102646. https://doi.org/10.1016/j.ijhcs.2021.102646

75. de Jong T, Lazonder AW, Chinn CA, Fischer F, Gobert J, Hmelo-Silver CE, Zacharia ZC et al (2023) Let's talk evidence–The case for combining inquiry-based and direct instruction. Educ Res Rev 100536. https://doi.org/10.1016/j.edurev.2023.100536
19. Driscoll MP, Burner KJ (2021) Psychology of learning for instruction, 4th edn. Pearson
20. Durning SJ, Artino AR (2011) Situativity theory: a perspective on how participants and the environment can interact: AMEE guide no. 52. Med Teach 33(3):188–199. https://doi.org/10.3109/0142159X.2011.550965
21. Gencoglu B, Helms-Lorenz M, Maulana R, Jansen EP, Gencoglu O (2023) Machine and expert judgments of student perceptions of teaching behavior in secondary education: added value of topic modeling with big data. Comput Educ 193:104682. https://doi.org/10.1016/j.compedu.2022.104682
22. Goldie JGS (2016) Connectivism: a knowledge learning theory for the digital age? Med Teach 38(10):1064–1069. https://doi.org/10.3109/0142159X.2016.1173661
23. Graesser AC, Li H, Forsyth C (2014) Learning by communicating in natural language with conversational agents. Curr Dir Psychol Sci 23(5):374–380. https://doi.org/10.1177/0963721414540680
24. Graesser A, McNamara D (2010) Self-regulated learning in learning environments with pedagogical agents that interact in natural language. Educ Psychol 45(4):234–244. https://doi.org/10.1080/00461520.2010.515933
25. Guyon I, Elisseeff A (2006) An introduction to feature extraction. In: Feature extraction: foundations and applications, pp 1–25
26. Hattie J, Timperley H (2007) The power of feedback. Rev Educ Res 77(1):81–112. https://doi.org/10.3102/003465430298487
27. Hwang G-J, Chien S-Y (2022) Definition, roles, and potential research issues of the metaverse in education: an artificial intelligence perspective. Comput Educ Artif Intell 3:100082. https://doi.org/10.1016/j.caeai.2022.100082
28. Ip HHS, Li C, Leoni S, Chen Y, Ma KF, Wong CHT, Li Q (2018) Design and evaluate immersive learning experience for massive open online courses (MOOCs). IEEE Trans Learn Technol 12(4):503–515. https://doi.org/10.1109/TLT.2018.2878700
29. Johnson W L (2003) Using agent technology to improve the quality of web-based education. In: Web intelligence, Springer, pp 77–101
30. Jordan MI, Mitchell TM (2015) Machine learning: trends, perspectives, and prospects. Science 349(6245):255–260. https://doi.org/10.1126/science.aaa8415
31. Ke F, Shute V, Clark K, Erlebacher G (2019) Interdisciplinary design of the game-based learning platform: a phenomenological examination of the integrative design of game, learning, and assessment. Springer
32. Ke FF, Shute VJ (2015) Design of game-based stealth assessment and learning support. In: Loh C, Sheng Y, Ifenthaler D (eds) Serious games analytics. Springer, pp 301–318
33. Ke F, Dai Z, Dai CP, Pachman M, Chaulagain R, Yuan X (2020) Designing virtual agents for simulation-based learning in virtual reality. In: Zheng R (ed) Cognitive and affective perspectives on immersive technology in education. IGI Global, pp 151–170. https://doi.org/10.4018/978-1-7998-3250-8.ch008
34. Ke F, Yuan X, Southerland S (2021) Teaching practices with multiplayer mixed reality simulations and virtual students. National Science Foundation (Award Abstract # 2110777). https://www.nsf.gov/awardsearch/showAward?AWD_ID=2110777
35. Kolb DA (1984) Experiential learning experience as a source of learning and development. Prentice Hall
36. Kroeze KA, Van Den Berg SM, Veldkamp BP, De Jong T (2021) Automated assessment of and feedback on concept maps during inquiry learning. IEEE Trans Learn Technol 14(4):460–473. https://doi.org/10.1109/TLT.2021.3103331
37. LeCun Y, Bengio Y, Hinton G (2015) Deep learning. Nature 521(7553):436–444. https://doi.org/10.1038/nature14539
38. Lee SY, Rowe JP, Mott BW, Lester JC (2013) A supervised learning framework for modeling director agent strategies in educational interactive narrative. IEEE Trans Comput Intell AI Games 6(2):203–215. https://doi.org/10.1109/TCIAIG.2013.2292010

39. Lee-Cultura S, Sharma K, Giannakos M (2022) Children's play and problem-solving in motion-based learning technologies using a multi-modal mixed methods approach. Int J Child-Comput Inter 31:100355. https://doi.org/10.1016/j.ijcci.2021.100355
40. Littenberg-Tobias J, Borneman E, Reich J (2021) Measuring equity-promoting behaviors in digital teaching simulations: a topic modeling approach. AERA Open 7:1–19. https://doi.org/10.1177/23328584211045685
41. Liu Z, Moon J, Kim B, Dai C-P (2020) Integrating adaptivity in educational games: a combined bibliometric analysis and meta-analysis review. Educ Tech Res Dev 68(4):1931–1959. https://doi.org/10.1007/s11423-020-09791-4
42. Livieris IE, Drakopoulou K, Tampakas VT, Mikropoulos TA, Pintelas P (2019) Predicting secondary school students' performance utilizing a semi-supervised learning approach. J Educ Comput Res 57(2):448–470. https://doi.org/10.1177/0735633117752614
43. Lorenzo G, Pomares J, Lledo A (2013) Inclusion of immersive virtual learning environments and visual control systems to support the learning of students with Asperger syndrome. Comput Educ 62:88–101. https://doi.org/10.1016/j.compedu.2012.10.028
44. Luan H, Tsai CC (2021) A review of using machine learning approaches for precision education. Educ Technol Soc 24(1):250–266. https://www.jstor.org/stable/26977871
45. Mislevy RJ, Steinberg LS, Almond RG (2003) Focus article: on the structure of educational assessments. Measur Interdisc Res Perspect 1(1):3–62. https://doi.org/10.1207/S15366359MEA0101_02
46. Nye BD, Davis DM, Rizvi SZ, Carr K, Swartout W, Thacker R, Shaw K (2021) Feasibility and usability of MentorPal, a framework for rapid development of virtual mentors. J Res Technol Educ 53(1):21–43. https://doi.org/10.1080/15391523.2020.1771640
47. Ouhaichi H, Spikol D, Vogel B (2023) Research trends in multimodal learning analytics: a systematic mapping study. Comput Educ Artif Intell 4:100136. https://doi.org/10.1016/j.caeai.2023.100136
48. Philippe S, Souchet AD, Lameras P, Petridis P, Caporal J, Coldeboeuf G, Duzan H (2020) Multimodal teaching, learning and training in virtual reality: a review and case study. Virtual Reality Intell Hardw 2(5):421–442. https://doi.org/10.1016/j.vrih.2020.07.008
49. Psotka J (1995) Immersive training systems: virtual reality and education and training. Instr Sci 23(5–6):405–431. https://doi.org/10.1007/BF00896880
50. Querrec R, Buche C, Maffre E, Chevaillier P (2004) Multiagents systems for virtual environment for training. application to fire-fighting. Int J Comput Appl 1(1): 25–34
51. Roberts ME, Stewart BM, Tingley D (2019) Stm: an R package for structural topic models. J Stat Softw. https://doi.org/10.18637/jss.v091.i02
52. Rogers MP, DeSantis AJ, Janjua H, Barry TM, Kuo PC (2021) The future surgical training paradigm: virtual reality and machine learning in surgical education. Surgery 169(5):1250–1252. https://doi.org/10.1016/j.surg.2020.09.040
53. Shah D, Patel D, Adesara J, Hingu P, Shah M (2021) Exploiting the capabilities of blockchain and machine learning in education. Augment Human Res 6:1–14. https://doi.org/10.1007/s41133-020-00039-7
54. Shepard LA (2000) The role of assessment in a learning culture. Educ Res 29(7):4–14. https://doi.org/10.3102/0013189X029007004
55. Shepard LA, Penuel WR, Pellegrino JW (2018) Using learning and motivation theories to coherently link formative assessment, grading practices, and large-scale assessment. Educ Meas Issues Pract 37(1):21–34. https://doi.org/10.1111/emip.12189
56. Shepard LA (2021) Ambitious teaching and equitable assessment: a vision for prioritizing learning, not testing. Am Educ 45(3):28–37. https://files.eric.ed.gov/fulltext/EJ1321974.pdf
57. Shute VJ (2008) Focus on formative feedback. Rev Educ Res 78(1):153–189. https://doi.org/10.3102/0034654307313795
58. Shute V, Rahimi S, Smith G, Ke F, Almond R, Dai CP, Sun C et al (2021) Maximizing learning without sacrificing the fun: stealth assessment, adaptivity and learning supports in educational games. J Comput Assist Learn 37(1):127–141. https://doi.org/10.1111/jcal.12473

59. Shute VJ (2011) Stealth assessment in computer-based games to support learning. In: Tobias S, Fletcher JD (eds) Computer games and instruction. Information Age Publishers, pp 503–524
60. Shute VJ, Ke F (2012) Games, learning, and assessment. In: Ifenthaler D, Eseryel D, Ge X (eds) Assessment in game-based learning: foundations, innovations, and perspectives. Springer, pp 43–58
61. Siemens G (2005) Connectivism: a learning theory for the digital age. Int J Instr Technol Distance Learn 2(1). http://www.itdl.org/
62. Sinatra AM, Pollard KA, Files BT, Oiknine AH, Ericson M, Khooshabeh P (2021) Social fidelity in virtual agents: impacts on presence and learning. Comput Hum Behav 114:106562. https://doi.org/10.1016/j.chb.2020.106562
63. So C, Khvan A, Choi W (2023). Natural conversations with a virtual being: how user experience with a current conversational AI model compares to expectations. Comput Animat Virtual Worlds e2149. https://doi.org/10.1002/cav.2149
64. Stenudd S (2010) Using machine learning in the adaptive control of a smart environment. Master's thesis. The VTT Technical Research Centre of Finland in the Software Architectures and Platforms Knowledge Centre
65. Thormundsson B (2022) Machine learning—statistics & facts. Statista. https://www.statista.com/topics/9583/machine-learning/#topicOverview
66. Valdez D, Picket AC, Young BR, Golden S (2021) On mining words: the utility of topic models in health education research and practice. Health Promot Pract 22(3):309–312. https://doi.org/10.1177/1524839921999050
67. Vaughan N, Gabrys B, Dubey VN (2016) An overview of self-adaptive technologies within virtual reality training. Comput Sci Rev 22:65–87. https://doi.org/10.1016/j.cosrev.2016.09.001
68. Vygotsky L (1978) Mind in society: the development of higher psychological process. Harvard University Press
69. Weizenbaum J (1966) ELIZA—a computer program for the study of natural language communication between man and machine. Commun ACM 9(1):36–45. https://doi.org/10.1145/365153.365168
70. Westera W, Dascalu M, Kurvers H, Ruseti S, Trausan-Matu S (2018) Automated essay scoring in applied games: reducing the teacher bandwidth problem in online training. Comput Educ 123:212–224. https://doi.org/10.1016/j.compedu.2018.05.010
71. Winkler-Schwartz A, Yilmaz R, Mirchi N, Bissonnette V, Ledwos N, Siyar S, Del Maestro R et al (2019) Machine learning identification of surgical and operative factors associated with surgical expertise in virtual reality simulation. JAMA Netw Open 2(8):e198363–e198363.https://doi.org/10.1001/jamanetworkopen.2019.8363
72. Wollny S, Schneider J, Di Mitri D, Weidlich J, Rittberger M, Drachsler H (2021) Are we there yet? A systematic literature review on chatbots in education. Front Artif Intell 4:654924. https://doi.org/10.3389/frai.2021.654924
73. Wu JY, Hsiao YC, Nian MW (2020) Using supervised machine learning on large-scale online forums to classify course-related Facebook messages in predicting learning achievement within the personal learning environment. Interact Learn Environ 28(1):65–80. https://doi.org/10.1080/10494820.2018.1515085
74. Zahabi M, Abdul Razak AM (2020) Adaptive virtual reality-based training: a systematic literature review and framework. Virtual Reality 24:725–752. https://doi.org/10.1016/j.cosrev.2016.09.001

Machine Learning in Educational Research

Machine Learning for Analyzing the Relationship Between Well-Being, Academic Performance with Large-Scale Assessment Data

Chong Ho (Alex) Yu, Zizhong Xiao, and Janet Hanson

Abstract Educational researchers have been utilizing large-scale assessment data from cross-cultural studies. However, the complexity of these big data poses several challenges to traditional statistical models, despite the benefits they provide to decision support and pedagogical practice. First, the sample size and the statistical power of these archival data are so huge that any trivial effect might be misidentified as significant. Consequently, this overfit model has low generalizability, contributing to the replication crisis. Second, in addition to academic performance test scores, these large-scale assessments also collect data about students' family backgrounds, living environment, educational institution characteristics, and other aspects associated with learning, such as motivation, engagement, and well-being. This curse of high dimensionality often results in multicollinearity, thus yielding an unstable model. To rectify the situation, machine learning methods, such as neural networks, bagging, gradient boosting, and XGBoost can localize errors by partitioning the data into subsets and generating numerous submodels. The final model is a synthesis of repeated analyses and therefore big data can be processed effectively. Further, through regularization machine learning penalizes complexity, preventing irrelevant predictors from entering the model. An analysis of the relationship between well-being and academic performance based on 2018 PISA data is presented in this chapter as an example of how machine learning can be applied to educational research.

Keywords Boosting · Bagging · Neural networks · Data science · PISA

C. H. (Alex) Yu (✉)
Hawaii Pacific University, Honolulu, HI 96813, United States
e-mail: cayu@hpu.edu

Z. Xiao
University of Maryland, College Park, MD 20742, USA

J. Hanson
University of North Texas at Dallas, Dallas, TX 75241, USA

M. S. Khine (ed.), *Machine Learning in Educational Sciences*,
https://doi.org/10.1007/978-981-99-9379-6_13

1 Purposes

This study used machine learning (ML) to analyze the relationship between student well-being in the educational context and their academic performance in math and science, based on data collected from the 2018 Program for International Student Assessment (PISA). The purposes of this investigation were to: (1) provide an example of how ML approaches can be applied to the interpretation of high-dimensional datasets and avoid issues arising from overfitting and convergence found with the use of classical statistical analysis methods, (2) demonstrate how the use of ML can inform the analysis of "big data", and (3) discuss the importance of ML and digital literacy for students, teachers, and researchers in the context of Education 4.0 and Industry 4.0.

A review of the relevant literature on the topics suggests the academic context of the twenty-first Century will focus on assessing Literacy 4.0 core skills using "standardized scalable testing…and individual scores aggregated to the population achievement….[with] testing profoundly influenc[ing] the curriculum" (Farrell et al. [33], pp. 905 & 909). Literacy 4.0 refers to "contemporary workplace literacy…the literacy practices generated and demanded by the forms of automation associated with Industry 4.0" (p. 899). What makes this study important is its position in connecting the contribution of ML to "identifying and evaluating important predictors of educational achievement" (Hilbert et al. [47], p. 3), such as student well-being, set in the context of Education 4.0 and Industry 4.0, aka the 4th Industrial Revolution (4IR). 4IR refers to "the digitization and the application of digital technologies, associated with new emerging jobs in the circular economy, artificial intelligence, cloud computing, development, sales of products and services and the focus on human resources" (Matúšová and Kollár [60], p. 1). Hilbert et al. [47] explained that ML can evaluate vast quantities of data that has been "mined" from educational sources. For example, online education learning management systems (LMS), massive open online courses (MOOCs), large-scale assessments such as the PISA, and Trends in International Mathematics and Science Study (TIMSS) are sources of "big data" that are captured, without a specific research question in mind. Instead, big data methods are used for their potential to be used in future analysis. In this situation, classical statistics are limited while ML seeks meaningful patterns from educational data mining with an eye toward using automated processes to support changes in the current educational frameworks. Finally, ML techniques have been shown useful in "develop[ing] advanced and scalable educational processes that account for individual needs of learners and complement and support the existing learning infrastructure" (p. 2).

Therefore, the following sections of this paper provide a discussion of the theoretical framework, methods used to perform the ML analysis, demonstration of the use of the relevant ML techniques, discussion of the results, and the importance of students, teachers, and researchers to develop an understanding of ML and digital literacy in the context of Education 4.0 and Industry 4.0. Conclusions will be drawn and recommendations for the next steps will be provided.

2 Theoretical Framework

ML provides models of input–output relationships that are especially useful in modeling complex relationships with high-dimensional datasets. These complex relationships are often encountered in modern data with many variables, cases, and potentially nonlinear effects [47]. ML approaches to data analytics are ideal for use with big data collected in large-scale education assessments. In the current era, data are viewed as the "new oil" (Caspari-Sadeghi [17], p. 1), and thus new tools are needed for complex analysis. ML approaches to statistical analysis have been shown to avoid problems, such as overfitting and convergence that can be encountered in traditional statistical analysis.

Educational researchers have been utilizing large-scale assessment data from cross-cultural studies since their availability. The complexity of these big data poses several challenges to traditional statistical models, despite the benefits they provide to decision support and pedagogical practice. First, the sample size and the statistical power of these archival data are so huge that any trivial effect might be misidentified as significant. Consequently, this overfit model has low generalizability, contributing to the "replication crisis" [21, 67]. Second, in addition to academic performance test scores, these large-scale assessments also collect data about students' family backgrounds, living environment, educational institution characteristics, and other aspects associated with learning, such as motivation, engagement, and well-being. This curse of high dimensionality often results in multicollinearity, thus yielding an unstable model.

Taking all the above limitations into account, as well as the increasing importance of scalable standardized testing in the future, it is necessary to employ modern ML to rectify the situation. ML methods, such as neural networks, bagging, and boosting, can localize errors by partitioning the data into subsets and generating numerous submodels. The final ML model is a synthesis of repeated analyses, and therefore, big data can be processed effectively. Further, through regularization, ML penalizes complexity, preventing irrelevant predictors from entering the model.

This present study examines the extent to which ML maximizes interpretable outcomes of data analysis of large, multidimensional data sets. We further discuss the contributing factor of Education 4.0 in encouraging educators, students, and researchers to become data literate in the areas of ML to better prepare for Workplace 4.0, a convergence of "human workers, devices, data collection, analysis applications, cloud platforms and robots" (Farrell et al. [33], p. 899).

3 Background of Well-Being in the Educational Context

3.1 Global Education Measurement

As globalization quickly spread, there has been a growing interest in measuring students' educational attainment worldwide [3, 59]. On average, most children, regardless of country, will have completed a majority of their compulsory education by the age of 15. Thus, to capture children's academic performances and capabilities, the Organization for Economic Cooperation and Development (OECD) developed the Programme for International Student Assessment (PISA) to collect data on how each participating country's children stood academically [84]. Traditionally, PISA's three key assessment elements consist of reading, mathematics, and science. Additionally, PISA consistently measured students' family backgrounds, living environment, educational institution characteristics, and other aspects associated with learning (i.e., motivation and engagement). Recently, PISA added new subsections that include technology competence, career aspirations, parental and teacher characteristics, along students' well-being. For this current study, we are primarily interested in the PISA's newly developed well-being subsection, due to its important link to academic success.

3.2 Well-Being and Academic Achievement

Various studies regarding the relationship between well-being and academic performance were conducted with participants in different academic levels, ranging from grade school [55], high school [30], to higher education [19], as well as in different countries, such as China [95], Pakistan [15], and the United Kingdom [81]. Well-being has been established as a strong predictor of educational attainment, especially among adolescent children participating in compulsory education (e.g., Korhonen et al. [49], Nordlander and Stensöta [65])

The causal mechanism between well-being and academic performance is conceptualized in different ways. First, according to the *broaden-and-build theory* [34–38], well-being consists of a set of positive emotions, including joy, interest, contentment, and love. Each of these positive emotions could broaden an individual's momentum for positive actions. Specifically, joy leads to active engagement, interest drives exploration, contentment sparks the urge to integrate, and love enhances social relationships. As a result, these heightened psychological states facilitate the build-up of academic performance, because those states broaden one's cognitive awareness that identifies potential solutions to solve academic challenges [86]. Similarly, Datu and King [27] found that well-being is associated with a wide range of adaptive psychological functions in the school setting, including active academic engagement, which results in higher academic performance.

Another prominent theory about the relationship between well-being and academic performance is based on *self-efficacy* [6, 7]. According to Bandura's theory of self-efficacy, one's positive perception of his or her own ability can lead to a successful accomplishment of a task or fulfillment of a goal, including academic achievement. This theory has received much attention in the field of educational psychology [92] and significantly contributed to the development of positive psychology, which focuses on and nurtures positive aspects of human life, such as happiness and well-being ([5, 25, 78]). Additionally, positive psychology interventions (e.g., practicing positive thinking) were found to be effective in helping high-school students with poor academic achievement by boosting their motivation to study [64].

Although those research studies yielded insightful findings, they heavily concentrated on subjective well-being and abstract constructs. For example, the adolescent well-being scale used by many researchers consists of six dimensions of subjective perceptions, such as present life satisfaction, present positive affect, present negative affect, hopeful future expectations, positive affect toward future life, and negative affect toward future life [87]. The Rosenberg Self-Esteem Scale (RSE; [72]), which measures the perception of self-worth, is another example of subjective well-being. Even though these scales were psychometrically validated, it is crucial to point out that participants from diverse cultures might interpret and rate their subjective well-being and life satisfaction differently. For example, White students reported the highest level of happiness in activities important to them at that moment while Asians tended to enjoy activities that were perceived as important in the future [29]. Further, Western cultures emphasized personal satisfaction, whereas Asian cultures aimed to achieve the greatest happiness collectively [31].

3.3 A Holistic Approach to Well-Being

To rectify the conceptual differences in well-being, the OECD [70] defines well-being as a multi-faceted construct consisting of both material, objective components, and psychological, subjective perception. Specifically, factors of well-being include the individual dimension (self-health and psychological functioning), the school environment (social connections and schoolwork), and the out-of-school environment (social connection and material conditions). In addition, there are at least three main aspects (but not limited to) of learners' life that could influence how they feel, namely, life at school, their relationships with their parents, and body image perception [70]. It is crucial to note that although these dimensions are considered distinct, they are not independent. Specifically, students' happiness is tied to social and economic well-being. For example, OECD [68] found that disadvantaged students were less likely than advantaged students to report higher levels of life satisfaction. Moreover, disadvantaged students and first-generation immigrant students were also less likely to enjoy a sense of belonging at school than other students. Based on the preceding framework, the OECD developed both subjective and objective items in the well-being survey. Additionally, psychological and sociological factors, to some extent, are non-separable. Based on 2018 PISA data, Kuhn et al. [50] found that the strength of personal relationships and sense of belonging were strongly related to perceived life satisfaction and well-being.

The 2018 PISA findings and previous studies signified the importance of well-being in predicting academic achievement. Therefore, these results are a critical step and imply the need to further advance the research on how well-being contributes to student success. A 2018 PISA finding emphasized that among the OECD participating countries, one in ten students from economically disadvantaged backgrounds were able to score within the top quarter of academic performance in their country despite their economic circumstances [70]. This implies that poverty was not the sole determinant. The corroboration is that students with higher well-being levels, despite low socioeconomic status, have more academic resilience, parental support, and self-efficacy. Previous studies [10, 45, 58] further support this finding: well-being factors such as social belongingness and positive affect moderate the relationship between poverty and academic achievement. By including both objective and subjective well-being measures, PISA data provide researchers with opportunities to obtain a more holistic view of the issue.

In addition, PISA focuses on concrete scenarios rather than abstract constructs. For example, the Rosenberg [72] self-esteem scale used in other well-being studies asked respondents to give ratings to statements like "I feel that I am a person of worth, at least on an equal plane with others," and "I feel I do not have much to be proud of." On the contrary, PISA survey items are more specific, such as "I consider myself to be attractive," "How often do you worry about how much money your family has?" and "How did you feel the last time you spent time outside your home with your friends? Are you happy?" (OECD [69], p. 6, 15, 22). It is our conviction that the OECD's data are more accurate and informative due to their concreteness.

3.4 Student Well-Being in the Context of Education 4.0

According to the OECD [70], improvements in students' education, and ultimately in their academic outcomes, can be developed from a better understanding of student well-being in the educational context. The PISA 2018 "aims to provide information about the ability of 15-year-old students to face the challenges in their future lives" (Govorova et al. [43], p. 4).

Govorova et al. [43] performed a network analysis based on elements of student well-being from both the PISA 2015 and 2018 to develop a model of student well-being and test the interrelationships between them. Current research includes a focus on school effectiveness in areas that "[go] beyond students' learning of 'knowledge and skills' into the social dimensions of learning, such as student well-being" (p. 1). Govorova et al. [43] found the student well-being dimensions of the 2018 PISA scale "form[ed] a solid welfare construct in the educational context" with self-efficacy (operationalized on PISA as "resilience") a key aspect of the well-being construct (p. 14). However, the influence of school factors was found to be generally low except for the influence from the school's context, where educators and administrators can play an indirect role, especially in reducing bullying. Govorova et al. [43] provided a review of definitions of student well-being in the educational context from a variety of sources. Appendix A provides a chart comparing the dimensions and trends in the literature reviewed on the topic of well-being. Of note, the social dimension is present in each contributor's operationalization of the elements of student well-being. This is important because it aligns well with key aspects of Education 4.0 and Industry/ Economy 4.0 (4IR), where the social skill dimension will become a foundational attribute for successful workplace skills, in the near future. For example, Farrell et al. [33] contended that in Workplace 4.0 the role of the human worker will be drastically transformed by a "convergence of people and technologies" (p. 899). Routine cognitive skills, such as those skills previously promoted in degree attainment, will be replaced by AI. The human worker (emphasized "human" due to the entry of the robotic, AI, and computational/system workers) will be required to provide "higher order skills of analytical, creative, communication and interpersonal skills" (p. 904). This is the new Literacy 4.0, which requires the ability to negotiate successful relationships especially in online environments where individuals may be separated by time and geographic location. To keep pace and align with dynamic changes resulting from the transitions brought on by the 4IR, new research and analytical methods for the changing landscape of large-scale, data-driven analyses are being employed. Therefore, teachers, administrators, and researchers engaged in Education 4.0 can better develop applied curricular approaches through project explorations using ML methods to identify replicable and reliable indicators that are malleable to administrator and stakeholder influence; leading to improvements in the systems and quality of life for students in the context of Education 4.0 as well as outside of the classroom where practical.

4 A Machine Learning Approach to Big Data

In the past, educational research that examined the relationship between well-being and academic performance over-relied on using traditional statistics with small data sets (e.g., Bonhauser et al. [10], Malecki and Demaray [58]). However, when Pearson, Fisher, and other statistical theorists developed their methodologies during the late nineteenth and early twentieth centuries, these methods were intended for small-sample studies [99]. Additionally, traditional statistics tend to impose artificial boundaries with minuscule purposes: there are minimal differences between p-value(s) of 0.05 (significant) and 0.06 (non-significant) [26]. After all, the p-value does not inform us of the underlying pattern of the data [83, 90].

To rectify the limitations of traditional statistics, research has evolved to use data science and ML methods, especially big data analytics (BDA) [88]. BDA has several advantages. First, data science researchers could draw conclusions based on the pattern of big data at hand, rather than making inferences by inquiring about the probability of observing the test statistics based on a theoretical sampling distribution [99]. Following this line of reasoning, BDA does not depend on conventional cut-off points (e.g., p-values) to determine whether findings are significant. Further, BDA often utilizes millions of observations that include multiple cultures and countries. Therefore, the results are more representative and decrease its proneness to the replication crisis. BDA can further resolve the issue of replication crisis by resampling, meaning that multiple models are built with numerous subsets of the original dataset, so that the problem of overfitting due to a single analysis is alleviated [82].

Since most research on students' well-being and academic achievement was conducted using traditional statistics, this current study pushed the boundaries by utilizing BDA. The 2018 PISA data set is the source for this specific BDA. The PISA student well-being survey items are the potential predictors. The science and math test scores were extracted as the criterion variables. This analysis is composed of three modules: (1) ensemble methods, (2) model comparison, and (3) data visualization. First, four ML methods, including bagging, boosting, XGBoost, and neural networks, were utilized to analyze the relationship between the PISA well-being variables and the academic achievement—math and science scores. Second, the model comparison was conducted to identify the best model. Last, we employed data visualization to examine the relationship between the outcome and the important predictors [100].

5 Method

5.1 Participants

In the past, educational researchers tended to recruit participants from westernized, educated, industrialized, rich, and democratic (W.E.I.R.D.) countries, resulting in a non-representative sample [46]. In contrast, the PISA 2018 data set consists of hundreds of thousands of students participating globally. However, only nine countries/regions completed the well-being section: Bulgaria, Georgia, Hong Kong, Ireland, Mexico, Panama, Serbia, Spain, and the United Arab Emirates. Thus, the entire participant pool for this study is trimmed down to 97,878. Nevertheless, unlike many other studies that only rely on a limited and Caucasian-dominated sample pool, the observations from the nine countries/regions still reduce some likelihood of the representative bias.

5.2 Items of Concrete Scenarios as Independent Variables

Since we are only interested in well-being variables as the predictors, we extracted all the variables PISA categorized as well-being (e.g., "How is your health" and "How often did you smile the past day"). To see a complete list of PISA well-being variables, please refer to OECD [69]. It is also available from the authors upon request. In total, there were 82 PISA well-being variables serving as the independent variable [69, 70]. Because each item measures different aspects of well-being using specific scenarios, we entered the variables individually into the model building process rather than utilizing any data reduction techniques, such as principal component analysis or factor analysis. Although the psychometric attributes of the PISA instrument have been examined by the OECD [71, 89], using individual scale items as independent variables is warranted based on the findings of Govorova et al. [42]. According to Govorova et al. [42], the "proposed structure of the well-being concept could not be confirmed" by a confirmatory factor analysis (CFA) of the data on the PISA 2015; e.g., "some of the domains were not unidimensional and some of the indicators included in the domains were not relevant to student welfare" (p. 3). They reported a limitation of the study as a need for additional research on the definition of student well-being in the educational context due to the complexity of operationalizing the construct.

5.3 *Survey Items of Momentary Feeling*

It is important to note that many well-being items in the PISA survey are concerned with the most recent events, which is an approach to remediate the inaccuracy of self-report data. The reliability of self-report data is an Achilles' heel of survey research. Cook et al. [22] have pointed out that very often participants (a) tend to report what they believe the researcher expects to see, or (b) report what reflects positively on their own abilities, knowledge, or opinions. Another concern about such data centers on whether subjects are able to accurately recall past behaviors. Psychologists have warned that human memory is fallible [54, 77]. Sometimes people "remember" events that never happened. Thus, the reliability of self-reported data is tenuous.

To counteract this problem, some researchers suggested collecting data related to the momentary thoughts or feelings of the participant, rather than asking him or her to recall remote events [24, 35]. The following are two examples used in the 2018 PISA: "Did you smile or laugh a lot yesterday?" "Did you learn or do something interesting yesterday?" [69].

5.4 *Plausible Value as Dependent Variable*

As for the dependent variable, we aimed for a single variable that defined math and science scores. However, there were multiple scores embedded within the PISA report. Therefore, we used the plausible value (PV) to convert multiple scores into one score. It is common to have multiple test scores per single student on a global education assessment such as the PISA. Since having every student take a standardized test would be time-costly, researchers resort to calculating students' propensity scores [94, 98]. Considering the different pre-performed assessments by the individual countries, OECD researchers calculated a set of standardized performance scores for reading, math, and science (OECD, 2018). To select an individual PV out of the provided set of PISA subject performance scores to act as dependent variables, a randomized value was used. Although it is tempting to average all PVs to obtain a single test score, this method is inappropriate. In contrast to the true score theory, modern psychometrics, like Bayesianism, view the outcome as a distribution rather than a fixed score. If the mean score is used, then this score is nothing more than a point estimate, thus destroying the purpose of PV. Indeed, each score is equally as good as others. Hence, according to [94], "even if we only use one plausible value per student to estimate population characteristics, we still have unbiased estimates, in contrast to using each student's EAP estimates (mean of plausible values for each student) and getting biased estimates. Hence, the apparent paradox is that using one random draw (PV) from the posterior distribution is better than using the mean of the posterior, in terms of getting unbiased estimates" (p. 977).

5.5 Big Data Analytics

Although ordinary least squares (OLS) regression has been widely used by many social scientists, this method was considered inappropriate for this study for several reasons. First, the variance of an OLS regression model could be inflated when some of the variables are highly correlated (multicollinearity), resulting in model instability [1]. Furthermore, traditional regression cannot handle the gargantuan sample size like the 2018 PISA data set, because the high statistical power tends to misidentify trivial effects as significant. Put bluntly, the model might be overfitted when both the numbers of observations and predictors are large. Last, unlike data science methods, case-wise deletion happens when there are missing data in regression modeling. Only 4,700 participants answered all question items related to well-being in this study. Using the OLS regression method would lead to a substantive information loss. To rectify the situation, four ML methods were employed as described below. All the following procedures were implemented in JMP Pro 17 (SAS [74]).

Bagging

Bagging, also known as the bootstrap forest, stands for Bootstrap Aggregation. This is a machine learning method that uses an algorithm to create multisets of additional training data from the original sample repeatedly [13, 16]. The bagging method is an ensemble of many decision trees obtained from repeated sampling with replacement from the same data set. Afterward, the multitude of results are then combined to form a converged conclusion. Since the validity of one single analysis may be weak due to the possibility of overfitting, this bootstrapping method can produce stronger validity because the end results are based on many sub-samples. In addition, bagging increases the size of these generated data and effectively minimizes the variance by decreasing the influence of extreme scores [63].

Boosting

Boosting is also known as the gradient boosted tree—a sequential machine learning method. Initially, the algorithm assigns all observations equal weight before producing a statistical model. If the model fails to classify some of the observations correctly, then these observations will be assigned a heavier weight, so that they are more likely to be selected in the subsequent model. As this process continues, each model is revised and updated constantly in order to successfully classify all or most of the observations. Unlike bagging, which requires multiple independent models to reach a solution convergence, the boosting method produces a solution after a few analytical repetitions.

XGBoost

XGBoost, which stands for E**x**treme **G**radient **Boost**ing, was introduced in 2014 as an enhancement of the original gradient boosting method. In order to search for the optimal solution, XGBoost can perform various boosting techniques, including classical gradient boosting, regularized boosting, and stochastic boosting [28]. In addition to handling missing data efficiently, XGBoost can also prevent overfitting through regularization, a technique of penalizing model complexity. XGBoost is also scalable in all situations. Specifically, the method can run billions of records on a distributed system or on a single computer. In traditional regression modeling, if the algorithm attempts to find the best subset of predictors by exhausting all possible combinations of an extremely big data set, the computer would be frozen. With the help of an advanced algorithm known as the exact greedy algorithm, XGBoost can enumerate all possible splits for every variable in the decision tree [20].

Neural Networks

Neural network, also known as artificial neural network, is an ML technique inspired by cognitive psychology and neuroscience that mimic human neural networks. A typical neural network model consists of the input, hidden, and output layers [76]. An input layer has a neuron for each input variable without any adjustable parameters (weights). It simply passes the positive or negative value to the next layer. A hidden layer performs a transformation of the weighted and summed input activations without the intervention of the analyst. In traditional linear regression, the researcher might try different transformations of the predictors, interactions between predictors, or both. But in neural networks, these are automatically processed in the hidden layer. Without the hidden layer, a neural network just looks like a regression model. An output layer combines the nonlinear hidden layer activation values in the final presentation. When there are multiple hidden layers in the network, it becomes a deep learning model [99]. Although the neural network modeling technique is immensely powerful, the price of this solution is complexity and low interpretability. For this reason, usually, the result yielded from neural networks is called a black box model [73].

To avoid overfitting to a specific sample, all four preceding machine learning techniques were run with fivefold cross-validation, which is a resampling procedure [51]. In cross-validation, the sample is randomly divided into several subsets. The first subset is used for training and proposing a model whereas the others are utilized for validating the result.

Model Comparison

After generating all four models, we can choose between model selection and model averaging to construct the final model. In model selection, the best model is retained based on predictive accuracy and error rates, such as R^2 and Root Absolute Squared Error (RASE). The R^2 is the variance explained whereas RASE is similar to Root Mean Squared Error (RMSE), except that RMSE adjusts for degrees of freedom while RASE does not. By doing so, the second best and other models are totally discarded. On the other hand, in model averaging information from all models is utilized in several ways, such as averaging the prediction estimates from all models, selecting the highest estimates of all models, and returning the proportion of the models that can substantially contribute to the outcome. It is our conviction that model selection is the most cost-effective way because prior research found that in most cases model averaging and model section lead to comparable results [66]. Although model averaging can yield more accurate results than model selection when researchers study complex models, such as nonlinear mixed-effect models [14], in this study, data visualization revealed that the relationships between input and target variables are linear, and thus it is unnecessary to employ model averaging.

Data Visualization

We used dynamic data visualization techniques for pattern seeking to efficiently explore and display insightful findings until a plausible story emerged [39, 97]. Median smoothing is a data reduction technique for visualization [83, 101]. When there are 90,000+ observations, the data pattern could be hidden by the noise, and this problem is known as "over-plotting." To overcome over-plotting, the median of y on the boxplot is displayed at each level of x, and hence the analyst can look at the trend by visually connecting the median. Mihalisin et al. [62] extended the preceding idea by using the mean rather than the median. However, using the median is recommended because the median is more resistant to extreme scores. When the distributions are fairly normal, mean smoothing and median smoothing yield identical results. The violin plot was superimposed on the boxplot to check the distributions at each level of x. The violin plot consists of two-mirrored frequency polygons or density curves. In other words, if the violin plot is cut in the middle, the display is a regular density curve depicting the distribution [97].

Table 1 Model comparison of validation sets using math scores as the target variable

Method	R^2	Mean RASE	SD RASE
Boosted Tree	0.3973	79.746	0.3920
Neural Boosted	0.3943	78.123	2.4596
XGBoost	0.3687	81.619	0.4833
Bootstrap Forest	0.3344	83.804	0.2493

Table 2 Model comparison of validation sets using science scores as the target variable

Method	R^2	Mean RASE	SD RASE
Boosted Tree	0.4222	75.921	0.7360
XGBoost	0.3912	77.930	0.6887
Neural Boosted	0.3871	67.467	2.8017
Bootstrap Forest	0.3576	80.055	0.6221

6 Results

6.1 Best Model: Boosting

Tables 1 and 2 indicate that for predicting both math and science test performance, the best model is boosting. Because the results of fivefold cross-validation were taken into account, the tables show the means of R^2 and RASE, as well as the standard derivation of RASE, of all validated results. Although in this analysis the original gradient boosting method outperformed the enhanced XGBoost approach and the powerful neural networks, these results are not surprising. While scalable XGBoosting is suitable for analyzing extremely large data sets (e.g., n = billions) and neural networks aim to deal with complicated nonlinear relationships, gradient boosting can function well with a moderately large sample and a less complicated problem. For predicting math performance, the boosting model yielded the largest R^2 (0.3973) and the second lowest error rate (79.746), second to the neural networks (78.123). However, the variation of boosting is much narrower than that of the neural networks (0.3920 vs. 2.4596). For predicting science test performance, boosting is considered the best in terms of R^2.

6.2 Math Scores: Boosting Results

Table 3 highlights the best well-being predictors of math test performance by boosting. There are three major indicators of predictor importance: the number of splits, the sum of squares, and the portion. The number of splits indicates how many times the variable is selected as a crucial predictor to split a tree across all repeated samples whereas the portion shows the share of the SS value attributed to the variable.

Table 3 Relationship between PISA well-being predictors and math performance score (Boosting)

Variable	# of splits	Level of agreement	*n*	Math score: Mean	Math score: SD
How did you feel the last time you spent time outside your home with your friends? Bored	400	1, 2, 3 4	95,450 2428	459.36 404.60	102.46 99.04
How did you feel the last time you spent time outside your home with your friends? Nervous or tense	412	1, 2, 3 4	94,323 3555	459.58 416.09	102.50 99.95
Now think of the last time you had a break between classes at school. How did you feel? Nervous or tense	400	1, 2, 3 4	94,333 3545	459.51 418.00	102.55 99.48

Although there is no consensus on which criterion is the best, the number of splits is the most intuitive and easiest to explain. Simply put, the number of splits is a form of voting. After repeated analyses are performed, the most frequent recurring variables are considered more important than those less popular predictors. The three most important predictors are themed on how students felt when spending time outside the classroom either at home or during breaks with peers.

6.3 *Science Scores: Boosting Results*

Table 4 highlights the best three predictors of science test performance by boosting. As with previous analyses, the most important predictors are themed on how students felt when spending time outside the classroom either at home or during breaks with peers.

Had a regression analysis been run, 16 variables would have been identified as significant based on the criterion of alpha <=0.05. Not only is the risk of committing Type I errors so high that the conclusion would be erroneous, but also this complicated model might not lead to practical implications and actionable items, even if the conclusion is correct. It is important to point out that boosting is built upon a collection of individual decision trees. The primary objective of the decision tree approach is to develop a parsimonious model for decision support. In some situations, the number of most important predictors is as small as three [12].

Table 4 Relationship between PISA well-being predictors and science performance score (Boosting)

Variable	# of splits	Level of agreement	*n*	*Science* score: Mean	Science score: SD
How did you feel the last time you spent time outside your home with your friends? Bored	400	1, 2, 3 4	95,425 2453	455.58 397.3	99.50 97.96
How did you feel the last time you spent time outside your home with your friends? Nervous or tense	407	1, 2, 3 4	91,771 6107	455.23 427.48	99.86 98.68
Now think of the last time you had a break between classes at school. How did you feel? Nervous or tense	403	1, 2, 3 4	94,621 3617	455.70 413.20	99.65 97.06

6.4 Data Visualization: Median Smoothing

The preceding numbers alone might not reveal the whole story. To illustrate the patterns and the relationships between the most important well-being predictors and math test performance, the median-smoothing technique was employed, as shown in Figs. 1, 2 and 3. The violin plot superimposed on the boxplot indicates that the math scores are normally distributed at each level of the well-being variables. The relationships between those variables and the science test performance are similar.

7 Discussion of Well-Being and Academic Performance

7.1 Peer Engagement and Academic Success

Three PISA well-being predictors on social engagement and emotional state are identified by boosting as significant predictors of math and science test performance. Previous research (e.g., Wentzel [91]) justifies our study's findings that, out of many PISA well-being predictor variables, a sense of belongingness among peers and social acceptance emerged as the most significant predictors. There are several explanations as to why peer relationships are robust predictors of academic success. During school, students are guaranteed to experience setbacks (i.e., fail an exam). However, having peer support will buffer those setbacks and boost resilience to maintain stamina for school achievement. Conversely, a lack of social support contributes to low academic performance and motivation, which leads to dropping out [53]. Other studies pointed out that the relationship between school environment and academic performance hinges on peer harassment, which decreases the sense of

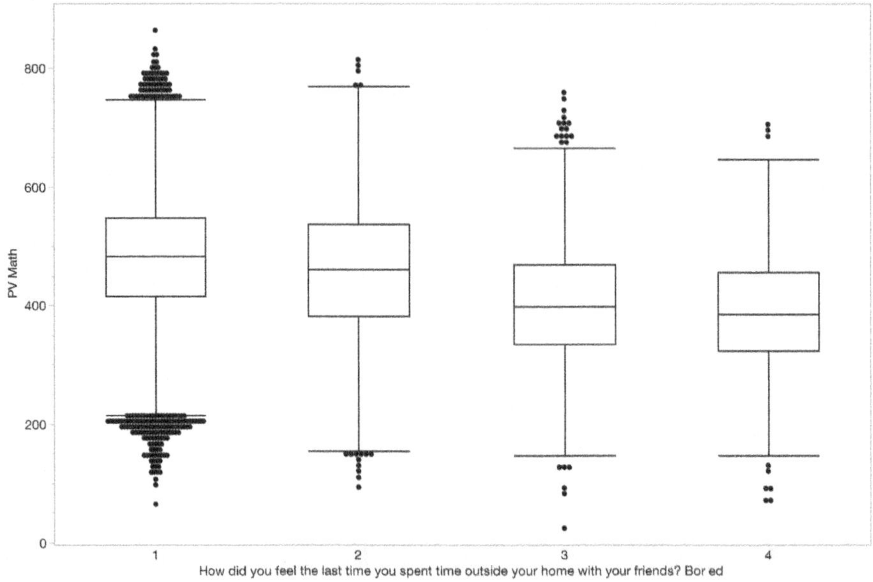

Fig. 1 The relationship between "how did you feel the last time you spent time outside your home with your friends? bored" and math scores. *Note:* This medium-smoothing plot depicts a negative linear relationship as noted by the downward trend of the median. The less boredom students experience with their friends outside the classroom (which translates to positive social engagement and peer environment), the better their math scores become

belongingness and social acceptance (e.g., Gloppen et al. [41], Graham and Juvonen [44]). For example, students who received peer harassment exhibited amotivation to attend school. Therefore, this further deprives them of the benefits of educational advancement [32]. These negative behavioral patterns then further prevent students from attaining their academic potential.

7.2 *Negative Emotion*

Although we cannot tell why some students feel nervous and tense during a break between classes at school, this negative emotion that negatively impacts their academic performance might be somehow related to the school environment. Previous research shows an association between teacher's willingness to support students and students' academic success (e.g., Khan et al. [48], Mazer [61]). Teachers often play a prominent role in maintaining a positive school environment [79]. Thus, an open communication channel between the teachers and students allows teachers to solve problems (i.e., peer conflicts and academic struggles). Another highly possible cause of anxiety at school is academic pressure. When schoolwork is demanding or expectations are high, students are likely to burn out, resulting in reduced academic

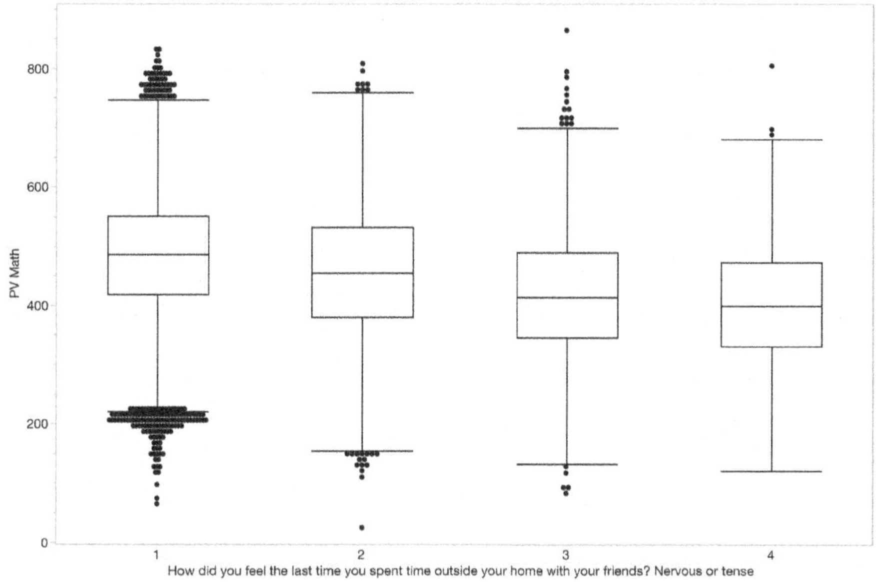

Fig. 2 The relationship between "how did you feel the last time you spent time outside your home with your friends? nervous or tense" and math scores. *Note:* This median smoothing depicts a negative linear relationship as noted by the downward trend of the median. The easier the students find to communicate with their friends, the better their math scores

performance [75]. This phenomenon is especially true among East-Asian cultures, such as Hong Kong, where students face extreme pressures to excel [52]. There is a widespread phenomenon about the Asian orientation toward over-achievement: Any grade lower than an A is technically an F [100]. Furthermore, students are more likely to fail when they are hyper-conscious of a particular task ranging from sports and academics—a phenomenon called *choking under pressure* [8, 40]. Indeed, Beilock and Carr [9] found that when pressure was overly high, scholastic capabilities such as solving math problems and working memory decreased compared to when pressure was reasonable. Neuroimaging research conducted by Lyons and Beilock [56, 57] also indicated that the subcortical regions of the brain implicated in motivation could lead to stress-induced choking in math performance.

Past works and our current study demonstrate the strong association between peer acceptance, sense of belongingness, emotion, and academic performance. However, this study is distinctive from most past research using traditional statistics because it demonstrated that, out of many potential PISA well-being predictors that can be related to educational attainment (e.g., at-home resources and physical health), social belongingness came out as the strongest variable. Additionally, we utilized numerous ratings of concrete scenarios and holistic well-being measures, rather than vague statements that are highly subject to interpretation. At the end, if teachers wish to strengthen students' academic achievement based on the well-being variables, then teachers should promote peer engagement, a sense of belongingness,

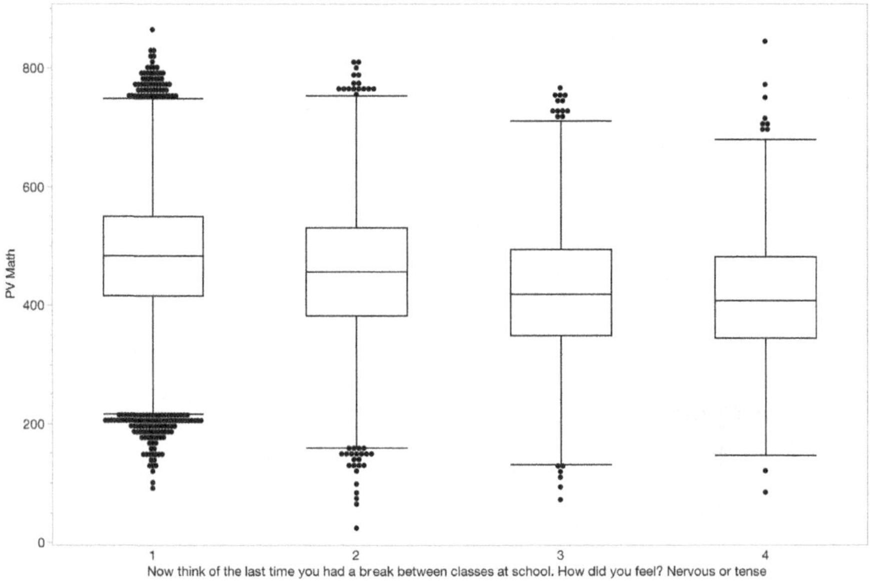

Fig. 3 The relationship between "now think of the last time you had a break between classes at school. How did you feel? nervous or tense" and math scores. *Note:* This medium-smoothing plot depicts a negative linear relationship as noted by the downward trend of the median. The less nervousness students experience at school, the better their math scores become

and positive emotion. These mental and behavioral attributes are not only important to test performance, but also to future job performance. Specifically, Education 4.0 identified social and collaborative skills as one of the most crucial skills for the jobs of the ever-evolving future workplace [93].

7.3 Limitations and Future Directions

Despite many strengths of BDA, this study is limited by the fact that only nine countries completed the PISA well-being survey. The reason for this lack of participation is due, perhaps, to the well-being section being relatively new. Because our result might not represent the global population, we hope more countries can complete the newly introduced well-being section in the future so we can perform follow-up studies. In addition, some PISA well-being questions are concerned with the momentary feelings of the participant (e.g., "Did you smile or laugh a lot yesterday?"). In this case, the response depends on what happened to the participant around that particular moment, which might not be typical. Specifically, even though the respondent did not smile or laugh a lot yesterday, it does not necessarily imply that the respondent is always unhappy. Nevertheless, none of those responses related to momentary feelings is identified as critical factors for predicting test performance. If these well-being variables are shown to be important in other studies, the results should be interpreted with caution.

8 Discussion of Machine Learning and Educational Research

This current study demonstrated how ML approaches can be applied to the interpretation of high-dimensional datasets to avoid issues arising from overfitting found with the use of classical statistical analysis methods. We encourage readers to begin an exploration of the novel methods presented here in spite of their unfamiliarity with them. As Meyers and Rose (2005) explained, it takes a long time for people to realize how to use technologies with radically new capabilities creatively and productively, especially when the new technologies drastically change the current system. A shift in viewpoint can only occur when users spend time experimenting with the new tools. Matúšová and Kollár [60] argued that in order to meet the challenges of Industry 4.0, new philosophies and tools are needed in education sciences and research. In a similar vein, Crawford et al. [23] asserted that new analytical methods like machine learning (ML), which support scalable, technology-driven instructional methods like artificial intelligence and "serious gaming" (p. 25), are contributing to the changing educational landscape in the Twenty-First Century/Information Age/ Digital Age/4IR/Education 4.0.

Soon, educational systems will be challenged to prepare students for success in the new generation of workplace practices, including Literacy 4.0. Therefore, it is important for students, teachers, and researchers to understand digital literacy tools such as ML [33]. Though the boundaries of this study do not extend to the debate on the efficacies of the "recurring theme" of the Industry 4.0/4IR, we recognize there is an "expectation of difficulties that can be overcome...[and] a predominant [theme of the] social benefit from the transformative power of technology" (Avis [4], p. 6). Through this lens, keeping an open mind to the changing terrain of education and research, the ML methods of analytics provided here could enhance educational research with large-scale assessment data, resulting in developing useful interventions and curricula to successfully prepare our students for the future.

Appendix A

Chart comparing and contrasting the dimensions of student well-being in the educational context-how they interact with and are influenced by school factors

	New south wales	Australia	International context						
Factors contributing to WB	Definition of			Interventions	Reasons to include in the PISA survey	Four General Principles for best interest of the child	Elements of	Five domains of-Elements common indicators related to students' school-lives which depend on their personal and family lives and are needed to live a happy and fulfilling life	Dimensions of - based upon PISA 2015 & 2018
	Being a moral actor in relation to oneself or behaves well toward others	Self-esteem	Dynamic state	Resilience	Human mind-Development and formation of; Educational achievement	Right to develop-ment	Mental health	Cognitive-related to knowledge and abilities in everyday life	Cognitive - growth mindset; self-efficacy and fear of failure
	Happiness or absence of sadness	Self-concept	Abilities and opportunity to fulfill personal goals…	Resilience		Right to development	Happiness and satisfaction in students' lives	Psychological perceptions of their own lives, school, and future	Psychological-meaning in life, feelings,

(continued)

(continued)

Positive class atmosphere; Teacher support regardless of intellect or temperament	Harmonious social relationships	Self-respect	…and social goals	Improving relationships; school climate reducing bullying	Socialization and social values	Non-discrimination-Respect for the views of the child	Socialization and interactions with peers and teachers	Social perceive their relationship in and out of school	Social-competition and cooperation, sense of belonging, parents emotional support, bullying
					Importance for health	Right to life, survival	Physical health	Physical health and habits related to sports and eating; Material-available resources to meet needs	
Govorova et al. [43]	Fattore et al. [34]	Anderson and Graham [2]	Borgonovi and Pail [11]	Slee and Skrzypiec [80]	Chapman [18])	UN General Assembly [85]	Govorova et al. [43]	OECD [68, 69]	Govorova et al. [43]

References

1. Althauser RP (1971) Multicollinearity and non-additive regression models. In: Blaock H (ed) Causal models in the social sciences. Aldine, pp 453–472
2. Anderson DL, Graham AP (2016) Improving student wellbeing: Having a say at school. Sch Eff Sch Improv 27(3):348–366. https://doi.org/10.1080/09243453.2015.1084336
3. Anderson JO, Lin HS, Treagust DF, Ross SP, Yore LD (2007) Using large-scale assessment datasets for research in science and mathematics education: Programme for International Student Assessment (PISA). Int J Sci Math Educ 5(4):591–614
4. Avis J (2020) Socio-technical imaginaries and the fourth industrial revolution: vocational education in the fourth industrial revolution. J Vocat Educ Train. https://doi.org/10.1007/978-3-030-52032-8_2
5. Bandura A (2008) An agentic perspective on positive psychology. In: Lopez SJ (ed) Positive psychology. Greenwood Publishing, pp 167–196
6. Bandura A (1986) Social foundations of thought and action: a social cognitive theory. Prentice Hall
7. Bandura A (1997) Self-efficacy: the exercise of control. Freeman Press
8. Baumeister RF (1984) Choking under pressure: self-consciousness and paradoxical effects of incentives on skillful performance. J Pers Soc Psychol 46(3):610–620
9. Beilock SL, Carr TH (2005) When high-powered people fail: working memory and "choking under pressure" in math. Psychol Sci 16(2):101–105
10. Bonhauser M, Fernandez G, Püschel K, Yañez F, Montero J, Thompson B, Coronado G (2005) Improving physical fitness and emotional well-being in adolescents of low socioeconomic status in Chile: results of a school-based controlled trial. Health Promot Int 20(2):113–122
11. Borgonovi F, Pál JA (2015) Framework for the analysis of student well-being in the PISA 2015 study: being 15 in 2015. OECD Publishing
12. Breiman L (1984) Classification and regression trees. Routledge
13. Breiman L (1996) Bagging predictors. Mach Learn 24(2):123–140
14. Buatois S, Ueckert S, Frey N, Retout S, Mentré F (2018) Comparison of model averaging and model selection in dose finding trials analyzed by nonlinear mixed effect models. Am Assoc Pharm Scientists J 20(3), Article 56. https://doi.org/10.1208/s12248-018-0205-x
15. Bukhari SR, Khanam SJ (2017) Relationship of academic performance and well-being in university students. Pak J Med Res 56(4):126–130
16. Bühlmann P, Yu B (2002) Analyzing bagging. Ann Stat 30(4):927–961
17. Caspari-Sadeghi S (2023) Artificial intelligence in technology-enhanced assessment: a survey of machine learning. J Educ Technol Syst 51(3):372–386. https://doi.org/10.1177/00472395221138791
18. Chapman A (2015) Wellbeing and schools: exploring the normative dimensions. In: Wright K, McLeod J (eds) Rethinking youth wellbeing. Springer, pp 143–159
19. Chattu VK, Sahu PK, Seedial N, Seecharan G, Seepersad A, Seunarine M, Singh A et al (2020) Subjective well-being and its relation to academic performance among students in medicine, dentistry, and other health professions. Educ Sci 10(9), Article 224. https://doi.org/10.3390/educsci10090224
20. Chen T, Guestrin C (2016) XGBoost: a scalable tree boosting system. In: Proceedings of the 22nd ACM SIGKDD international conference on knowledge discovery and data mining Knowledge discovery and data mining (KDD '16), pp 785–794. https://doi.org/10.1145/2939672.2939785
21. Colling L, Szucs D (2018) Statistical inference and the replication crisis. Rev Philos Psychol 12(1):121–147. https://doi.org/10.1007/s13164-018-0421-4
22. Cook TD, Campbell DT, Day A (1979) Quasi-experimentation: design & analysis issues for field settings. Houghton Mifflin
23. Crawford CM, Moffett NL, Wallace JY (2020) The C Class lesson plan framework: disrupting lesson strategies and intentions by designing a new style of lesson engagement that deconstructs and reconstructs the future of learning. Int J E-Learn 19(1):25–47. https://www.learntechlib.org/primary/p/173349/

24. Csikszentmihalyi M, Larson R (1987) Validity and reliability of the experience-sampling method. J Nerv Ment Dis 175:526–536
25. Csikszentmihalyi M (2014) Applications of flow in human development and education: the collected works of Mihaly Csikszentmihalyi. Springer Science. https://doi.org/10.1007/978-94-017-9094-9
26. Dahiru T (2008) P-value, a true test of statistical significance? A cautionary note. Ann Ib Postgrad Med 6(1):21–26
27. Datu JAD, King RB (2018) Subjective well-being is reciprocally associated with academic engagement: a two-wave longitudinal study. J Sch Psychol 69:100–110. https://doi.org/10.1016/j.jsp.2018.05.007
28. Dhaliwal SS, Nahid A, Abbas R (2018) Effective intrusion detection system using XGBoost. Information 9(7), Article 149. https://doi.org/10.3390/info9070149
29. Diener E, Oishi S, Lucas RE (2003) Personality, culture, and subjective well-being: emotional and cognitive evaluations of life. Annu Rev Psychol 54:403–425. https://doi.org/10.1146/annurev.psych.54.101601.145056
30. Diseth Å, Danielsen AG, Samdal O (2012) A path analysis of basic need support, self-efficacy, achievement goals, life satisfaction and academic achievement level among secondary school students. Educ Psychol 32(3):335–354. https://doi.org/10.1080/01443410.2012.657159
31. Eckersley R (2009) Population measures of subjective wellbeing: How useful are they? Soc Indic Res 94:1–12
32. Eisenberg ME, Neumark-Sztainer D, Perry CL (2003) Peer harassment, school connectedness, and academic achievement. J Sch Health 73(8):311–316
33. Farrell L, Newman T, Corbel C (2021) Literacy and the workplace revolution: a social view of literate work practices in industry 4.0. Discourse Stud Cultural Politics Educ 42(6), 898–912. https://doi.org/10.1080/01596306.2020.1753016.
34. Fattore T, Mason J, Watson E (2007) Children's conceptualisation(s) of their well-being. Soc Indic Res 80(1):5–29. https://doi.org/10.1007/s11205-006-9019-9
35. Finnigan KM, Vazire S (2018) The incremental validity of average state self-reports over global self-reports of personality. J Pers Soc Psychol 115(2):321–337. https://doi.org/10.1037/pspp0000136
36. Fredrickson BL (1998) What good are positive emotions? Rev Gen Psychol 2(3):300–319. https://doi.org/10.1037/1089-2680.2.3.300
37. Fredrickson BL (2001) The role of positive emotions in positive psychology: the broaden-and-build theory of positive emotions. Am Psychol 56(3):218–226. https://doi.org/10.1037/0003-066X.56.3.218
38. Fredrickson BL (2004) The broaden-and-build theory of positive emotions. Philos Trans R Soc London Ser B Biol Sci 359(1449):1367–1378. https://doi.org/10.1098/rstb.2004.1512
39. Friendly M, Chen CH, Hardle W, Unwin A (2008) A brief history of data visualization. In: Chen CH, Hardle WK, Unwin A (eds) Handbook of data visualization. Springer, pp 15–56
40. Gimmig D, Huguet P, Caverni JP, Cury F (2006) Choking under pressure and working memory capacity: when performance pressure reduces fluid intelligence. Psychon Bull Rev 13(6):1005–1010. https://doi.org/10.3758/BF03213916
41. Gloppen KM, Gower AL, McMorris BJ, Eisenberg ME (2017) Associations between peer harassment and school risk and protection profiles. J Sch Health 87(11):832–841. https://doi.org/10.1111/josh.12557
42. Govorova E, Benítez I, Muñiz J (2020a) How schools affect student well-being: a cross-cultural approach in 35 OECD Countries. Front Psychol 11, Article 431. https://doi.org/10.3389/fpsyg.2020.00431
43. Govorova E, Benítez I, Muñiz J (2020b) Predicting student well-being: network analysis based on PISA 2018. Int J Environ Res Public Health 17(11), Article 4014. https://doi.org/10.3390/ijerph17114014
44. Graham S, Juvonen J (2002) Ethnicity, peer harassment, and adjustment in middle school: an exploratory study. J Early Adolesc 22(2):173–199. https://doi.org/10.1177/0272431602022002003

45. Heaven P, Ciarrochi J (2008) Parental styles, gender and the development of hope and self-esteem. Eur J Pers 22(8):707–724. https://doi.org/10.1002/per.699
46. Henrich J, Heine SJ, Norenzayan A (2010) Most people are not WEIRD. Nature 466(7302):29–29. https://doi.org/10.1038/466029a
47. Hilbert S, Coors S, Kraus E, Bischl B, Lindl A, Frei M, Wild J, Krauss S, Goretzko D, Stachl C (2021) Machine learning for the educational sciences. Rev Educ 9, Article e3310. https://doi.org/10.1002/rev3.3310
48. Khan A, Khan S, Zia-Ul-Islam S, Khan M (2017) Communication skills of a teacher and its role in the development of the students' academic success. J Educ Pract 8(1):18–21
49. Korhonen J, Linnanmäki K, Aunio P (2014) Learning difficulties, academic well-being and educational dropout: a person-centered approach. Learn Individ Differ 31:1–10. https://doi.org/10.1016/j.lindif.2013.12.011
50. Kuhn L, Bradshaw S, Donkin A, Fletcher L, Liht J, Wheater R (2021) PISA 2018 additional analyses: What does PISA tell us about the wellbeing of 15-year-olds? National Foundation for Educational Research. https://files.eric.ed.gov/fulltext/ED612548.pdf
51. Kurtz AK (1948) A research test of the Rorschach test. Pers Psychol 1:41–53. https://doi.org/10.1111/j.1744-6570.1948.tb01292.x
52. Lee MY, Lee MK, Lee MJ, Lee SM (2020) Academic burnout profiles and motivation styles among Korean high school students. Jpn Psychol Res 62(3):184–195. https://doi.org/10.1111/jpr.12251
53. Legault L, Green-Demers I, Pelletier L (2006) Why do high school students lack motivation in the classroom? Toward an understanding of academic amotivation and the role of social support. J Educ Psychol 98(3):567–582. https://doi.org/10.1037/0022-0663.98.3.567
54. Loftus E (2005) Planting misinformation in the human mind: a 30-year investigation of the malleability of memory. Learn Mem 12(4):361–366
55. Lv B, Zhou H, Guo X, Liu C, Liu Z, Luo L (2016) The relationship between academic achievement and the emotional well-being of elementary school children in China: the moderating role of parent-school communication. Front Psychol 7, Article 948. https://doi.org/10.3389/fpsyg.2016.00948
56. Lyons LM, Beilock SL (2012) When math hurts: math anxiety predicts pain network activation in anticipation of doing math. PLoS ONE. https://doi.org/10.1371/journal.pone.0048076
57. Lyons LM, Beilock SL (2012) Mathematics anxiety: separating the math from the anxiety. Cereb Cortex 22(9):2102–2110. https://doi.org/10.1093/cercor/bhr289
58. Malecki CK, Demaray MK (2006) Social support as a buffer in the relationship between socioeconomic status and academic performance. Sch Psychol Q 21(4):375. https://doi.org/10.1037/h0084129
59. Martin AJ, Liem GA, Mok M, Xu J (2012) Problem solving and immigrant student mathematics and science achievement: multination findings from the Programme for International Student Assessment (PISA). J Educ Psychol 104(4):1054–1073. https://doi.org/10.1037/a0029152
60. Matúšová S, Kollár V (2023) Labour and education markets in Industry 4.0. Acta Educationis Generalis 13(1):1–25. https://doi.org/10.2478/atd-2023-0001
61. Mazer J (2013) Associations among teacher communication behaviors, student interest, and engagement: a validity test. Commun Educ 62(1):86–96. https://doi.org/10.1080/03634523.2012.731513
62. Mihalisin T, Timlin J, Schwegler J (1991) Visualization and analysis of multi-variate data: A technique for all fields. In: 1991 Proceeding visualization. IEEE Computer Society, pp 171–178. https://doi.org/10.5555/949607.949635
63. Miller PJ, Lubke GH, McArtor DB, Bergeman CS (2016) Finding structure in data using multivariate tree boosting. Psychol Methods 21(4):583–602. https://doi.org/10.1037/met0000087
64. Muro A, Soler J, Cebolla A, Cladellas R (2018) A positive psychological intervention for failing students: does it improve academic achievement and motivation? A pilot study. Learn Motiv 63:126–132. https://doi.org/10.1016/j.lmot.2018.04.002

65. Nordlander E, Stensöta H (2014) Grades–for better or worse? The interplay of school performance and subjective well-being among boys and girls. Child Indic Res 7(4):861–879. https://doi.org/10.1007/s12187-014-9233-y
66. Okoli K, Breinl K, Brandimarte L, Botto A, Volpi E, Di Baldassarre G (2018) Model averaging versus model selection: estimating design floods with uncertain river flow data. Hydrol Sci J 63(13–14):1913–1926. https://doi.org/10.1080/02626667.2018.1546389
67. Open Science Collaboration (2015) Psychology: estimating the reproducibility of psychological science. Science 349(6251), Article aac4716. https://doi.org/10.1126/science.aac4716
68. Organization for Economic Cooperation and Development (OECD) (2017a) PISA 2015 results. https://www.oecd.org/education/pisa-2015-results-volume-iii-9789264273856-en.htm
69. Organization for Economic Cooperation and Development (OECD) (2017b) Well-being questionnaire for PISA 2018. https://www.oecd.org/pisa/data/2018database/CY7_201710_QST_MS_WBQ_NoNotes_final.pd
70. Organization for Economic Cooperation and Development (OECD) (2019) PISA 2018 results. https://www.oecd.org/education/pisa-2018-results-volume-iii-acd78851-en.htm
71. Organization for Economic Cooperation and Development (OECD) (2020) PISA 2018 technical report. https://www.oecd.org/pisa/data/pisa2018technicalreport/
72. Rosenberg M (1989) Society and the adolescent self-image. Wesleyan University Press
73. Rudin C (2019) Stop explaining black box machine learning models for high stakes decisions and use interpretable models instead. Nat Mach Intell 1:206–215. https://doi.org/10.1038/s42256-019-0048-x
74. SAS Institute (2021) JMP Pro 17 [Computer software]. https://www.jmp.com/en_us/home.html
75. Salmela-Aro K, Kiuru N, Pietikäinen M, Jokela J (2008) Does school matter? The role of school context in adolescents' school-related burnout. Eur Psychol 13(1):12–23. https://doi.org/10.1027/1016-9040.13.1.12
76. Sarang P (2020) Artificial neural networks with TensorFlow 2: ANN architecture machine learning projects. Apress.
77. Schacter D (1999) The seven sins of memory: insights from psychology and cognitive neuroscience. Am Psychol 54(3):182–203. https://doi.org/10.1037/0003-066X.54.3.182
78. Seligman M, Csikszentmihalyi M (2000) Positive psychology: aN introduction. Am Psychol 55(1):5–14
79. She H, Fisher D (2000) The development of a questionnaire to describe science teacher communication behavior in Taiwan and Australia. Sci Educ 84(6):706–726. https://doi.org/10.1002/1098-237X(200011)84:6%3c706::AID-SCE2%3e3.0.CO;2-W
80. Slee P, Skrzypiec G (2016) Well-being, positive peer relations and bullying in school settings. Springer
81. Topham P, Moller N (2011) New students' psychological well-being and its relation to first year academic performance in a UK university. Couns Psychother Res 11(3):196–203. https://doi.org/10.1080/14733145.2010.519043
82. Tseng GC, Wong WH (2005) Tight clustering: A resampling-based approach for identifying stable and tight patterns in data. Biometrics 61(1):10–16. https://doi.org/10.1111/j.0006-341X.2005.031032.x
83. Tukey JW (1980) We need both exploratory and confirmatory. Am Stat 34(1):23–25. https://doi.org/10.1080/00031305.1980.10482706
84. Turner R, Adams RJ (2007) The programme for international student assessment: an overview. J Appl Measure 8(3):237–248. https://doi.org/10.1787/9789264048096-2-en
85. United Nations General Assembly (1989) Convention on the Rights of the Child. United Nations, Treaty Series 1577(3). https://www.refworld.org/docid/3ae6b38f0.html:1989
86. Valiente C, Swanson J, Eisenberg N (2012) Linking students' emotions and academic achievement: when and why emotions matter. Child Dev Perspect 6(2):129–135. https://doi.org/10.1111/j.1750-8606.2011.00192.x

87. Wang W, Gai X, Zhang Y, Wang G (2018) Promoting academic achievement via adolescents' subjective well-being: the mediating role of school engagement. Psychol Explor 39(3):260–266
88. Wang Y, Kung L, Byrd TA (2018) Big data analytics: understanding its capabilities and potential benefits for healthcare organizations. Technol Forecast Soc Chang 126:3–13. https://doi.org/10.1016/j.techfore.2015.12.019
89. Warm T (1989) Weighted likelihood estimation of ability in item response theory. Psychometrika 54(3):427–450. https://doi.org/10.1007/BF02294627
90. Wasserstein R, Lazar N (2016) The ASA statement on *p*-values: Context, process, and purpose. Am Stat 70(2):129–133. https://doi.org/10.1080/00031305.2016.1154108
91. Wentzel KR, Battle A, Russell SL, Looney LB (2010) Social supports from teachers and peers as predictors of academic and social motivation. Contemp Educ Psychol 35(3):193–202. https://doi.org/10.1016/j.cedpsych.2010.03.002
92. Williams DM, Rhodes RE (2016) The confounded self-efficacy construct: conceptual analysis and recommendations for future research. Health Psychol Rev 10(2):113–128. https://doi.org/10.1080/17437199.2014.941998
93. World Economic Forum (2023) Education 4.0: Here are 3 skills that students will need for the jobs of the future. https://www.weforum.org/agenda/2023/01/skillsets-cultivated-by-education-4-0-davos23/#:~:text=Education%204.0%20reimagines%20education%20as,acting%20as%20facilitators%20and%20enablers
94. Wu M (2005) The role of plausible values in large-scale surveys. Stud Educ Eval 31(2–3):114–128. https://doi.org/10.1016/j.stueduc.2005.05.005
95. Wu XJ, Gai XS, Wang W (2020) Subjective well-being and academic performance among middle schoolers: a two-wave longitudinal study. J Adolesc 84:11–22. https://doi.org/10.1016/j.adolescence.2020.07.011
96. Yu CH, Lee HS, Lara E, Gan S (2018) The ensemble and model comparison approaches for big-data analytics in social sciences. Pract Assess Res Eval 23(1), Article 17. https://scholarworks.umass.edu/cgi/viewcontent.cgi?article=1381&context=pare
97. Yu CH (2014) Dancing with the data: the art and science of data visualization. Lambert Academic Publisher
98. Yu CH (2016) How to randomly select a plausible value. http://www.creative-wisdom.com/computer/sas/PV_excel.html
99. Yu CH (2022) Data mining and exploration: from traditional statistics to modern data science. CRC Press
100. Yu CH, Lee HS (2020) Creating changes in science and math education: lessons from Hong Kong. Springer
101. Yu CH, Douglas S, Lee A, An M (2016) Data visualization of item-total correlation by median smoothing. Pract Assess Res Eval 21(1), Article 1. https://scholarworks.umass.edu/cgi/viewcontent.cgi?article=1334&context=pare

Chong Ho (Alex) Yu holds a Ph.D. in Educational Psychology with a focus on Measurement, Statistics and Methodological Studies, and also a Ph.D. in Philosophy with a specialization in History and Philosophy of Science. Both degrees are conferred by Arizona State University. He is a former Professor and Director of Data Analytics at Azusa Pacific University. Currently he is a Professor and the Program Director of Data Science at the Department of Mathematics, Hawaii Pacific University. He is a three-time winner of SAS faculty scholarship. In addition, he also won the Distinguished SAS Educator Award in 2021. He authored numerous books on data science and education and. His most recent books on these topics are "Data mining and exploration: From traditional statistics to modern data science," published by CRC Press, and "Creating changes in science and math education: Lessons from Hong Kong," coauthored with H. S. Lee and published by Springer.

Using Large Language Models to Probe Cognitive Constructs, Augment Data, and Design Instructional Materials

Fabian Kieser and Peter Wulff

Keywords Large language models · Problem solving · Machine learning · Natural language processing

1 Motivation

Not only since the dawn of electrical computers, humans sought to instruct machines to automate processes, and outsource capabilities. While electrical (and analogical) computers could solve simple arithmetic problems for quite some time, only during the mid-twentieth century they became increasingly capable of tackling some more complex problems such as games. While simple arithmetic problems can be solved by deterministic algorithms, complex problems such as games like checkers or chess cannot be efficiently solved by deterministic algorithms given the vast number of possible moves (state space) in such games. Arthur Samuel [1] famously enabled computers (i.e., machines) to learn from game playing and hence coined a term that pervaded many scientific disciplines until a couple of decades ago: machine learning (ML). One essential of Samuel's ML approach was an evaluation function that compared actual states to desired states, and utilized this information for learning. This approach was an early inception of the so-called reinforcement learning. Parallel to these advances in computer science, researchers in the field of cognitive and computer science advanced the development of artificial neural networks (ANNs) and presented early algorithms for how to train these networks through backpropagation in order to make them learn desired capabilities. As promising as these approaches were, given the limitations in computer processing

F. Kieser · P. Wulff (✉)
Physics and Physics Education Research, Heidelberg University of Education, Heidelberg, Germany
e-mail: peter.wulff@ph-heidelberg.de

M. S. Khine (ed.), *Machine Learning in Educational Sciences*,
https://doi.org/10.1007/978-981-99-9379-6_14

power, storage, and data sets, a so-called AI winter followed, where research on artificial intelligence (AI) ceased in funding.

This situation fundamentally changed when Hinton et al. [2], among others, showed that a generative ANN (here: a deep belief net) could be trained through a fast, greedy algorithm that could be trained to learn to classify pictures of handwritten digits. It became apparent that supervised deep learning, i.e., training ANNs through input–output pairs, is possible and achieves state-of-the-art performance on relevant problems [3]. This progress was soon supplemented by developing large data sets of labelled images, competitions, and ever-increasing ANNs with specialized architectures such as convolutions [4]. A similar landslide moment as for image processing was probably the invention of the transformer architecture for language processing [5]. While computers have been utilized for natural language processing (NLP) for quite some time [6, 7], the versatility of utilizing ANNs for language processing advanced the field for good [6, 8]. It was shown that transformer-based large language models (LLMs could be pretrained on large language data sets such as the Common Crawl of the Internet, the Books Corpus, or Wikipedia, and then further fine-tuned in specific tasks [9, 10]. Encoder and decoder architectures are differentiated, whereas the former (such as "bidirectional encoder representations from transformers," BERT) can be utilized optimally for classification and fine-tuning, and the latter (such as "generative pretrained transformers," GPT) for text generation of all kinds. Moreover, it became clear that even generally trained LLMs could solve many tasks off-the-shelf without further fine-tuning (zero-shot) [11]. These powerful tools now became available to the public in conversational AIs such as ChatGPT, which exceeded the expectations even of experts regarding task performance and popularity [12].

For educational fields such as science education, these LLMs provide exciting opportunities to process learning- and teaching-related complex data, generate guidance on the basis of this data, and even design instructional materials [13–15]. For once, researchers in science education utilized LLMs to analyze written reflections of students and automate assessment [16, 17], identify meaningful clusters in complex language data [18], and even combine analysis of image and language data to assess students' modelling capabilities [19]. Specifically with ChatGPT (based on either GPT-3.5, or GPT-4), science education researchers engaged in ad-hoc, feasibility studies. It was shown that ChatGPT could write up physics essays on par with students, be used for task development, solve concept inventories, and thus show basic physics understanding, augment educational data, and more generally, assist in writing in science education [14, 15, 20–23]. What is less clear, however, is to what extent conversational AIs such as ChatGPT can also be used to analyse problem-solving processes, and even generate novel problems. Science problem-solving is among the primary competencies for the twenty-first century [24], and played a major role in science education research throughout the years [25]. Hence, in this study, we seek to explore the capabilities of ChatGPT to solve a physics problem and generate physics problems.

2 Using AI to Assess Physics Problem-Solving

AI methods based on ML have been utilized extensively in science education to assess learner's data such as forced-choice responses or even constructed responses such as short-form essays [26, 27]. ML is a form of inductive problem-solving by computers [28]. Basically, ML seeks to identify patterns in complex data sets (unsupervised ML), or learn a mapping from input data to gold-standard outputs (supervised ML) [29]. Particularly with the development of LLMs (language models in essence seek to predict words in a sequence), more advanced forms of training, e.g., self-supervised learning, were introduced. Supervised ML was found to be too restrictive to allow LLMs to acquire the generic capabilities that they have [6]. In a paradigm called transfer learning, base LLMs (also called: "foundation models") are trained on large language data sets such as the Common Crawl of the Internet, the Book Corpus, or Wikipedia. This pretraining provides versatile base LLMs that can be used to gain dense, contextualized representations (called embeddings) of language data [10], which then can be further used as feature representations in downstream tasks such as classifying written reflections of students [16, 17]. With adequate preprocessing, it was then possible to enable quantitative reasoning (with formula and natural language) in LLMs [11]. Sophisticated learning procedures such as chain-of-thought prompting were devised that specifically enabled these LLMs to be able to outline their reasoning processes.

Pretrained and fine-tuned LLMs were found to be able to perform reasoning-like tasks and solve problems in domains such as mathematics, biology, chemistry, and physics [11]. Problem-solving in scientific disciplines was described as a structured, interactional process, where the problem solver starts with a given situation and works towards a goal state, while the means of attaining this goal state are initially unclear to the problem solver [30, 31]. Experts often approach problems in physics in a principle-first manner, where they work from the given situation by analyzing what key concepts and principles have to be applied, and what assumptions have to be made [32–34]. Experts often apply a strategy called forward chaining, where they can work from an appropriate formula towards the goal state [34]. Novice problem solvers oftentimes seek to work backwards from the goal state, and recursively apply formulas that contain given quantities, a strategy called backwards chaining (also: working backwards) or plug and chug [33, 35, 36]. Moreover, assumptions remain oftentimes unstated for novices, which prevents successful problem-solving [37]. Researchers devised process models of physics and scientific problem-solving. Typically, subsequent stages are differentiated, where (1) first the problem solver engages in analysis and representation of the problem situation, then (2) a strategy to solve the problem is selected, afterwards (3) the strategy is executed, and finally (4) the solution is evaluated by comparing it to the goal state [38–40].

Problem-solving skills can be assessed through think-aloud, cognitive interviews, or constructed responses [30]. Each form of assessment has its benefits and challenges. Whereas think-aloud protocols allow for an interactive assessment of reasoning processes, they are barely scalable to larger samples, which restricts evaluation of generalizability of findings. Constructed responses, on the other hand, are scalable in that they do not require a researcher to supervise the data acquisition process, however, principled and systematic analytical tools are required in order to reliably assess them. If human researchers are meant to perform much of the data processing and analysis, constructed responses are typically not scalable either. Interestingly, advances in NLP and ML research allow researchers to perform novel kinds of assessments, namely probing LLMs on reasoning and problem-solving performance. By being trained on large data sets that no human could have been trained on, these LLMs might exhibit totally new and unanticipated ways of solving physics problems. As such, LLMs might be used to investigate (A) problem-solving processes, (B) augment data on the existing problem-solving surveys, and even (C) generate instructional materials to be used to teach physics problem-solving.

3 Research Questions

As for (A), problem-solving processes are laid out throughout the Internet and LLMs gained generic and subject-specific problem-solving abilities by merely processing the existing data. Some of these abilities were explicitly trained into the LLMs through respective data sets and prompting strategies. Hence we inquire:

> RQ1: How does physics problem-solving by a LLM map to a problem-solving process model and resemble human problem-solving performance?

As for (B), LLMs were found to generate linguistically appropriate responses to questions. Specifically, LLMs could solve physics concept inventories, and adequate prompting of the LLMs could alter the outputs. As such, we inquire:

> RQ2: In what ways can specific prompting on problem-solving strategy use impact the output by the LLM and can be utilized to augment a physics problem-solving data set?

As for (C), while it is known that LLMs are proficient in replicating data that was seen in the training data set, it is of interest to what extent LLMs also have the capability to produce meaningful novel problems. Hence, we ask:

> RQ3: What kinds of physics problems does the LLM generate when explicitly prompted?

4 Method

4.1 ChatGPT as a Resource

Decoder LLMs such as GPT are pretrained models that can generate the next words in a sequence in an autoregressive manner. With recent advances in training decoder LLMs, they became largely capable of quantitative reasoning and scientific problem-solving. In particular, prompting strategies play a major role in getting from LLMs desired output and train them appropriately [9]. For example, Lewkowycz et al. [11] used a prompting strategy called chain-of-thought prompting in order to enable LLMs to provide more detailed reasoning on typical mathematics and science problems. They found that training their LLM with chain-of-thought prompting enabled it to achieve state-of-the-art performance on standardized mathematics and science problem-solving datasets. The model family Generative Pretrained Transformers (GPT) became particularly capable of problem-solving in a wide range of disciplines such as law, business, and many others, in particular, the version GPT-4 [9, 22]. The company OpenAI provides a browser-based front-end to access ChatGPT based on GPT-4 under: https://chat.openai.com/. Moreover, older models can be accessed through a dedicated API, e.g., via the open source software tool Python [41]. In this study, we will utilize the browser-based front-end, as we are interested in exploring the outputs of the most versatile and elaborate version, GPT-4 (as of December 2023). Hence, we investigate our RQs by inputting prompts that contain specific instructions for the LLMs in the browser. The responses will also be generated in the browser and can be copy-pasted into our analysis software.

4.2 A Textbook-Style Physics Problem

To prompt ChatGPT with an authentic physics problem, we leveraged a mechanics problem[1] that we utilized for research and teaching. The problem requires learners to figure out the initial heights for a (point) mass, given that it runs through a vertical loop (see Fig. 1). The task requires important assumptions such as that the movement takes place without friction, without air resistance, and that the mass is a point mass. Two physical concepts are essential to this physical problem-solving task: 1) The conservation of energy: The potential energy of the mass at the starting point is converted into kinetic energy and back. 2) The concept of centripetal force as the cause of circular motion. To obtain a solution, these two concepts must be combined. The solution must then be calculated from this approach using mathematical methods.

[1] All materials related to this physics problem solving test instrument can be accessed freely under the following url: https://osf.io/gsmvr/?view_only=dfc5c366675041e1a4dd4e9e40347fbe.

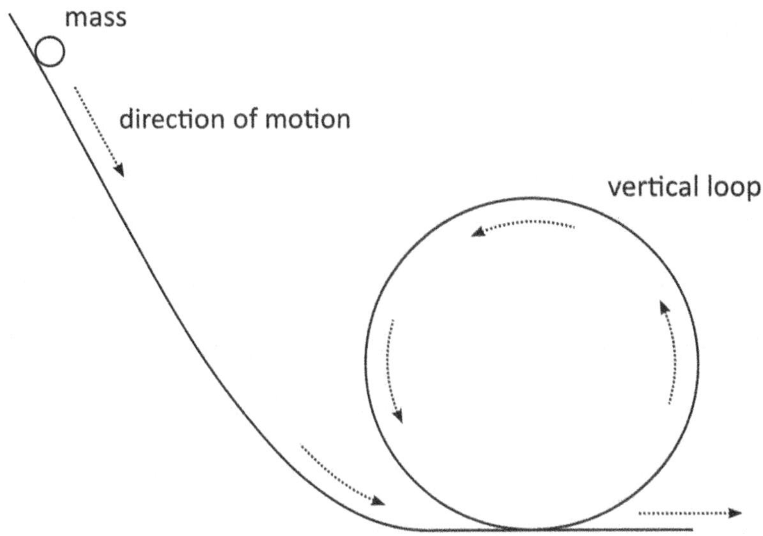

Fig. 1 Textbook-style physics problem of a mass running through a vertical loop (the mass is assumed to be a point mass that slides through the loop).

5 Findings

5.1 RQ1–ChatGPT's Problem-Solving

To address RQ1, we employ ChatGPT to solve the physics problem: mass in a vertical loop (see Fig. 1). We analyze the phases of the problem-solving process that ChatGPT implicitly or explicitly goes through. The object of the task is a loop. The question is to be answered at what? (minimal) height a mass should start in order not to fall down when moving through a loop. .

We find that ChatGPT solves the task correctly five times out of 10 attempts. In each correct solution attempt, the 1st phase of problem-solving process model (problem representation) and the 3rd phase (elaboration of solution) are correctly and explicitly implemented. The 2nd Phase is used implicitly in eight cases and is only used explicitly once. In no single attempt, the solution is evaluated (4th phase). The most frequent errors, with a rate of 4 out of 5 wrong answers, occur in the 3rd phase, the elaboration of the solution. Only one solution explicitly mentions the selection of a problem schema (Table 1).

Table 1 ChatGPT's problem solutions of the looping task

Number of the solution attempts	1. Problem representation	2. Selection of a problem scheme	3. Elaboration of a solution	4. Evaluation of a solution
1	Explicitly present and correct	Implicitly present	Explicitly present but incorrect	Not present
2	Explicitly present and correct	Implicitly present	Explicitly present and correct	Not present
3	Explicitly present but incorrect	Not present	Not present	Not present
4	Explicitly present and correct	Explicitly present and correct	Explicitly present but incorrect	Not present
5	Explicitly present and correct	Implicitly present	Explicitly present and correct	Not present
6	Explicitly present and correct	Implicitly present	Explicitly present and correct	Not present
7	Explicitly present and correct	Implicitly present	Explicitly present but incorrect	Not present
8	Explicitly present and correct	Implicitly present	Explicitly present and correct	Not present
9	Explicitly present and correct	Implicitly present	Explicitly present but incorrect	Not present
10	Explicitly present and correct	Implicitly present	Explicitly present and correct	Not present

5.2 *RQ2–Prompting ChatGPT to Use Different Strategies*

As we have seen in the examination of RQ1, not all relevant phases of the problem-solving process are explicitly mentioned in ChatGPT, if at all. For example, the last phase of the problem-solving process, the phase of evaluating the solution, does not appear in any solution proposal generated by ChatGPT. RQ2 now deals with the question of how the output can be changed through targeted promoting in such a way that these phases are also taken into account in the solution proposals generated by ChatGPT [15].

In a first attempt, the prompt for the looping task is supplemented with the addition: "In particular, describe whether your solution is plausible and and for what reasons you chose your solution" (cf. Appendix 1). In the reply generated by ChatGPT, the addition appears at the end: "This answer is plausible because it applies two well-established physical principles: the conservation of energy and the balance of forces. It demonstrates that for the mass to just complete the loop, it needs to start at a certain minimum height. If it starts any lower, it won't have enough speed to overcome gravity at the top of the loop, and it will fall. If it starts any higher, it will have more than enough speed to stay on the track" (cf. Appendix 1). ChatGPT tries to evaluate the generated solution by looking for arguments for the correctness of the solution. So ChatGPT tries to justify the solution rather than verify it. Nevertheless, the solution changes through targeted prompting and the phase for evaluating the solution appears in the problem-solving phase.

Another experiment that shows that targeted promoting can change the outcome of problem-solving is when ChatGPT tries to use the "means-ends analysis" problem-solving strategy to solve the looping task. ChatGPT is able to apply the important points of means-ends analysis to the task, in the sense that it identifies the initial state and the final state. It identifies the means that are important to solve the task (in our case the necessary physical concepts). Subgoals are also identified. In this prompting, the solution is not carried out but only described as to which path would lead to the solution (cf. Appendix 2). In a similar way, this also works with the "working backwards" problem-solving strategy. ChatGPT starts backwards at the looping high point and argues backwards how one would arrive at the solution using the "backwards method". (cf. Appendix 3).

However, problems arise when applying the plug and chug method to the looping task. It can be stated, however, that the method in which the solution is determined by inserting values is not really suitable for solving such conceptual problems, as in plug and chug learners would search for a given formula that includes the given physical quantities. This could be done for the heights, but there would not really be a suitable formula in this case, except for combined formulas. ChatGPT also recognizes this correctly (cf. Appendix 4a). If we explain more closely what this strategy means, then ChatGPT gives it a better try and arrives at the correct solution (cf. Appendix 4b). However, it is doubtful if this really resembles a novice using this strategy.

5.3 RQ3–Utilizing ChatGPT to Generate Instructional Materials

As for RQ3, we sought to utilize ChatGPT to generate novel learning materials, namely physics problems, through deliberative prompting. We particularly tested three prompting strategies: (A) generate similar physics problems to a given problem, (B) generate novel physics problems based on the underlying concepts required to solve them, and (C) generate problems in different subfields in physics.

In (A), we prompted the LLM the physics problem with the vertical loop (see methods section) as a template for novel problems. The LLM generated, as requested, two "similar physics problems" to the vertical loop problem[2]:

> Question: Generate two similar physics problems as the following problem: "A mass starts from a height and enters a vertical loop of a given radius R. At what height does the mass need to start from to move through the vertical loop?"

[2] See: https://chat.openai.com/share/3f0a299d-f1bc-499e-a8ba-832854fa0ee4, last access: June 2023.

> Answer: Problem 1: "A ball is released from a certain height and subsequently goes into a vertical circular motion along a track with a radius of R. What should be the initial height from which the ball must be released to complete the vertical circle?"
>
> Problem 2: "A car moves on a frictionless, vertical circular track of a given radius R. From what initial height does the car need to start in order to successfully complete the circular track without losing contact at any point?"

Problem 1 comprises a ball instead of a mass that runs along a similar (if not identical) track. Problem 2 comprises a car that drives (frictionless) on a "vertical circular track", which is an elaborate description of a loop. In essence, the LLMs picked up on the surface features of the loop problem and transformed it into two novel problems. However, the innovativeness is limited, given that it only—from a physics stance—switched the objects and language, rather than the underlying physical setting. Both problems resemble the original problem insofar as exactly the same physical variables, namely radius R, are given. Interestingly, in problem 2 ChatGPT adds in the information on the assumptions to be made, which is not explicitly given in the original problem. It is quite interesting that this assumption is added and not other relevant assumptions, such as point mass, or negligence of air resistance.

In (B), we then turned to the underlying concepts that are required to solve the vertical loop problem, namely conservation of energy and force balance. To evaluate to what extent ChatGPT could utilize this information and generate novel problems, we prompted it explicitly to utilize these concepts to generate two physics problems[3]:

> Question: Generate two physics problems where learners have to use the energy conservation principle, and the force concept"
>
> Answer: Problem 1: "A 5 kg block is sliding down a frictionless incline plane of 30 degrees. If the block starts from rest at a height of 10 meters, find (a) the speed of the block when it reaches the bottom of the incline, and (b) the force exerted by the incline on the block at the bottom. Consider the acceleration due to gravity to be 9.8 m/s^2."
>
> Problem 2: "A pendulum with a mass of 2 kg swings from a height of 1 m above its lowest point. Calculate (a) its speed when it is at its lowest point and (b) the tension in the string at the lowest point if the string length is 1.5 m. Assume there is no friction or air resistance, and the acceleration due to gravity is 9.8 m/s^2."

Problem 1 of the generated problems is a typical inclined plane problem, where a block slides down the plane, and quantities such as velocity (first part) or normal force (second part) had to be calculated. Problem 2 is also a classic in physics, namely the pendulum. Here, velocity and tension in the string had to be calculated. In both problems, assumptions are explicitly stated (frictionless movement, no air resistance, constant acceleration due to gravity). Technically, no air resistance is also an assumption in problem 1, however, it is not explicitly stated. Maybe this is due to the fact that air resistance is oftentimes only considered for falling or swinging movements.

---

[3] See: https://chat.openai.com/share/3f0a299d-f1bc-499e-a8ba-832854fa0ee4, last access: June 2023.

The generated problems in fact require the abovementioned concepts for solving them. What was unsatisfactory at this point is that the two concepts had to be used independently of each other. It would be quite interesting to generate a problem where both concepts had to be used in conjunction in order to reach a solution. We thus further prompted ChatGPT to devise a problem where both concepts had to be used in conjunction in order to solve it[4]:

> Question: Generate two physics problems where learners have to use the energy conservation principle, and the force concept, where you have to combine the two concepts
>
> Answer: Problem 1: "A skier of mass 70 kg starts from rest at the top of a frictionless ski slope of height 50 m. Halfway down the slope, she encounters a patch of rough ground 10 m long with a coefficient of kinetic friction of 0.3. (a) Calculate her speed at the bottom of the slope. (b) Calculate the force exerted by the friction on the skier as she passes over the rough ground."
>
> Problem 2: "A block of mass 2 kg is at rest on a frictionless horizontal surface. The block is connected to a spring with a spring constant of 100 N/m. The block is then pulled by a force of 50 N, stretching the spring by a distance of 0.5 m from its equilibrium position, then released. (a) Determine the speed of the block when it passes through the equilibrium position. (b) Calculate the force exerted by the spring when the block is 0.2 m from the equilibrium position."

Again, ChatGPT generated two problems where conservation of energy and force had to be used again, and both approaches had to be integrated slightly more. However, closer inspection reveals that problem 1 is not a well-defined task. In order to solve the two parts of the problem, information about the angle of the slope is required, but not given in the problem statement by ChatGPT. Note that slight variations in framing the problem (e.g., "... have to be used in conjunction to find a solution/solve it.") did not change the fact that, again, both concepts had to be used quite independently from each other. Moreover, it was interesting that all generated problems were in mechanics, which was not part of the instruction. Hence, we finally sought to generate problems in other subfields such as electricity-magnetism.

In (C), we instructed ChatGPT to devise problems in electricity-magnetism where the same concepts, conservation of energy and the force concept, have to be used to solve a problem:

> Question: Generate two physics problems where learners have to use the energy conservation principle, and the force concept in electromagnetism.
>
> Answer: Problem 1: "A proton with a mass of 1.67 x 10^-27 kg is released from rest in a uniform electric field of strength 2 x 10^5 N/C. The electric field directs towards the positive x-axis. (a) What is the kinetic energy of the proton after it has moved 0.02 m in the direction of the field? (b) What is the velocity of the proton at this point?–"
>
> Problem 2: "A 500-turn solenoid carries a current of 4 A and has a length of 0.3 m and radius of 0.01 m. A single electron enters the solenoid perpendicularly to the axis of the solenoid with a velocity of 2 x 10^6 m/s. (a) What is the magnetic force acting on the electron due to

[4] See: https://chat.openai.com/share/3f0a299d-f1bc-499e-a8ba-832854fa0ee4, last access: June 2023.

the magnetic field of the solenoid? (b) If the electron continues to move inside the solenoid for 0.001 s, what will be the change in its kinetic energy?"

Both problems now could be classified as electricity-magnetism problems, rather than mechanics problems. This was due to the fact that the objects of interest were a proton and a coil (solenoid), but also because electrical fields (hence: electrical forces) and magnetic fields (hence: magnetic forces) were constitutive for the problem. In a way, however, a similar issue was apparent as in the mechanics' problems, namely, the concept use was quite independent of each other in the sub-problems. An overarching goal, as in the vertical loop problem where it was of interest at which height the mass will pass the loop, was not designed into these problems. Moreover, it is implicitly assumed that the force of gravity is to be neglected. This is also often an unstated assumption in physics textbooks for such problems in electricity-magnetism, and conincides with the observation that in any such problems much world-knowledge is assumed to be shared knowledge, which might or might not be a valid assumption.

6 Discussion

Not only since the advent of generative AI such as LLMs it is likely that research on teaching and learning processes can be enhanced with such tools. Essentially, because they allow for scalable distribution of knowledge, such as designing learning materials and guiding learners in the process of engaging with these materials. However, educational researchers have to explore the quality of the generated learning materials. In our study, we explored three capabilities of a state-of-the-art LLM (GPT-4): (1) investigating problem-solving processes, (2) augmenting educational data, and (3) generating instructional materials. We found that ChatGPT, as our test case for a decoder LLM, could well solve physics problems much in alignment with shared problem-solving process models (RQ1). However, we also found that the final phase (evaluation of solution) was never performed by ChatGPT in our trials. This coincides with the fact that evaluation of solution is barely performed in practice, after all, and LLMs can be seen as advanced "stochastic parrots" that—on some level—only mirror back what they encountered in the training data [42]. Be that as it may, it is still impressive that such LLM then compress the large training data and be capable to output meaningful solutions to our problem. In RQ2, we found that targeted prompting can be used to modify the answers to our problem in such a way as to selectively extend a dataset to physical solutions to problems, e.g., with learners that adhere to different problem-solving strategies. This would allow researchers to augment their data [15]. Finally, ChatGPT was found capable of generating physics problems, based on our prompts (RQ3). We prompted it to (A) generate similar problems to a given problem, (B) devise problems based on underlying concepts that are necessary to solve it, and (C) generate problems in different subfields in physics. ChatGPT generated well-defined, typical textbook-style problems which are quite common in physics instruction [43, 44]. Interestingly, ChatGPT used mechanics problems by default, likely because they are most prevalent in the training data sets. While the problems were well posed and solvable, they were also quite arcane without any reference to personally relevant contexts, which could enhance motivation to solve problems. This might be imposed through prompting, to raise the motivation of learners to engage in the problems. ChatGPT could well adapt to subfields and underlying concepts. As such, it provides a valuable resource for instructors to design test items in all kinds of assessments.

All of these capabilities of LLMs will likely have wide-ranging consequences for research on teaching and learning processes, and implementing effective and efficient teaching environments: (1) Researchers have at their fingertips tools to inquire about reasoning and thinking processes. This can help them to validate their own process models or frameworks for all sorts of cognitive processes such as reflective thinking, problem-solving, or argumentation [45–47]. (2) Researchers gain tools to quasi-empirically (via simulations) probe their devised test instruments, surveys, questionnaires, concept inventories, etc. Piloting surveys is a resource-consuming process and it is unclear to what extent the collected samples provide representative responses. With LLMs such as GPT-4, researchers can use prompting strategies to collect synthetic data for their surveys. Prompt engineering will play a crucial role in adequately piloting surveys with the help of LLMs. It is important that prompts are as specific as possible in order for the LLMs to provide meaningful output. Finally, (3), LLMs can help instructors come up with textbook problems and tasks in order to facilitate students' learning processes. This can spare instructors valuable resources that they can use for other things such as engaging in conversations with students on their conceptual understanding.

It is important to be wary of the challenges that come along with utilizing LLMs. For example, we found that they express certain misconceptions on their own. This can be expected in domains such as physics, as they are trained on unfiltered data from the Internet where all sorts of established misconceptions and preconceptions are presented. An interesting strategy to improve the LLMs is to explicitly instruct them with the relevant principles and concepts that they should apply for a certain problem. As such, instructors can even gain access to the effectiveness of their own instruction, given that the LLMs should have learned the materials afterwards. Even more concerning are the ethical and gender biases present in many LLMs [6, 48]. Researchers have to ensure that they probe the effectiveness of their LLM-generated instructional approaches and materials for relevant subgroups as they might be biased towards certain subgroups.

LLMs such as GPT-4, we argue, can help augment the tandem between human researchers and machines [49]. They can facilitate a bootstrapping program where researchers can use the LLMs to critically probe their materials such as surveys,and the researchers can instruct (or: prompt) the LLM to become capable of solving certain relevant tasks for their research—even label data according to their coding rubrics and similar. This potentially extends the capabilities of educational researchers, however, it has to be assured that LLMs become open source and publicly accessible, given that they are trained on public data in the first place. These LLMs might then be extended to the so-called foundation models [6], which also utilize other modalities such as vision and sensory data to enrich the self-supervised learning process and become knowledgeable in a richer sense.

Funding Information This research has been partly supported by the Klaus-Tschira-Stiftung (grant number: 00.001.2023).

Appendices

Appendix 1: Evaluation of the Solution

Question: A mass runs frictionless through a track with a loop. The mass starts from a height above the highest point of the loop. Determine the minimum starting height that is necessary so that the mass does not fall down when passing through the loop. Assume that the mass rolls along the track without sliding and assume that the mass is point-shaped. Describe in a comprehensible way how you would proceed with the solution. In particular, describe whether your solution is plausible and for what reasons you chose your solution. Try to write in complete sentences.

Answer: To answer this question, we need to understand a few fundamental concepts from physics: the conservation of energy and the concept of centripetal force.

Firstly, the conservation of energy principle states that the total energy in an isolated system remains constant. In this context, there are two forms of energy that matter: kinetic energy and potential energy. The kinetic energy, given by the formula $(1/2)mv^2$ (where m is mass and v is velocity), is the energy the mass has due to its motion. The potential energy, given by mgh (where m is mass, g is gravitational acceleration, and h is height), is the energy the mass has due to its position in a gravitational field.

The principle of conservation of energy states that the sum of kinetic and potential energy at any point of the path is equal to the sum at any other point. This implies that if we know the mass starts from rest, its total energy at the start is fully potential energy and will be transformed into kinetic energy as it falls.

Secondly, the mass must not fall when at the top of the loop. This means that the centripetal force, which keeps the mass moving in a circle, must be equal to or greater than the force of gravity pulling the mass down. The centripetal force is given by mv^2/r (where m is mass, v is velocity, and r is the radius of the loop), and the gravitational force is mg.

The minimum starting height necessary to complete the loop is found when the centripetal force is just equal to the gravitational force at the top of the loop. So we have:

$$mv^2/r = mg$$

This simplifies to:

$$v^2 = r * g$$

Applying the conservation of energy (remembering that all potential energy will be converted into kinetic energy at the top of the loop):

$$\text{mgh} = (1/2)\text{mv}^2 + \text{mg}^2 * \text{r}$$

We substitute v^2 from the above equation:

$$\text{mgh} = (1/2)\text{mrg} + \text{mg}^2\text{r}$$

This simplifies to:

$$\text{h} = (1/2)\text{r} + 2\text{r} = 5/2 * \text{r}$$

So the minimum height is 2.5 times the radius of the loop.

This answer is plausible because it applies two well-established physical principles: the conservation of energy and the balance of forces. It demonstrates that for the mass to just complete the loop, it needs to start at a certain minimum height. If it starts any lower, it won't have enough speed to overcome gravity at the top of the loop, and it will fall. If it starts any higher, it will have more than enough speed to stay on the track. This solution was chosen because it accurately applies these principles to solve the problem.

Appendix 2: Chat History on the Means-Ends Strategy

Question: A mass runs frictionlessly through a track with a loop. The mass starts from a height above the highest point of the loop. Determine the minimum starting height that is necessary so that the mass does not fall down when passing through the loop. Assume that the mass rolls along the track without sliding and assume that the mass is point-shaped. Describe in a comprehensible way how you would proceed with the solution if you were to use the "Means-ends analysis" strategy. Try to write in complete sentences.

Answer: Means-ends analysis is a problem-solving strategy that involves identifying the current state, the goal state, and the means to get from one to the other.

To solve this problem with means-ends analysis, you'd break down the overall goal into a series of subgoals. The overall goal is to determine the minimum starting height necessary to ensure the mass does not fall off the track when it goes through the loop.

Here are the steps you would take:

1. Identify the Current State: You start with a mass at a certain height above a loop in a frictionless track. The mass is point-shaped and rolls along the track without sliding.
2. Identify the Goal State: The goal is to find the minimum height at which the mass can start and still successfully make it around the loop without falling.
3. Identify the Means to Reach the Goal State: This problem is a classic physics problem that involves concepts like energy conservation, the force of gravity, and centripetal force.

   - Energy Conservation: At the starting point, the mass has potential energy that's converted into kinetic energy as it moves along the track. At the top of the loop, it will have a combination of potential and kinetic energy.
   - Gravity: The force of gravity pulls the mass downwards, and this force is what gives it the potential energy at the start and is constantly working against its motion as it moves along the track.
   - Centripetal Force: At the top of the loop, the only force acting on the mass is the force of gravity, which must be equal to or greater than the centripetal force needed to keep the mass moving in a circular path.

4. Develop Subgoals to Apply the Means:

   - Subgoal 1: Use the principle of conservation of energy. At the start, the potential energy of the mass is mgh (mass x gravity x height). As it descends, this potential energy gets converted into kinetic energy, and by the time it reaches the top of the loop, this kinetic energy must be sufficient to keep it in the loop.
   - Subgoal 2: Apply the understanding of gravity and centripetal force. At the top of the loop, the gravitational force acting on the mass (mg) must be equal to the centripetal force (mv^2/r, where v is the velocity of the mass and r is the radius of the loop) necessary to keep it in the loop.

5. Evaluate and Iterate: Once the subgoals are identified, you can perform the necessary calculations to find the height. Evaluate the results and iterate the process if necessary, adjusting the assumptions and calculations until you have a satisfactory answer that achieves the goal state.

By using means-ends analysis, you can systematically work your way through the problem, breaking it down into manageable chunks and using physics principles to find a solution.

Appendix 3: Chat History on the "Working Backwards " Strategy

Question: A mass runs smoothly through a track with a loop. The mass starts from a height above the highest point of the loop. Determine the minimum starting height necessary so that the mass does not fall down while running through the loop. Assume that the mass rolls along the track without sliding and assume that the mass is point-like. Describe in an understandable way how you would solve the problem if you used the "working backwards" strategy. Try to write in complete sentences.

Answer: Sure! This is a classic problem in physics, and it's a wonderful application of the concept of energy conservation, specifically gravitational potential energy and kinetic energy. To solve it using the "working backwards" strategy, we'll start from the point we're most interested in (the top of the loop), and calculate what conditions need to be met at that point to keep the mass from falling down. Then, we'll work our way back to the starting point.

Here are the steps:

1. Determine the Necessary Speed at the Top of the Loop: At the top of the loop, the mass needs to have enough speed so that the gravitational force pulling it downward is equal to the centripetal force necessary to keep it moving in a circle. This can be expressed as $mg = mv^2/r$, where m is the mass, g is the acceleration due to gravity, v is the velocity, and r is the radius of the loop. From this, we find that the necessary speed at the top of the loop is sqrt(gr).
2. Determine the Necessary Kinetic Energy at the Top of the Loop: Using the speed from step 1, we can calculate the kinetic energy at the top of the loop using the equation $K = 1/2\ mv^2$. This gives us $K = 1/2\ m * (gr)$.
3. Determine the Necessary Total Energy at the Start: At the top of the loop, the mass has both kinetic and potential energy. So, the total energy at the start needs to be equal to the sum of these two quantities. The potential energy at the top of the loop is given by $U = mgh$, where h is the height above the ground, which in this case is 2r (the diameter of the loop). So, the total necessary energy at the start is $U + K = mgh + 1/2\ m * (gr) = mg(2r) + 1/2\ m * (gr) = 2.5\ mgr$.
4. Determine the Necessary Starting Height: We know the mass starts with all its energy as potential energy, which is given by $U = mgh$, where h is the starting height. Setting this equal to the necessary total energy from step 3, we get $mgh = 2.5\ mgr$. Solving for h, we get $h = 2.5r$.

So, the minimum starting height necessary for the mass to not fall down while running through the loop is 2.5 times the radius of the loop. This "working backwards" strategy is effective because it starts by identifying the conditions necessary at the critical point (the top of the loop), and then uses the principle of conservation of energy to relate these conditions to the initial situation.

Appendix 4a: Chat History on the "Plug and Chug"-Strategy

Question: A mass runs smoothly through a track with a loop. The mass starts from a height above the highest point of the loop. Determine the minimum starting height necessary so that the mass does not fall down while running through the loop. Assume that the mass rolls along the track without sliding and assume that the mass is point-like. Describe in an understandable way how you would solve the problem if you used the "Plug and Chug" strategy. The plug and chug strategy is a technique of entering values to get a numerical answer without requiring any problem-solving skills. Try to write in complete sentences.

Answer: In this physics problem, we are interested in finding the minimum starting height required for a mass to make it through a loop without falling. This situation involves concepts from classical mechanics, specifically the conservation of mechanical energy. The "plug and chug" strategy you referred to is more straightforward in a scenario where we have specific numerical values to substitute, but since this problem is set up in a more conceptual way, we will still have to perform a bit of problem-solving to set up the equation that we will use.

1. Understand the Physics Concepts:

   Energy conservation is the fundamental principle we're working with here. The idea is that the total mechanical energy (the sum of kinetic and potential energy) is conserved in an isolated system if there are no non-conservative forces doing work, such as friction or air resistance. Here, we're given that the mass rolls without sliding, which means we can consider this a frictionless scenario and thus energy is conserved.

2. Identify the Relevant Energies:

   The relevant energies here are gravitational potential energy and kinetic energy. Gravitational potential energy is associated with the mass's height off the ground and is given by U = mgh, where m is the mass, g is the gravitational acceleration (approx. 9.8 m/s^2), and h is the height. Kinetic energy is associated with the mass's speed and is given by K = 1/2 mv^2, where v is the velocity.

3. Set up the Energy Conservation Equation:

   For the mass to just barely make it through the loop, it must have zero kinetic energy at the top of the loop. All its energy would be in the form of potential energy. Therefore, the energy at the starting height (all potential energy) must be equal to the energy at the top of the loop (also all potential energy).

   So we can say mgh1 = mgh2, where h1 is the initial height and h2 is the height at the top of the loop (which is twice the radius of the loop if we assume the loop to be a perfect circle).

   The mass m and gravitational acceleration g cancel out on both sides, leaving us with h1 = h2.

   However, since the mass has to maintain contact with the track at the top of the loop without falling, it must have some non-zero velocity at the top, and therefore some kinetic energy. To maintain contact, the gravitational force at the top of the loop must equal the centripetal force required to move in a circle of the loop's radius, r. This gives us mg = mv^2/r, where v is the velocity of the mass at the top of the loop. Solving for v gives us v = sqrt(gr), which can be substituted back into the kinetic energy expression to give the kinetic energy at the top of the loop: K = 1/2 m * (gr), or K = mgr/2.

We need to account for this extra energy requirement in our conservation of energy equation. The adjusted equation is mgh1 = mgh2 + K, or h1 = h2 + r/2.

4. Conclusion:

Therefore, to ensure the mass can just make it around the loop, the starting height must be at least equal to the height at the top of the loop plus half the radius of the loop. Note that we made some simplifying assumptions to derive this result, including that the mass is point-like and rolls without sliding.

When you're given specific values for the radius of the loop, you can "plug" these values into the final formula to "chug" out your answer. Remember, the heights and radius should be in the same unit when you substitute values.

Appendix 4b: Chat History on the "Plug and Chug"-Strategy

Question: A mass runs smoothly through a track with a loop. The mass starts from a height above the highest point of the loop. Determine the minimum starting height necessary so that the mass does not fall down while running through the loop. Assume that the mass rolls along the track without sliding and assume that the mass is point-like. Describe in an understandable way how you would solve the problem if you used the "Plug and Chug" strategy. The plug and chug strategy is a technique that problem solvers use where they seek to identify relevant physical quantities in the problem description and find matching equations where they can plug in the quantities and find the desired quantity.

Answer: Sure, this problem is a classic one in introductory physics, especially within the subject of energy conservation. We can approach it using the "Plug and Chug" strategy, with several steps:

1. Identify Relevant Quantities: In our case, the quantities that seem relevant include the gravitational potential energy (PE), kinetic energy (KE), and the height of the loop (H).
2. Identify Relevant Concepts: The main concept here is conservation of energy, which states that the total energy of an isolated system remains constant if no non-conservative forces are involved. Another relevant concept is the centripetal force needed to keep the mass moving in a circular path at the top of the loop.
3. Find Matching Equations: There are three key equations here:

   - PE = m * g * h (gravitational potential energy)
   - KE = 1/2 * m * v^2 (kinetic energy)
   - For centripetal force at the top of the loop, we have F_c = m * v^2 / r (where r is the radius of the loop). This force must be at least equal to the gravitational force acting on the mass at the top of the loop, m * g.

4. Plug in Quantities:

   - Start by setting the gravitational potential energy (PE) at the starting height equal to the kinetic energy (KE) plus the gravitational potential energy at the top of the loop. This uses the principle of conservation of energy: PE_start = KE_top + PE_top.
   - Convert these to equations: m * g * h_start = 1/2 * m * v_top^2 + m * g * 2*r* *(we use* 2r for the height at the top of the loop because the mass has to travel the diameter's length of the loop to reach the top)
   - Cancel the mass (m) from both sides (since it's a common factor), and plug in the centripetal force condition at the top of the loop, v_top^2 = g * r, to get g * h_start = 1/2 * g * r + 2 * g * r

5. Solve the Equation: Finally, solve the equation for the starting height:

   - Combine like terms on the right side to get g * h_start = 5/2 * g * r
   - Divide both sides by g to isolate h_start: h_start = 5/2 * r

This equation gives the minimum height from which the mass needs to start so that it will not fall off the track at the top of the loop. The answer is that the initial height should be 2.5 times the radius of the loop, which implies the height of the loop itself *(2r) plus an additional 0.5*r. This result ensures the mass will have enough kinetic energy at the top of the loop to counteract the force of gravity and maintain the necessary centripetal force.

References

1. Samuel AL (1959) Some studies in machine learning using the game of checkers. IBM J
2. Hinton G, Osindero S, Teh Y-W (2006) A fast learning algorithm for deep belief nets. Neural Comput 18:1527–1554
3. Géron A (2018) Praxiseinstieg machine learning mit scikit-learn und tensorflow: konzepte, tools und techniken für intelligente systeme (K Rother, Trans.). Animals. O'Reilly. http://nbn-resolving.org/urn:nbn:de:bsz:31-epflicht-1303476
4. Mitchell M (2020) Artificial Intelligence: a guide for thinking humans. Pelican Books.
5. Vaswani A, Shazeer N, Parmar N, Uszkoreit J, Jones L, Gomez AN, Kaiser U, Polosukhin I (2017) Attention is all you need. In: Guyon I, Von Luxburg U, Bengio S, Wallach H, Fergus R, Vishwanathan S, Garnett R (eds) Advances in neural information processing systems. Curran Associates, Inc. https://proceedings.neurips.cc/paper/2017/file/3f5ee243547dee91fbd053c1c4a845aa-Paper.pdf
6. Manning CD (2022) Human language understanding & reasoning. Daedalus 151(2):127–138. https://doi.org/10.1162/daed_a_01905
7. Manning CD, Schütze H (1999) Foundations of statistical natural language processing. MIT Press
8. Goldberg Y (2017) Neural network methods for natural language processing. Synthesis lectures on human language technologies. Morgan and Claypool
9. Brown TB, Mann B, Ryder N, Subbiah M, Kaplan J, Dhariwal P, Neelakantan A, Shyam P, Sastry G, Askell A, Agarwal S, Herbert-Voss A, Krueger G, Henighan T, Child R, Ramesh A, Ziegler DM, Wu J, Winter C, Amodei D (2020) Language models are few-shot learners. ArXiv.

10. Devlin J, Chang M-W, Lee K, Toutanova K (2018) BERT: pre-training of deep bidirectional transformers for language understanding. ArXiv 1810:04805
11. Lewkowycz A, Andreassen A, Dohan D, Dyer E, Michalewski H, Ramasesh V, Slone A, Anil C, Schlag I, Gutman-Solo T, Wu Y, Neyshabur B, Gur-Ari G, Misra V (2022) Solving quantitative reasoning problems with language models. ArXiv.
12. Wolfram S (2023) What is ChatGPT doing and why does it work? Wolfram Media
13. Kasneci E, Sessler K, Küchemann S, Bannert M, Dementieva D, Fischer F, Gasser U, Groh G, Günnemann S, Hüllermeier E, Krusche S, Kutyniok G, Michaeli T, Nerdel C, Pfeffer J, Poquet O, Sailer M, Schmidt A, Seidel T, Kasneci G (2023) ChatGPT for good? On opportunities and challenges of large language models for education. Learn Individ Differ 103:102274. https://doi.org/10.1016/j.lindif.2023.102274
14. Küchemann S, Steinert S, Revenga N, Schweinberger M, Dinc Y, Avila KE, Kuhn J (2023) Physics task development of prospective physics teachers using ChatGPT. ArXiv.
15. Kieser F, Wulff P, Kuhn J, Küchemann S (2023) Educational data augmentation in physics education research using ChatGPT. Phys Rev Phys Educ Res 19(2):1–13. https://doi.org/10.1103/PhysRevPhysEducRes.19.020150
16. Carpenter D, Geden M, Rowe J, Azevedo R, Lester J (2020) Automated analysis of middle school students' written reflections during game-based learning. In: Bittencourt II, Cukurova M, Muldner K, Luckin R, Millán E (eds) Artificial intelligence in education. Springer International Publishing, pp 67–78
17. Wulff P, Mientus L, Nowak A, Borowski A (2022b) Utilizing a pretrained language model (BERT) to classify preservice physics teachers' written reflections. Int J Artif Intell Educ https://doi.org/10.1007/s40593-022-00290-6
18. Odden TOB, Marin A, Rudolph JL (2021) How has science education changed over the last 100 years? An analysis using natural language processing. Sci Educ 105(4):653–680. https://doi.org/10.1002/sce.21623
19. Zhai X, He P, Krajcik JS (2022) Applying machine learning to automatically assess scientific models. J Res Sci Teach
20. Kortemeyer G (2023) Could an Artificial-Intelligence agent pass an introductory physics course? Phys Rev Phys Educ Res 19(1):010132. https://doi.org/10.1103/PhysRevPhysEducRes.19.010132
21. Wang J (2023) ChatGPT: a test drive. Am J Phys 91(4):255–256. https://doi.org/10.1119/5.0145897
22. West CG (2023) Advances in apparent conceptual physics reasoning in GPT-4. ArXiv
23. Zhai X (2023) ChatGPT user experience: implications for education. SSRN
24. Csapó B, Funke J (eds) (2017) The nature of problem solving: Using research to inspire 21st century learning. OECD Publishing. https://doi.org/10.1787/9789264273955-en
25. Odden TOB, Marin A, Caballero MD (2020) Thematic analysis of 18 years of physics education research conference proceedings using natural language processing. Phys Rev Phys Educ Res 16(1). https://doi.org/10.1103/PhysRevPhysEducRes.16.010142
26. Wulff P, Buschhüter D, Westphal A, Mientus L, Nowak A, Borowski A (2022a) Bridging the gap between qualitative and quantitative assessment in science education research with machine learning—a case for pretrained language models-based clustering. J Sci Educ Technol https://doi.org/10.1007/s10956-022-09969-w
27. Zhai X, Yin Y, Pellegrino JW, Haudek KC, Shi L (2020) Applying machine learning in science assessment: a systematic review. Stud Sci Educ 56(1):111–151. https://doi.org/10.1080/03057267.2020.1735757
28. Rauf IA (2021) Physics of data science and machine learning. CRC Press. https://doi.org/10.1201/9781003206743
29. Marsland S (2015) Machine learning: an algorithmic perspective (Second edition). Chapman & Hall/CRC machine learning & pattern recognition series. CRC Press. http://proquest.tech.safaribooksonline.de/9781466583283
30. Docktor JL, Dornfeld J, Frodermann E, Heller K, Hsu L, Jackson KA, Mason A, Ryan QX, Yang J (2016) Assessing student written problem solutions: a problem-solving rubric with

application to introductory physics. Phys Rev Phys Educ Res 12(1):10130. https://doi.org/10.1103/PhysRevPhysEducRes.12.010130
31. Hsu L, Brewe E, Foster TM, Harper KA (2004) Resource letter RPS-1: research in problem solving. Am J Phys 72(9):1147–1156. https://doi.org/10.1119/1.1763175
32. Fortus D (2009) The importance of learning to make assumptions. Sci Educ 93(1):86–108. https://doi.org/10.1002/sce.20295
33. Larkin JH, McDermott J, Simon DP, Simon HA (1980) Expert and Novice performance in solving physics problems. Science 208:1335–1342
34. Sweller J (1988) Cognitive load during problem solving: effects on learning. Cogn Sci 12:257–285
35. de Jong T, Ferguson-Hessler MG (1996) Types and qualities of knowledge. Educ Psychol 31(2):105–113
36. Walsh LN, Howard RG, Bowe B (2007) Phenomenographic study of students' problem solving approaches in physics. Phys Rev Spec Top Phys Educ Res 3(2). https://doi.org/10.1103/PhysRevSTPER.3.020108
37. Savelsbergh ER, de Jong T, Ferguson-Hessler MG (1997) The importance of an enhanced problem representation: instructional technology memorandum series. Univ. of Twente, Enschede, Netherlands
38. Bransford J, Stein BS (1984) The ideal problem solver: a guide for improving thinking, learning, and creativity. W. H. Freeman and Company
39. Polya G (1945) How to solve it. Princeton University Press
40. Reif F (1995) Millikan Lecture 1994: understanding and teaching important thought processes. Am J Phys 63(1):17–31
41. Python Software Foundation (2020) Python language reference: version 3.8. http://www.python.org
42. Bender EM, Gebru T, McMillan-Major A, Shmitchell S (2021) On the dangers of stochastic parrots. FAccT 610–623. https://doi.org/10.1145/3442188.3445922
43. Kim E, Pak S-J (2002) Students do not overcome conceptual difficulties after solving 1000 traditional problems. Am J Phys 70(7):759–765. https://doi.org/10.1119/1.1484151
44. Ogilvie CA (2009) Changes in students? problem-solving strategies in a course that includes context-rich, multifaceted problems. Phys Rev Spec Top Phys Educ Res 5(2):139. https://doi.org/10.1103/PhysRevSTPER.5.020102
45. Wulff P, Buschhüter D, Nowak A, Westphal A, Becker L, Robalino H, Stede M, Borowski A (2020) Computer-based classification of preservice physics teachers' written reflections. J Sci Educ Technol Adv https://doi.org/10.1007/s10956-020-09865-1
46. Xing W, Lee H-S, Shibani A (2020) Identifying patterns in students' scientific argumentation: content analysis through text mining using Latent Dirichlet allocation. Educ Technol Res Dev https://doi.org/10.1007/s11423-020-09761-w
47. Li Y, Sha L, Yan L, Lin J, Raković M, Galbraith K, et al (2023) Can large language models write reflectively. Comput Educ: Artif Intell 4:100140. https://doi.org/10.1016/j.caeai.2023.100140
48. Caliskan A, Bryson JJ, Narayanan A (2017) Semantics derived automatically from language corpora contain human-like biases. Science (New York, N.Y.) 356(6334):183–186. https://doi.org/10.1126/science.aal4230
49. Sherin B (2013) A computational study of commonsense science: an exploration in the automated analysis of clinical interview data. J Learn Sci 22(4):600–638. https://doi.org/10.1080/10508406.2013.836654

Machine Learning Applications for Early and Real-Time Warning Systems in Education

Mariel F. Musso and Eduardo C. Cascallar

Abstract During the last decade, research has shown the accuracy and robustness of machine-learning methods for modelling nonlinear and complex relationships among several variables and for different prediction and classification purposes in the educational field. In this chapter we will present the main highlights coming from several studies that we carried out to predict a wide range of educational outcomes across different domains (reading readiness, mathematics, language, writing, general academic performance, and complex problem solving) using artificial neural networks. They have encompassed all educational levels from primary school to higher education, in different developed and developing countries. Furthermore, this approach was also useful to model key educational outcomes in academic trajectories such as retention, degree completion, and online education satisfaction in higher education. The main objective is to provide a comprehensive, integrated, systematic, and evidence-based framework for the application of machine learning systems in the educational field for improved prediction of performance, better understanding of factors influencing such performance, and improved policy decision-making regarding these kinds of problems. In addition, we will analyze and discuss the main specific patterns found in all these educational outcomes.

Keywords Machine learning · Educational outcomes · Prediction · Artificial neural networks · Self-regulated learning

M. F. Musso (✉)
University of Granada, Granada, Spain
e-mail: mariel.musso@hotmail.com

CIIPME-CONICET, Buenos Aires, Argentina

E. C. Cascallar
KU Leuven, Leuven, Belgium

M. S. Khine (ed.), *Machine Learning in Educational Sciences*,
https://doi.org/10.1007/978-981-99-9379-6_15

1 Early Beginnings

The application of machine learning algorithms to several problems in the educational field has become an emerging interdisciplinary area during the last two decades. Although this methodology had already been used since the 1980s in other fields, such as economics, meteorology, and engineering [1–6], the first articles using artificial neural networks on educational measurement issues, date back to 1994 [7–10]. Additionally, pioneer applications using machine learning techniques appeared in the self-regulation assessment field [11]. At the same time, two independent but related areas, Educational Data Mining (EDM) and Learning Analytics (LA) have grown exponentially since the first book about these areas was published in 2006 and two important conferences were carried out in 2008/2011 [12]. EDM concentrates on advancing and utilizing data mining methods to extract knowledge from educational settings, to address critical questions in this domain [12, 13]. LA, on the other hand, places a significant emphasis on the measurement, collection, analysis, and reporting of data pertaining to learners and their environments, with the aim of gaining insight and enhancing learning and its context [14]. Although they focus on diverse aspects, both EDM and LA converge in their focus on how to use educational data to benefit the science of learning and education [12].

The growing interest in EDM and LA has been documented in recent reviews showing an application of a wide range of tools, frameworks, and methods for solving educational problems, such as visualization, clustering, prediction, and association analysis techniques, in different environments (traditional, computer-based educational systems, and blended learning) (e. g. [12, 15]). Prediction is one of the most widely used approaches within data mining in the educational field. This approach addresses inferring a target attribute (dependent outcome) of the data from other variables named predictors or input. There are several types of prediction methods such as classification (categorial target), regression (continuous target), or density estimation (probability density function).

The present chapter aims to present some highlights regarding the methods used and results from several studies carried out by the authors, following a predictive classification approach, and specifically using artificial neural networks (ANN). They have been developed for several purposes: forecasting student performance, detecting students at risk and undesirable behavioral outcomes, early learning recommendations, and for scientific research objectives in the cognitive and learning fields. These studies have encompassed the predictive classification of a wide range of educational outcomes across different domains and educational levels from pre-school to higher education, in both developed and developing countries, using machine learning techniques. Table 1 summarizes the general and specific objectives for each one of the domains across every educational level.

This chapter is organized as follows: the first section introduces some general science frameworks about how we conceptualize modelling and information in science. The second section describes one procedure called Structured Neural

Table 1 Predictive classification studies and objectives

Domain/ educational content	Educational levels			
	Pre-school	Primary school	Secondary school	Higher education
Executive attention, self-regulated behaviour and IQ	Modeling-scientific research in cognition Early detecting risk groups			
Fluid intelligence/ WM training gains		Student modeling-scientific research in cognition & genetics area		
Reading readiness		Provide course adaptation and learning recommendations		
Math performance				Student modeling-scientific research in cognition and learning recommendations
Writing			Early learning recommendations	
GPA				Early learning recommendations Grouping students
Retention				Detecting undesirable student behaviors
Degree completion				Planning and scheduling
Complex problem solving				Student modeling-scientific research in cognition
Online education satisfaction				Learning recommendation

Network technique (SNN) [16] which we followed for modelling the different educational problems in order to address a common criticism to neural networks, namely being a "black box" method. The SNN technique led us to present in this section a general theoretical framework of academic performance, and specifically some key constructs for the learning field. The third section will focus on methodological aspects of ANN in order to understand both the general procedures and specific measures needed to interpret the most important neural networks results. The fourth section of this chapter will summarize the most relevant results of ANN applications. Finally, a discussion about certain patterns found across all the studies will be presented, and some potential new trends for future research will be drawn.

2 Theoretical Underpinnings of the Scientific Framework for Predictive Analysis in Education

2.1 Model-Dependent Realism

A general science framework called model-dependent realism has guided the conceptualization of our overall research. Simply put, this approach states that there is no point in assuming the existence of a "real" or final model, rather, models are considered as long as they are useful in predicting the phenomenon or outcome of interest, to be replaced by a more successful model whenever one is developed. Considering this viewpoint, models that explain the same phenomenon should be considered validated if and only if there are no inconsistencies in the overlapping areas where they are applicable.

It is assumed that ultimately, every mental state has a physical correlate. These models are "emerging", not fundamental. It is a deterministic and physical conceptualization, but not reductionist. Emerging models have their own validity at the corresponding level of analysis, but they can never contradict the most basic physical model in their logic and formalisms. Each model is a set of interlocking descriptions that correspond to and form a stable model within the domain of applicability of the system constituted by the model.

When examining the role and significance of modeling in science, as elucidated by [17], it can be observed that scientific models can perform two distinct representational functions, which are not mutually exclusive. These functions are the representation of an aspect or selected part of the world, referred to as the "target system," and the representation of a theory, encompassing its rules, laws, and axioms. It is evident that Artificial Neural Networks (ANNs) contribute to the development of more comprehensive representational models, specifically in the form of "models of data" [18]. This contribution is based on extensive research that has been crucial in establishing the connection between ANNs' representations and their relationship to the resulting outputs [19].

2.2 Information Theory Perspective

According to information theory, as interpreted according to the holographic principle, the maximum information content of any conceptual space is defined by its surface area. Each bit of information of an event or element is represented on the surface of the conceptual sphere around it [20]. A consequence of this assumption is that because there is a limit to the number of bits of information that can be represented on a surface, there is also a limit to the amount of information content in the event or element space [21]. Any arbitrary section of a holographic representation enables the reconstruction of the whole information of an event or element, with distortions decreasing gradually with the increase in the size of the holographic representation (information) available in a given surface area, and there are no restrictions on the order in which this data or information is considered [22]. Thus, if we conceptualize any event or element that we wish to understand or classify (in our case, a certain educational performance level) as surrounded by a conceptual sphere of information, any portion of which contains the full representation we wish to predict or understand, it will be the density of the information considered that will determine the degree of "distortion" or error with which we can predict the event or element in question. The sampling of this information, either all in one given surface area or as a sum of sampling elements of information from various points in the surface will determine the degree to which we can specify or describe the event we wish to classify, its precision being directly related to the density of the information collected from the conceptual surface [22]. We will be presenting neural network models based on previous evidence and educational theory in order to achieve a high density of information related to the event or educational target.

3 Conceptual Bases for Developing Neural Networks.

An ANN is a computational system that processes information through interconnected processing units known as neurons [23, 24]. The neuron, which serves as the basic building block of any ANN, enables the mapping of various inputs to an output by implementing a weighted sum operation and an activation function. The weights represent the strength of the connections between neurons and the information utilized by the network to minimize the discrepancy between the predicted and actual outputs [25]. Additionally, each neuron has an internal state, that is, an activation function which depends on the inputs the neuron receives. Therefore, a neural network can be described in terms of: a its topology or architecture, which refers to the organization of connections between neurons, b the learning or training algorithm employed to establish the strength or weights of these connections; and c its activation functions: the weighted sum of the input information, transformed into an output from a node or nodes in a layer of the network.

3.1 Structure Neural Network (SNN)

Traditional ANN have been applied as empirical models without considering theoretical frameworks underlying their data. This called "black box" approach has aimed to predict specific outcomes rather than provide insights about the relationships among input data and the targets. Although we recognize the complexity inherent in educational phenomena, we take into consideration underlying theoretical frameworks for this modeling problem in each one of our studies as it has been suggested by previous studies [16]. They have demonstrated that a SNN technique could be useful when theoretical frameworks are well defined. [16] proposed to build a neural network based on contextual and theoretical knowledge. In our studies, the structure of inputs related to the educational targets were designed based on solid theoretical constructs derived from previous learning and self-regulation models (e.g. [26] and classical analyses on the same data [27, 28]). The objective was to avoid the problem of blind data mining techniques [29]. Therefore, the interpretation of the potential weight of each variable to the classification problem was more feasible.

3.2 Theoretical Framework of Academic Performance

A large body of research has shown that school and academic achievement depend on multiple factors at different levels (e.g., [30, 31]). Based on several classifications proposed in the literature [32, 33], we suggested the following categories of factors into two levels: student and environmental levels. On the one hand, the student level has involved (a) cognitive factors, (b) demographics (i.e., age, gender), and (c) self-regulated learning factors. In addition, other categories of variables were included in primary school studies: (a) child physical factors & health, (b) child preschool and school trajectory, and (c) child routines and habits. On the other hand, the environmental level for higher education has included (a) background variables and socio-economic status, and (b) institutional characteristics. Factors related to the teacher and school categories were additionally analyzed for predicting educational outcomes in primary and secondary school.

Student level

A broader definition of self-regulation has guided the designing of the ANN models presented in this chapter. The concept of self-regulation involves the ability to monitor and control one's cognitive, emotional, and behavioral processes with the objective of accomplishing personal goals and/or adjusting to the social and cognitive demands of specific situations [34]. The complex educational activities require an integrative theoretical framework of self-regulated learning (SRL) as proposed by Boekaerts's model [26]. In this model, with knowledge as a central construct, SRL is divided in six components: (1) domain-specific knowledge and skills, (2) cognitive strategies, (3) cognitive regulatory strategies, (4) meta-cognitive knowledge and motivational beliefs, (5) motivation strategies, and (6) motivational regulatory strategies. The components are grouped into two mechanisms, cognitive and affective-motivational self-regulation, and at three levels: (a) the domain-specific level, which is the bottom level, (b) the strategic level, which is considered the intermediate level, and (c) the goal level, which represents the top level. In addition, the Dual

Processing self-regulation model [35] pointed out three purposes of self-regulation during the learning process: 1 to expand knowledge and skills (top-down strategies, 2 to preserve well-being and caring the self and loss of resources (bottom-up strategies, and 3 to protect one's promises redirecting attention from well-being to the mastery/growth pathway. These purposes are triggered by positive or negative appraisals of the students regarding the learning situation and the self. Boekaerts and collaborators (e.g., [36–38]) have analyzed different appraisals, such as perceived relevance of the task, task attraction, subjective competence, perceived difficulty, perception of control, and success expectation. In addition, emotions and the strategies to regulate them play a key role in this model [27]. Motivational beliefs included in Boekaerts' model are judgments and opinions of the students regarding a specific aspect of the learning context: (a) values, beliefs, and attitudes about a curricular content and/or task, (b) beliefs related to the self-capacity coping a specific content or task; and (c) goal orientations [39]. According to this model, when students face a learning task, attributions take place as knowledge structures to interpret the cause of a successful or failure learning outcome [27]. Additionally, as academic situations could be as stressful for a student, coping strategies were included in our models. Coping strategies are conceptualized as cognitive, emotional, and behavioral efforts that are developed to handle stressors [40]. Previous research has also demonstrated the relevance of perceived social support for students during their transition to college [41], and also predicting academic results [42].

Our ANN also incorporated Learning Strategies (LS), which encompass any cognitive or behavioral processes that facilitate students' acquisition and integration of new information into their existing knowledge [43, 44]. As delineated by Weinstein and Palmer [43], LS comprises three fundamental components: skill, will, and self-regulation. The "skill component" of the learning strategies construct involves behaviors and processes related to "identifying, acquiring and constructing meaning for important new information, ideas and procedures, and how the students prepare for and demonstrate their new knowledge on tests or other evaluative procedures" ([43], p. 5). The "self-regulation component" of the LS construct encompasses the management and control of the entire learning process, with the goal of optimizing time utilization, "focusing the attention and maintaining the concentration, self-testing, and using study supports such as review sessions, tutors or special features of a textbook" ([43], p. 5). Finally, the "will component of strategic learning" as mentioned by Weinstein and Palmer [43] encompasses attitudes and interest in a learning situation, concern for academic performance, receptiveness to learning, effort, self-discipline, and the management of anxiety.

The literature has shown that SRL is influenced by cognitive processes such as working memory capacity (WMC) and attentional control [45, 46]. WMC is a limited cognitive system that enables the temporary storage and manipulation of information [47]. Individual differences in the capacity and efficiency of the storage and processing components of WMC predict the performance in reasoning, reading, math, and complex problem solving [48–51]. Individuals with high WMC are better in the active maintenance of information and apply effectively learning strategies for a specific learning goal [52–54]. In addition, a three-attentional systems model was considered based on large empirical evidence, including orienting, alerting, and executive control systems [55, 56]. Orienting allows the selection of information from stimuli entering into the system. Alerting refers to the maintaining of an alert state. The Executive system controls the interference between competitive stimulus maintaining the focus on a goal [55, 57–59].

Environment Level

Literature reviews have also identified several environmental variables predicting academic achievement and drop-out, such as sociodemographic background and socio-economic status [60, 61]. Background factors involve socio- demographic information (gender, age), and socioeconomic status (parental education, income, and occupation of parents). There are some controversial results regarding the predictive contribution of background factors depending on the academic level, the factors included in the model, and the method used [61]. However, modelling educational outcomes at pre-school and primary school requires considering the crucial role of the environment. It is widely acknowledged that family socioeconomic status has a substantial influence on early attention and self-regulation development, as evidenced by numerous cross-sectional and longitudinal studies [62–65].

4 A Systematic Procedure

The application of machine learning techniques for predicting in educational field must follow systematic procedures according to the scientific standards established in the literature. These procedures have to do with well-known methodological processes which guarantee the reproducibility and transparency of science [66–68]. We have considered the six stages proposed by Alyahyan and Düştegör [69], but we also propose a seventh phase for a more comprehensive framework in the implementation of ANN to educational prediction problems: (1) Data collection, (2) Initial preparation of data, (3) Preliminary statistical analyses, (4) Data preprocessing, (5) Implementation of ANN, (6) Evaluation of the models, and (7) Analysis of predictive patterns. Several decisions have to be made at various stages of the process. Here we focus on the implementation and evaluation of the models as these phases include most of the design decisions to be made in the overall process.

(1) *Data collection.* As it was noted in the previous sections, a "good model" begins with a clear and specific question about the educational outcome to be predicted. Whether this question derives from an empirical or research need, it should lead to a review and systematic study of the literature in order to determine which are the most important factors associated with the outcome. As we have already pointed out, from the information theory perspective, we need high density of information, so collecting elements of information related to the educational target is a very crucial step of the process. Therefore, which are the most influential variables for educational achievement at the different levels of performance

under consideration? These factors were specified in Sect. 2.2 according to the comprehensive SRL model we have chosen, and basic cognitive processes which appear in the relevant scientific literature. These theoretical models have guided us in the selection of several measures and in the application of various validated instruments. As indicated in this table, the information which feeds the ANN can be extracted from multiple sources. Sociodemographic data of students (i.e., age, gender, ethnicity), socioeconomic status and other background/family indicators can easily be retrieved from general student surveys that institutions collect online. On the other hand, cognitive factors need to be obtained administering specific computerized tests (i.e., AOSPAN for working memory, ANT for attentional networks, IQ), and psychological data requires students to complete valid and reliable psychometric scales (i.e., scales for learning and coping strategies, personality, etc.).

(2) *Initial preparation of data.* The original data set needs to go through selection and cleaning, and some new variables are obtained from collapsing other data into scales, which need to be checked for their psychometric properties. This preparation involves the deletion of cases or imputation of missing data, deleting of incorrect, or duplicate data, and the scoring of scales and cognitive tests (see [70, 71], Data analyses section, p. 5).

(3) *Preliminary Statistical Analyses.* This step is necessary to obtain descriptive analyses of the data, the identification of possible outliers, and determine the distribution of each variable. In addition, we can identify the relationships between predictors and the educational target. In this phase we can generate new complex variables based on exploratory and/or confirmatory factor analysis.

(4) *Data preprocessing.* One critical decision in this step is to define the categories of the educational target, depending on the empirical or research interest. Most of our studies were conducted to identify a low or high level of educational achievement considering the centiles 33 and 66 respectively (see [70–73). In these cases, we labeled all the students classified as belonging or not belonging to each performance group. The next step within this stage is to standardize any independent continuous variable as they were measured with different scales. In addition, categorical variables have to be dummy coded [74].

(5) *Implementation of ANN.* This phase addresses systematically training and testing ANN models to classify student groups. One of the first decisions to be made in this stage is the selection of algorithms to build the models. A multilayer perceptron neural network (MLP), a type of neural network consisting of nonlinear units, was employed in our studies. These units compute their activation levels by aggregating weighted activations received from other units. Subsequently, the network transforms its activation into a response through a nonlinear transfer function that establishes a relationship between inputs and the weights assigned to them. To optimize its weights, the MLP utilizes an optimization function, with gradient descent being the choice in most of our studies to minimize the error from the mean squared error function. During the training phase, these systems evaluate the impact of weight patterns on the accuracy of their classification outputs and adjust those weights in a recursive

manner until they maximize the precision of the resulting classifications, using a backpropagation algorithm.

A second decision in this stage has to do with the adjusting of three main hyperparameters of the backpropagation algorithm: (a) learning rate, (b) momentum, and (c) stopping rule. Learning rate serves as an indicator of the network's efficiency in adjusting its weights and biases during the training process. The learning rate ultimately determines the extent to which the network's predictions improve with each iteration. The integration of momentum in the training process helps to avoid the network getting trapped in a suboptimal solution (local minimum) by adding a fraction of the previous weight update to the current one. Stopping rules refer to the moment when the network should stop "learning" to avoid over-fitting the current sample. Other specifications must be determined in this stage regarding the architecture of the ANN: number of hidden layers and nodes. Although a trial-and-error approach is frequently used by non-expert users [75], we suggest the training of ANN through systematic changes of the parameters' values and the analysis of error curves during the training and cross-validation sets [74]. Analyzing when the error curve started to increase has provided useful information on when to stop the training in order to prevent overfitting [74]. In addition, in this implementation stage we need to split randomly the total sample into training and testing sets. Sometimes there is a valid concern regarding the number of cases available for training purposes, but case generation methods and alternative means of training (i.e., leave one-out cross-validation) are available, and applicable to events or problems in which large data sets area not feasible. One suggested criterion for determining the adequacy of the training data is that the number of training inputs should be at least 5–10 times the number of input and middle layer neurons in the network [76]. It is also recommended that approximately 2/3 or 3/4 of the available data set be used for the training phase, to ensure that the network includes a representative set of patterns (as represented by the vector for each case). The remaining 1/3 or 1/4 of the data should be reserved for the testing phase of the network.

(6) *Evaluation of the models.* In the previous stage several models are obtained, so we need to evaluate them and select the most successful one, using different performance measures derived from the confusion matrix of the classification [77]. This matrix allows to calculate the rates for True Positive (TP), True Negative (TN), False Positive (FP), and False Negative (FN) outcomes, as products of the ANN analysis. In addition, certain summative evaluative algorithms have been developed in this field to assess overall quality of the predictive system. These overall measures are: "Recall, which represents the proportion of correctly identified targets, out of all targets presented in the set, and is represented as: Recall $=$ TP/(TP $+$ FN),and Precision, which represents the proportion of correctly identified targets, out of all identified targets by the system, and is represented as: Precision $=$ TP/(TP $+$ FP)." [11], p. 304). The characteristics of the system's detection sensitivity have been reported using two measures. Sensitivity, which is similar to Recall, represents the proportion of correctly identified

targets out of all targets presented in the set, and is expressed as Sensitivity = TP/(TP + FN). The other measure is Specificity, which is the proportion of correctly rejected targets from all the targets that should have been rejected by the system, and is expressed as Specificity = TN/(TN + FP). In addition, the performance of the ANN is evaluated using a summative measure that accounts for the complementary relationship between Precision and Recall. This measure, known as F1, is defined as F1 = (2*Precision*Recall)/(Precision + Recall) and assumes equal weights for Precision and Recall. However, this assumption can be modified to favor either Precision or Recall depending on the utility and cost/benefit ratio of outcomes favoring either Precision or Recall for any given predictive circumstance. Other algorithms of machine learning and classical predictive techniques (i.e., logistic regression, discriminant analysis, etc.) were used in our studies in order to compare their performance measures against those obtained by the ANN.

(7) *Analysis of predictive patterns.* In order to understand the contribution of the predictors for the classification output, a useful method for sensitivity analysis has been used in our studies. This approach provides not only a variable selection method but also information to interpret a model addressing the "black box" problem [78, 79]. This method, which is based on the relative importance, improves on Garson's [76] use of relative importance weights and incorporates sensitivity analysis to determine the causal importance of input variables on outputs. Sensitivity is a measure of the increase in the error of the predicted value when each variable is excluded from the model, and it systematically demonstrates the degree of influence on the network weights of each participating variable. Another approach used in our studies was to group the input variables into more general categories based on theoretical models, with the importance of each category being calculated by adding the relative importance of all the variables within that category. Finally, a comparison across the different performance level groups allowed us to analyze the commonalities and differences between the predictors for each level, and even to compare across the different educational outcomes (i.e., GPA, drop-out, degree completion, different domains).

5 Results

Predicting and understanding key educational outcomes in higher education

Findings of our predictive classification studies have found a high level of accuracy (90 to 100%) for all the predictive classifications of different levels in mathematics, general academic performance, retention, and degree completion in higher education [19, 71–74]. The results indicated that learning strategies had the most significant contribution in predicting GPA, with coping strategies being the top predictors for the prediction of degree completion. Background information was found to be the most important factor in identifying students who are likely to drop out of their university programs [70, 71].

Artificial neural networks applied in these studies were more robust and performed significantly better than other classical techniques. Regarding mathematics performance, we found specific patterns of predictors for different levels of achievement [73]. Among the student groups with the lowest 30% of math performance, the main determinants of performance appear to be basic cognitive processing variables, which indicate the degree to which they represent the areas of relative weakness in the group and they showed to be more discriminating from the rest of the students. On the other hand, among the student groups with the highest 30% of math performance, the main determinants of performance appear to be self-regulation and background variables (particularly, how interested students were in the task and some social indicators such as parents' occupation). The prediction pertaining to the middle 30% level of math performance among students demonstrates a discernible pattern that incorporates learning strategies, self-efficacy as critical motivational beliefs, and attentional resources as significant predictors. Furthermore, it appears that working memory does not enhance the prediction of performance for this particular group of students, suggesting that their mathematical performance is more strongly influenced by self-regulated learning processes, such as learning strategies, motivational beliefs, and attention, which are more closely associated with environmental, instructional, and training constructs rather than basic cognitive processes like working memory [73].

We have also found consistent results predicting general academic performance (Grade Point Average) in higher education in other previous studies [19, 72]. Among the student group with the lowest 33% of GPA, the two main predictors were learning strategies components, specifically cognitive resources/cognitive processing and time management. When considering the relative importance of each variable, it is apparent that working memory and other cognitive resources play a significantly more significant role in identifying lower GPA students compared to high-performance groups [19, 72].

Recently, we have applied several machine learning techniques to classify students belonging to low satisfaction with online university instruction during the COVID pandemic [81]. This study has collected data from 560 university students about their working memory capacity, attentional networks, self-regulation strategies, coping strategies, and socioeconomic status. Deep learning models and Gradient boosted trees had a very high accuracy (88 to 90%) in the identification of the student group with low satisfaction. Both types of models have found similar patterns of predictors: the two most important predictors were self-efficacy-expectations regarding the online modality (self-regulation strategies), and the social dimension of the online experience (for example: feeling part of the group, the possibility of online peer interaction, etc.). Both models have shown that the number of rooms in the home, father's educational level, working memory, and attentional alertness were among the first 10 factors that contributed to the classification.

Predicting and understanding the students' attributions in secondary school

We have developed an ANN model predicting causal attribution of writing performance in secondary school with data from 1130 vocational high-school students of both genders [27]. ANN were utilized to provide insights regarding the multiple interactions between the measured domain-specific variables, situation-specific appraisals and emotions, trace data, and background variables. ANN were able to accurately identify (with precision ranging from 87.5 to 100%) students who attributed their success or failure to their writing skills, affect, and self-regulation strategies on the basis of domain variables, appraisals, emotions, and performance indicators. The ANN model also detected significant differences in the factors that appear to underlie the students' causal attributions [27]. On the one hand, "students who can rely on stable, domain-dependent motivation beliefs are inclined to consider themselves competent to start the learning activity and are willing to invest effort" ([27], p. 18). On the other hand, students who anticipate a poor result have lower metacognition but their appraisals and emotions are different depending on specific attributions (deficit of writing skills or affective states) [27].

Predicting and understanding key educational outcomes in primary school

Our previous studies also involve the application of algorithms of machine learning to predict mathematics and language performance in primary school, using large datasets [70, 71]. We worked with a sample of 1,138 Vietnamese students from the Young Lives Studies survey (www.younglives.org.uk) including 194 variables which had data from birth to 15 years of age (individual physical and cognitive variables, socio-economic factors and environmental variables). The aforementioned models attained exceptionally high accuracy rates ranging from 95 to 100%. A consistent pattern has been established for both Math and Vietnamese language, applicable to both high and low levels of performance. The individual cognitive characteristics, physical factors, and daily routines/activities of the child serve as critical predictive factors of academic performance, as reflected in the student's final grade 5 math test and grade 5 Vietnamese language test results, respectively. Furthermore, parental expectations, pre-school attendance, and the school trajectory of the students also plays a notable role in the classification.

Predicting and understanding the foundation of key cognitive processes for learning

Currently, we have developed highly accurate predictive models for the gains in working memory and fluid intelligence that result from executive attention training in children, utilizing genetic markers, gender, and age variables [81]. ANN models reached high overall accuracy in their predictive classification. Furthermore, variations in genes associated with dopamine and norepinephrine neurotransmission have been found to impact children's susceptibility to benefit from executive attention training, a pattern that is consistent with previous research [81].

Incorporating both attentional and environmental variables as input in one recent study [82], ANN models were able to identify 100% of the children belonging to both low and moderate/high self-regulation (SR) groups. These more inclusive models achieved higher sensitivity and specificity compared to ANN models that involved either attentional or environmental predictors. Models using only attentional inputs achieved good sensitivity and correctly classified 75% of low SR children. However, models including only environmental factors were unable to correctly classify both groups of children simultaneously, achieving relatively low accuracy values for both groups. Father's education and correct anticipations were the top two predictors with the most significant importance in classifying between low SR and moderate/high SR. The inclusive model (attentional + environmental factors) was able to correctly identify both groups, determining an interaction among attentional and other socio-economic variables such as the education of the mother, SES, father's and mother's occupation, and complex correct anticipations. These predictors contributed more than 60% of the predictive weight of the variables for reaching a correct predictive classification. However, it is important to observe that all variables contribute to the prediction in relatively small proportions, and it is the joint effect of many contributing variables that influences SR development.

In recent studies, we are using this type of machine learning approach to predict different risk outcomes for the study of the development of attention and autism spectrum problems, at 36 months-old, from very early profiles of cognitive, non-cognitive, and environmental factors. Preliminary results are promising about the detection of these children who might be at risk, as early as 6 to 9 months of age [81].

6 Discussion

The advantages of the application of machine learning -artificial neural networks among them- in the educational field have been widely demonstrated in numerous studies over the last 10 years. The complexity of the non-linear relationships among multiple factors at different micro and macro-levels can be captured by this type of methodology, so it outperforms classical statistical approaches. All our studies have shown these advantages across several educational outcomes from pre-school to higher education. However, there is still a long way to go in relation to efforts of implementation of machine learning outcomes for decision-making in educational policy and organizations.

A major observation derived from our findings is that variables have a relatively minor impact on predictions, and it is the collective effect of multiple contributing variables that produces significant changes in performance. The educational outcome for each case is determined by a comprehensive set of variables. In essence, there is no single factor or "magic bullet" that solely contributes to the outcome, but rather the accumulation of effects from numerous sources that leads to significant improvements in outcomes [83]. Therefore, we establish that classical studies, which only account for a limited number of variables in simple designs and do not account for complex interactions among numerous factors, can and frequently produce incongruous results. This is not an insignificant issue for the understanding of various effects and phenomena in every scientific discipline [35] As corollary, many of the apparently contradictory results obtained by simpler traditional designs have to do with the uncontrolled effect of full vector information. Another issue refers to the directionality of the relationship between predictors and the educational outcome. For a specific student, a variable (e.g., socioeconomic status) could negatively affect his/her performance while the same variable could have a positive impact on the performance for another student, because its effect depends on the full vector in which this variable is immersed [83].

In general, our results suggest that the application of machine learning methods, such as ANN, would help to design more targeted interventions at individual and/ or group level. Moreover, a validated result from an ANN that has been properly constructed and tested could provide valuable diagnostic information for early-warning systems tailored to the specific needs. Specifically, the ANN developed for early prediction during a child's development suggests that environmental factors play a crucial role in determining learning outcomes, which are outside the scope of traditional educational interventions. While economists typically focus on school-level decisions, such as those involving teachers and principals, their scope is limited if they do not consider the child, family, and social milieu as important levels of analysis. Our ANN developed with the Vietnam data suggests that resources should be allocated in four priority areas: (a) promoting healthy cognitive and physical development of children from very early in life, (b) stimulating positive parental attitudes toward education and expectations/interest regarding their children's educational trajectories, (c) identifying children and families at risk using indicators such as health at birth, and d) selecting and training teachers to maximize those teacher characteristics that better predict high student performance, such as teacher attitudes to promote a positive school environment for learning.

References

1. Al-Deek HM (2001) Which method is better for developing freight planning models at seaports Neural networks or multiple regression? Transp Res Rec 1763:90–97
2. Detienne KB, Detienne DH, Joshi SA (2003) Neural networks as statistical tools for business researchers. Organ Res Methods 6:236–265
3. Laguna M, Martí R (2002) Scatter search methodology and implementations in C. Kluwer
4. Neal W, Wurst J (2001) Advances in market segmentation. Market Res 13:14–18
5. Nguyen N, Cripps A (2001) Predicting housing value: a comparison of multiple regression and artificial neural networks. J Real Estate Res 22:313–336
6. White H, Racine J (2001) Statistical inference, the bootstrap, and neural network modelling with application to foreign exchange rates. IEEE Trans Neural Netw 12:657–673
7. Everson HT, Chance D, Lykins S (1994) Exploring the use of artificial neural networks in educational research. Paper presented at the Annual meeting of the American Educational Research Association, New Orleans, Louisiana.
8. Hardgrave BC, Wilson RL, Walstrom KA (1994) Predicting graduate student success: a comparison of neural networks and traditional techniques. Comp Operat Res 21(3):249–263
9. Perkins K, Gupta L, Tamanna (1995) Predict item difficulty in a reading comprehension test with an artificial neural network. Lang Test 12:34–53

10. Wilson RL, Hardgrave BC (1995) Predicting graduate student success in an MBA program: regression versus classification. Educ Psychol Measur 55(2):186–195
11. Cascallar EC, Boekaerts M, Costigan TE (2006) Assessment in the evaluation of self- regulation as a process. Educ Psychol Rev 18:297–306
12. Romero C, Ventura S (2020) Educational data mining and learning analytics: an updated survey. WIREs Data Min Knowl Discovery. https://doi.org/10.1002/widm.1355
13. Bakhshinategh B, Zaiane OR, ElAtia S et al (2018) Educational data mining applications and tasks: a survey of the last 10 years. Educ Inf Technol 23:537–553. https://doi.org/10.1007/s10639-017-9616-z
14. Lang C, Siemens G, Wise A, Gasevic D (2017) Handbook of learning analytics. In: SOLAR, society for learning analytics and research. SOLAR, New York, NY
15. Shafiq DA, Marjani M, Habeeb RAA, Asirvatham D (2022) Student retention using educational data mining and predictive analytics: a systematic literature review. IEEE Access 10:72480–72503. https://doi.org/10.1109/ACCESS.2022.3188767
16. Lee C, Rey T, Mentele J, Garver M (2005) Structured neural network techniques for modeling loyalty and profitability. In: Proceedings of the thirtieth annual SAS® users group international conference. SAS Institute Inc., Cary, NC
17. Hartmann S, Frigg R (2006) Models in science. In: Zalta EN (ed) The Stanford encyclopaedia of philosophy. Summer 2006 Edition. Stanford University Press, Stanford, CA
18. Suppes P (1962) Models of data. In: Nagel E, Suppes P, Tarski A (eds) Logic, methodology and philosophy of science: proceedings of the 1960 International Congress. Stanford University Press, Stanford, pp 252–261
19. Cascallar E, Musso MF, Kyndt E, Dochy F (2015) Modelling for understanding and for prediction/classification the power of neural networks in research. Frontline Learn Res 2(5):67–81. https://doi.org/10.14786/flr.v2i5.135
20. Bruckstein AM, Holt RJ, Netravali AN (1998) Holographic representations of images. IEEE Trans Image Process 7(11):1583–1597. https://doi.org/10.1109/83.725365
21. Susskind L (1997) Black holes and the information paradox. Sci Am 276(4):52–57
22. Bekenstein JD (2003) Information in the holographic universe. Sci Am *289*(2):58–65
23. Garson GD (2014) Neural network models. Statistical Associates Publishers
24. Haykin SS (2009) Neural networks and learning machines, vol. 3. Pearson, Upper Saddle River, NJ, USA
25. Fausett LV (1994) Fundamentals of neural networks. Prentice-Hall
26. Boekaerts M (1997) Self-regulated learning: a new concept embraced by researchers, policy makers, educators, teachers, and students. Learn Instr 7(2):161–186. https://doi.org/10.1016/s0959-4752(96)00015-1
27. Boekaerts M, Musso MF, Cascallar EC (2022) Predicting attribution of letter writing performance in secondary school: a machine learning approach. Front Educ 7:1007803. https://doi.org/10.3389/feduc.2022.1007803
28. Musso MF, Boekaerts M, Segers M, Cascallar EC (2019) Individual differences in basic cognitive processes and self-regulated learning: their interaction effects on math performance. Learn Individ Differ 71:58–70. https://doi.org/10.1016/j.lindif.2019.03.003
29. Hand D, Mannila H, Smyth P (2001) Principles of data mining. The MIT Press, Cambridge, MA
30. Astin AW, Oseguera L (2012) PreCollege and institutional characteristics on degree attainment
31. Tross SA, Harper JP, Osher LW, Kneidinger LM (2000) Not the usual cast of characteristics: using personality to predict college performance and retention. J Coll Stud Dev 4(3):323–334
32. McKenzie K, Schweitzer R (2001) Who succeeds at university? Factors predicting academic performance in first year Australian university students. High Educ Res Dev 20(1):21–33. https://doi.org/10.1080/07924360120043621
33. Richardson M, Abraham C, Bond R (2012) Psychological correlates of university students' academic performance: a systematic review and meta-analysis. Psychol Bull 138(2):353–387. https://doi.org/10.1037/a0026838

34. Berger A, Kofman O, Livneh U, Henik A (2007) Multidisciplinary perspectives on attention and the development of self-regulation. Prog Neurobiol 82(5):256–286. https://doi.org/10.1016/j.pneurobio.2007.06.004
35. Boekaerts M, Corno L (2005) Self-regulation in the classroom: a perspective on assessment and intervention. Appl Psychol 54(2):199–231. https://doi.org/10.1111/j.1464-0597.2005.00205.x
36. Boekaerts M, Niemivirta M (2000) Self-regulated learning: finding a balance between learning goals and ego-protective goals. In: Boekaerts M, Pintrich PR, Zeidner M (eds) Handbook of self-regulation. Academic Press, San Diego, CA, pp 417–451. https://doi.org/10.1016/b978-012109890-2/50042-1
37. Crombach MJ, Boekaerts M, Voeten MJM (2003) Online measurement of appraisals of students faced with curricular tasks. Educ Psychol Meas 63:96–111. https://doi.org/10.1177/0013164402239319
38. Seegers G, Boekaerts M (1996) Gender-related differences in self-referenced cognitions in relation to mathematics. J Res Math Educ 27(2):215. https://doi.org/10.2307/749601
39. Boekaerts M (2006) Self-regulation and effort investment. In Sigel E, Renninger KA (eds) Handbook of child psychology, vol 4. Wiley, New York, pp 345–377. https://doi.org/10.1002/9780470147658.chpsy0409
40. Lazarus RS, Folkman S (1986) Cognitive theories of stress and the issue of circularity. In Trumbull R, Appley MH (eds) Dynamics of stress. The Plenum series on stress and coping. Springer, Boston
41. Fisher S, Hood B (1987) The stress of transition to university: a longitudinal study of psychological disturbance, absent-mindedness and vulnerability to homesickness. Br J Psychol 78:425–441
42. Scott DM, Spielmans GI, Julka DC (2004) Predictors of academic achievement and retention among college freshmen: a longitudinal study. Coll Stud J 38(1):66–80
43. Weinstein CE, Palmer DR (2002) LASSI: user's manual, 2nd edn. H&H Publishing Company Inc., Clearwater, FL
44. Weinstein CE, Schulte AC, Cascallar EC (1982) The learning and studies strategies inventory (LASSI): initial design and development. Technical Report, US Army Research Institute for the Social and Behavioural Sciences, Alexandria, VA
45. Bull R, Espy KA (2006) Working memory, executive functioning, and children's mathematics. In: Pickering SJ (ed) Working memory and education. Elsevier, San Diego, USA, pp 93–123
46. Lee H, Grosse R, Ranganath R, Ng AY (2009) Convolutional deep belief networks for scalable unsupervised learning of hierarchical representations. In: Proceedings of the 26th annual international conference on machine learning, Montreal, June 14–18. ACM, New York, pp 609–616
47. Conway ARA, Kane MJ, Bunting MF, Hambrick DZ, Wilhelm O, Engle RW (2005) Working memory span tasks: a methodological review and user's guide. Psychon Bull Rev 12:769–786. https://doi.org/10.3758/BF03196772
48. Colom R, Escorial S, Shih PC, Privado J (2007) Fluid intelligence, memory span, and temperament difficulties predict academic performance of young adolescents. Pers Individ Differ 42(8):1503–1514. https://doi.org/10.1016/j.paid.2006.10.023
49. Conway ARA, Cowan N, Bunting MF, Therriault DJ, Minkoff SRB (2002) A latent variable analysis of working memory capacity, short-term memory capacity, processing speed, and general fluid intelligence. Intelligence 30:163–183. https://doi.org/10.1016/S0160-2896(01)00096-4
50. Engle RW, Kane MJ (2004) Executive attention, working memory capacity, and a two-factor theory of cognitive control. Psychol Learn Motiv Adv Res Theory 145–199. https://doi.org/10.1016/S0079-7421(03)44005-X
51. Unsworth N, Redick TS, Heitz RP, Broadway J, Engle RW (2009) Complex working memory span tasks and higher-order cognition: a latent variable analysis of the relationship between processing and storage. Memory 17:635–654

52. Dunlosky J, Kane MJ (2007) The contributions of strategy use to working memory span: a comparison of strategy assessment methods. Quart J Exp Psychol 60(9):1227–1245. https://doi.org/10.1080/17470210600926075
53. Dunlosky J, Thiede KW (2004) Causes and constraints of the shift-to-easier-materials effect in the control of study. Mem Cognit 32:779–788
54. Dunning DL, Holmes J (2014) Does working memory training promote the use of strategies on untrained working memory tasks? Mem Cognit 42(6):854–862. https://doi.org/10.3758/s13421-014-0410-5
55. Fan J, McCandliss BD, Sommer T, Raz A, Posner MI (2002) Testing the efficiency and independence of attentional networks. J Cogn Neurosci 14(3):340–347
56. Posner MI, Petersen E (1989) The attention system of the human brain. ONR technical report #89–1. Oregon
57. Checa P, Rueda MR (2011) Behavioral and brain measures of executive attention and school competence in late childhood. Dev Neuropsychol 36(8):1018–1032. https://doi.org/10.1080/87565641.2011.591857
58. Kane MJ, Conway ARA, Hambrick DZ, Engle RW (2008) Variation in working memory capacity as variation in executive attention and control. In: Conway ARA, Jarrold C, Kane MJ, Miyake A, Towse JN (eds) Variation in working memory. Oxford, NY, pp 21–48
59. Posner MI, Rothbart MK, Sheese BE, Voelker P (2014) Developing attention: behavioral and brain mechanisms. Adv Neurosci 2014:1–9. https://doi.org/10.1155/2014/405094
60. Jun J (2005) Understanding dropout of adult learners in e-learning
61. Kovacic JC (2010) Early prediction of student success: mining students enrolment data. 647–665. https://doi.org/10.28945/1281
62. Brandes-Aitken A, Braren S, Swingler M, Voegtline K, Blair C (2019) Sustained attention in infancy: a foundation for the development of multiple aspects of self-regulation for children in poverty. J Exp Child Psychol 184:192–209. https://doi.org/10.1016/j.jecp.2019.04.006
63. Conejero Á, Rueda MR (2018) Infant temperament and family socio-economic status in relation to the emergence of attention regulation. Sci Rep 8(1):11232. https://doi.org/10.1038/s41598-018-28831-x
64. Lipina SJ, Evers K (2017) Neuroscience of childhood poverty: evidence of impacts and mechanisms as vehicles of dialog with ethics. Front Psychol 8(January):1–13. https://doi.org/10.3389/fpsyg.2017.00061
65. Musso MF, Richaud MC, Cascallar EC (2015) Self-regulation and executive functions: understanding learning and school performance. In: Huaire Inacio EJ, Elgier AM, Maldonado Paz G (eds) Cognitive psychology and learning processes. Universidad Nacional de Educación Enrique Guzmán y Valle Lima, Perú, pp 25–47
66. Ahmad F, Ismail NH, Aziz AA (2015) The prediction of students' academic performance using classification data mining techniques. Appl Math Sci 9(129), 6415–6426. https://doi.org/10.12988/ams.2015.53289
67. Huang S (2011) Predictive modeling and analysis of student academic performance in an engineering dynamics course, All Grad. Theses Diss.
68. Pittman K (2008) Comparison of data mining techniques used to predict student retention. ProQuest Diss Publ 3297573
69. Alyahyan E, Düştegör D (2020) Predicting academic success in higher education: literature review and best practices. Int J Educ Technol High Educ 17:1–21
70. Musso MF, Cascallar EC, Bostani N, Crawford M (2020a) Identifying reliable predictors of educational outcomes through machine-learning predictive modeling. Front Educ 5. https://doi.org/10.3389/feduc.2020.00104.Article104
71. Musso MF, Hernández CFR, Cascallar EC (2020b) Predicting key educational outcomes in academic trajectories: a machine-learning approach. High Educ 80:875–894. https://doi.org/10.1007/s10734-020-00520-7
72. Musso MF, Kyndt E, Cascallar EC, Dochy F (2013) Predicting general academic performance and identifying the differential contribution of participating variables using artificial neural networks. Frontline Learn Res 1(1):42–71. https://doi.org/10.14786/flr.v1i1.13

73. Musso M, Kyndt E, Cascallar E, Dochy F (2012) Predicting mathematical performance: the effects of cognitive processes and self-regulation factors. Educ Res Int 2012:1–13. https://doi.org/10.1155/2012/250719
74. Rodriguez-Hernandez CF, Musso MF, Kyndt E, Cascallar EC (2021) Artificial neural networks in academic performance prediction: systematic implementation and predictor evaluation. Comp Educ Artif Intell 2:100018. https://doi.org/10.1016/j.caeai.2021.100018
75. Ruano MV, Ribes J, Sin G, Seco A, Ferrer J (2010) A systematic approach for fine-tuning of fuzzy controllers applied to WWTPs. Environ Model Softw 25(5):670–676
76. Garson GD (1998) Neural networks. An introductory guide for social scientists. Sage Publications Ltd., London
77. Saito T, Rehmsmeier M (2015) The precision-recall plot is more informative than the ROC plot when evaluating binary classifiers on imbalanced datasets. PLoS ONE 10(3):e0118432. https://doi.org/10.1371/journal.pone.0118432
78. Kewley R, Embrechts M, Breneman C (2000) Data strip mining for the virtual design of pharmaceuticals with neural networks. IEEE Trans Neural Netw 11(3):668–679. https://doi.org/10.1109/72.846738
79. Yeh IC, Cheng WL (2010) First and second order sensitivity analysis of MLP. Neurocomputing 73(10–12):2225–2233. https://doi.org/10.1016/j.neucom.2010.01.011
80. Kyndt E, Musso M, Cascallar E, Dochy F (2015) Predicting academic performance: the role of cognition, motivation and learning approaches. A neural network analysis. In: Donche V, De Maeyer S, Gijbels D, van den Bergh H (eds) Methodological challenges in research on student learning. Garant, Antwerp, Belgium
81. Musso MF, Cómbita LM, Cascallar EC, Rueda MR (2022) Modeling the contribution of genetic variation to cognitive gains following training with a machine learning approach. Mind Brain Educ 16:300–317. https://doi.org/10.1111/mbe.12336
82. Musso MF, Moyano S, Rico-Picó J, Conejero Á, Ballesteros-Duperón MÁ, Cascallar EC, Rueda MR (2023) Predicting effortful control at 3 years of age from measures of attention and home environment in infancy: a machine learning approach. Children 10:982. https://doi.org/10.3390/children10060982
83. Cascallar EC (2022, September) Personal communication
84. Seidman (Ed.). College student retention. Rowman and Littlefield Publishers, Inc., Lanham, MD

Text Identification for Questions Generation According to Bloom's Taxonomy Using Natural Language Processing

Pallavi Asthana, Sumita Mishra, and Bramah Hazela

Abstract Adaptivity is actively pursued for testing in education as it provides alternative paths of action for the students. Particularly in engineering education adaptive assessments provide an estimation of students' competence to provide novel solutions and accurate evaluation of knowledge. This chapter discusses the question identification based on Bloom's taxonomy for the design of Computerized adaptive assessment. Indicators selected for classification are pre-determined verbs that represent the Bloom's stages indicating the level of course expertise. Questions are labelled as per Blooms taxonomy. Natural Language Processing is a Machine learning based method to classify the text, as it easily converts the text into tokenized format that can be used for classification. Classification models are developed using the Bidirectional Long Short-Term Memory networks and Convolutional Neural Network. Deep Learning toolbox of MATLAB 2023a is used to train and test the model. The CNN model has provided a validation accuracy of 85.71% for the labeled questions.

Keywords Machine learning · Deep learning · Natural processing language · BiLSTM · Recurrent neural network

1 Introduction

Adaptive assessments are designed in a manner where questions are not randomly generated but instead follow a meaningful pattern. It is based on certain parameters that are known as concepts. Concept maps (CM) have been employed to organize the relationship between concepts that simplify the adaptive assessment process. Bloom's taxonomy provides a proper morphological way of creating questions based on learning stages ([22], p. 527). Bloom's taxonomy has been accepted to map the learning stages with different learning outcomes, as question complexity is mapped with learning stages. Bloom's taxonomies are framed in a structured manner where

P. Asthana (✉) · S. Mishra · B. Hazela
Amity School of Engineering and Technology, Amity University, Lucknow, India
e-mail: pasthana@lko.amity.edu

M. S. Khine (ed.), *Machine Learning in Educational Sciences*,
https://doi.org/10.1007/978-981-99-9379-6_16

verbs clearly define Bloom's stage, so they help generate automated concept maps faster. They can also identify the cognitive level of the students ([16], p. 1872).

Concept maps are automatically generated through the process known as Concept Map Mining (CMM) ([15], p. 248). CMM is used in the generation of adaptive assessments where words are extracted from the questions and then they are mapped with the stages of Bloom's taxonomy [21].

Natural Language Processing (NLP) is the widely Machine learning based tool used in the text extraction in CMM [8]. It accomplishes the task by breaking large blocks of text into smaller ones creating words, phrases, and sentences through tokenization. The process aids in comprehending the NLP model by evaluating the series of sentences to read the context of language. Tokenization can be done using different techniques and databases using libraries such as NLTK, Genism, and Keras. The use of lookup tables for Bloom's taxonomy is also suggested due to the better accuracy in the data retrieval. NLP utilizes computational linguists to identify the structure of the sentences and model knowledge and reasoning to accomplish a task [19]. To complete these tasks, it has algorithms, data structures, formal models of representation, and reason. Mining of CM is strongly supported by methods present in NLP.

Using NLP to identify Bloom's stages and classify the questions to create meaningful relations between concepts based on learning stages would help measure the achieved outcomes directly ([25], p. 12709). The method is even more helpful in engineering education, where a hierarchical knowledge structure exists. The method will provide a real-time meta-cognitive tool to assess students' learning. In this chapter, we propose to develop the Computerized adaptive assessment where questions generation is based on Bloom's taxonomy.

2 Automatic Generation of Concept Map

Adaptivity in assessment describes modification in questions in response to students' scores, and it needs to comply with specific rules that correspond to various parameters ([11], p. 297). Computerized adaptive Tests (CATs) are already present that change the difficulty of questions based on students' responses. Adaptive assessment systems are getting popular as they can challenge the students' learning and test the epistemological growth of students ([17], p. 256).

Adaptivity in the assessment system must be defined unambiguously and should be mapped with the desired outcomes. Adaptivity is always based on specific parameters to determine the path of the following questions [23]. Concept maps create the path of the following questions, and they are trained through Natural language processing to identify the keywords from different courses and fetch the best matched question. The primary task is creating a database where questions will be developed based on Bloom's taxonomy for defining a path for the adaptivity of the questions ([2], p. 41).

Each question is converted to suit the requirement of key words used in Bloom's taxonomy. Concept Map Mining (CMM) has achieved some promising results, where concept extraction from text is done. Moreover, based on the text extraction, automatic CM is generated using the required tools [7]. Here, we propose the method to create a concept map with five stages:

(1) Preprocessing: In this stage, resources are utilized for the text extraction process.
(2) Concepts extraction: Extraction of potential candidates for the CM based on rule-based extraction. Since the data is structured text customized for the CMM process, rule-based extraction will be a suitable technique for information extraction from the question.
(3) Creation of propositions: to search the relationship between concepts to be linked.
(4) Ranking: To discover and rank the most relevant information based on concepts.
(5) Visualization: Concept maps can be visualized using any tool like Cmap tools by translating them into concept map extensible language.

3 Natural Language Processing (NLP) for Automated Concept Map Generation

Bloom's taxonomy provides a proper morphological way of creating questions based on learning stages ([13], p. 106). Bloom's taxonomy is framed in a structured manner where verbs clearly define Bloom's stage, so they help in generation of automated concept maps faster. They can also identify the cognitive level of the students. Table 1 shows the cognitive level of Bloom's taxonomy and verbs associated with each level. For labelling the questions, it is necessary that verb is defined in the question. Bloom's has taxonomy has total of six stages namely remembering, understanding, applying, analysing, evaluation and creation. In the designing of adaptive assessment, we have considered the five stages as creativity cannot be easily assessed using Multiple choice questions ([22], p. 530).

Table 1 Bloom's taxonomy

Bloom's level	Expected learning	Verbs to define bloom's level
Remembering	Recall associated with memory	List, name, state, define, describe, identify, outline, recall, reorganize
Understanding	Understanding for explanation	Identify, explain, describe, generalize, paraphrase, restate, summarize, discover
Applying	Use of knowledge or information	Apply, use, solve, compute, demonstrate, express, investigate, predict, discover
Analysing	Establishing connection between concepts	Analyze, compare, classify, characterize, correlate, dissect, maximize, synthesize, relate
Evaluating	Formation of judgement or critique	Evaluate, appraise, conclude, determine, support, validate, interpret

NLP accomplishes the task by breaking large blocks of text into smaller ones creating words, phrases, and sentences through tokenization. The process aids in comprehending the NLP model by evaluating the series of sentences to read the context of language. Tokenization can be done using different techniques and databases using libraries such as NLTK, Genism, and Keras. The use of lookup tables for Bloom's taxonomy is also suggested due to the better accuracy in the data retrieval.

Using NLP to identify Bloom's stages and classify the questions to create meaningful relations between concepts based on learning stages would help measure the achieved outcomes directly. The method is even more helpful in engineering education, where a hierarchical knowledge structure exists ([12], p. 10). The method will provide a real-time meta-cognitive tool to assess students learning by enhancing the difficulty of questions ([5], p. 731).

Table 1 shows the verbs associated with different Bloom's stages to represent the knowledge of students.

As shown in Fig. 1, Natural language processing has the following steps:

(1) Input: Reading of the raw data file from a source. This file must be composed of the sub-topics that need to be extracted.
(2) Segmentation of Text and Syntactic Analysis: The question is retrieved from the database and divided into text consisting of each word in the questions.
(3) Syntactic analysis: Analysis of the phrases and words of the question to create a dictionary-like semantics from the text.
(4) Tokenization and Morphological analysis: These two processes are performed simultaneously. First, the text is divided into tokens, and each part of speech

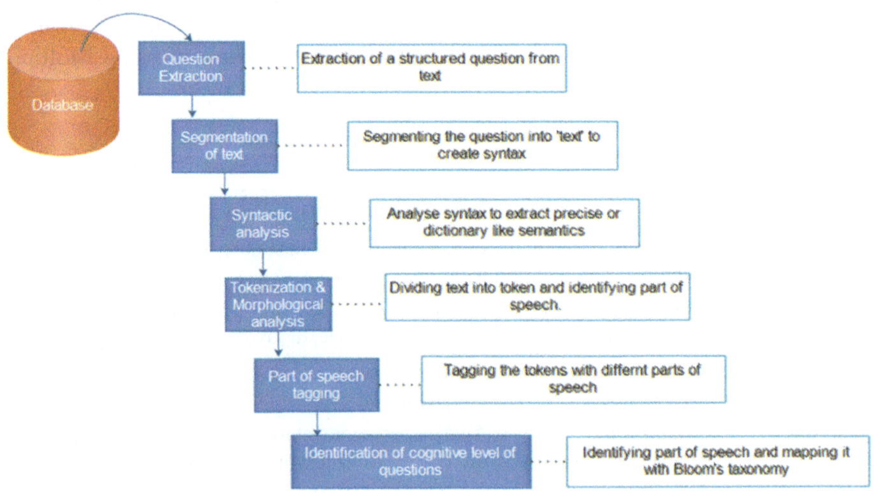

Fig. 1 Steps in natural language processing

is identified. Tokenization is a crucial process in NLP because these tokens are tagged for pattern recognition

(5) Part of speech (PoS) tagging: Each token is tagged as part of speech that contains the tag as a Noun (NN), Plural Noun (NNS), singular proper Noun (NNPS), Adjective (JJ), Cardinal number (CD), or verb (VB). Many tags are used in different applications, but for the adaptive assessment, questions are structured so that identification must have minimum ambiguity. Words that have no information are called stop words. These words are removed during this phase, and with deeper linguistic knowledge techniques, each token is associated with the tags using PoS tagging.

4 Experiment and Methodology

Nowadays, a tremendous number of Multiple choice-based questions are available online, but these questions cannot be used in the present form, questions need to be specifically designed or remodeled to suit the semantics of the automated generation with adaptivity ([9], p. 1355).

4.1 Framing of Questions Based on Bloom's Taxonomy

The design of the data bank is crucial for an adaptive assessment system. Questions are retrieved from the database by defining a path. Questions designed by course experts to represent Bloom's stages are defined in Table 1. Each question must have a '*verb*' to determine Bloom's level. Also, the trajectory of questions is created for a single '*noun*' to test the concept. Each question is specified for a single verb defining Bloom's level and one noun to be assessed. Questions are Multiple choice based and have one correct answer.

4.2 Mapping of Course Learning Outcomes with Bloom's Levels

In this work, adaptive assessment is designed to calculate the learning outcomes of a trajectory of courses that forms the hierarchy of courses. It is considered that knowledge of pre-requisite courses is essential for the better understanding of course, so if adaptive assessment is designed in such a way that it can evaluate the knowledge of the three courses simultaneously, then the students' performance can become a measure of Course Learning Outcomes (CLOs) as shown in Table 2. CLOs are the expected learning that must occur after the completion of a course.

Table 2 Mapping of the selected courses with the course learning outcomes and bloom's taxonomy

Course learning outcomes	Courses	Stages of bloom's taxonomy	Level of taxonomy
CLO1	MMI (PR1), TAA (PR2)	Memorization	Level 1
CLO2	TAA (PR2)	Understanding	Level 2
CLO3	PPI (C1)	Applying	Level 3
CLO4	PPI (C1)	Analyzing	Level 4
CLO5	PPI (C1)	Evaluation	Level 5

For this work, we have sample questions from three courses taught in the undergraduate engineering program of Electrical and Electronics Engineering (EEE) at Amity university, Lucknow Campus. For designing the questions, it is essential to follow the sequence of courses where a lower level of Bloom's taxonomy (Memorization and understanding) is mapped with the pre-requisite courses and upper levels (Applying, analyzing, and evaluating) are mapped with the current course, not necessarily in the same order for assessment. The illustration trajectory of the courses is Power Plant Instrumentation (PPI, C1), Transducers and Applications (TAA, PR1), and Measurement and measuring instruments (MMI, PR2). ([3], p. 3).

4.3 Significance of Course Expert for Framing Questions

It has been widely recognized that during the design of classification model for adaptive assessment, unstructured text may lead to ambiguity. Therefore, the experts will design or restructure questions for suitability and easy retrieval. Here, we have shown an example where questions have been replaced with suitable verbs and nouns for the proper tokenization. These questions will be used to generate automated concept maps.

The first step in NLP is extraction of structured text from the questions. The questions need to be labelled for proper identification. The questions do not contain the keywords needed in Bloom's taxonomy, so this question is restructured as shown in the table. Analyze is the verb mapped with the *Level 4* of Bloom's taxonomy and Temperature is the noun.

In this system, information extraction is done to classify the questions based on Bloom's stages. The system must be able to classify questions based on the keywords that have been defined to specify the stages. Classification is suitably done with semantic and syntactic approach using WordNet and Stanford Parser with Universal Dependency. Questions are exclusively labelled according to the verb in Bloom's Taxonomy, in the supervised learning, this is multiclass classification where classes are defined as the Levels of Bloom's stages. For this work, questions have been restructured as shown in Table 3.

Table 3 Modification of question for PoS tagging

Original question:	Which of the characteristics is not valid for 'RTD', being considered as a sensor for temperature measurement in case of fuel oil measurement?			
Modified question	*'Analyze'* the false characteristic for *'RTD'* when it measures *'temperature'* in fuel oil measurement			
PoS tagging of modified question	*Analyze*	**Verb** represents the Bloom's Level	*Temperature*	**Noun** groups the questions of three courses for common topic defining the CLOs to make a sequence

4.4 Block Diagram of Proposed System

As a first step, the classification of cognitive level has been discussed using a hybrid syntactic rule-based and semantic approach for the correct results. Keyword identification is a significant task. It is done with Stanford Parser Universal dependency to measure the semantic similarity between Bloom's keywords and keywords of the question; WordNet similarity is used to measure semantic similarity, which is used to classify the questions. Figure 2 shows the Block diagram of the proposed system ([6], p. 81).

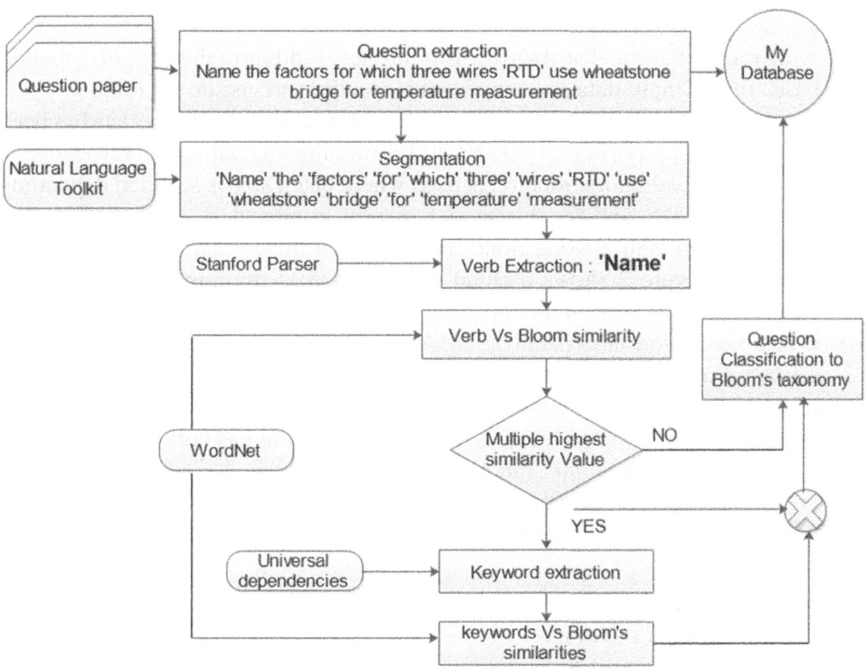

Fig. 2 Block diagram of the proposed system in question classification

5 Data Preprocessing, Training, and Validation

The total dataset consists of 100 questions belonging to various levels of Bloom's taxonomy. Before running the data through classifier, preprocessing is done that involves many stages such labelling data, importing data, categorization of classification level, segmentation and tokenization.

5.1 *Data Preprocessing in MATLAB*

In this section, these stages have been discussed. Deep Learning Toolbox of MATLAB 2023a is used to design and run the Model.

(i) Import Data: Here data is already labeled and available in.csv format, this file is imported that contains textual descriptions of the questions. A few samples of the labeled questions are shown here where 'verb', 'noun', and Level of Bloom's taxonomy is mentioned in the imported data.
Input data is labeled to create five classes namely: *'Level 1', 'Level 2', 'Level 3', 'Level 4', and 'Level 5'*. Categorical function is used as shown below.
df.("Blooms level") = categorical(df.("Blooms level"));
df.("Blooms level").
This command to distribute the Bloom's level and show the class distribution based on the input data and map is with specific verb as shown Table 4.
(ii) Next step after categorization is the training and validation of data, data has been split into the proportion of 80 and 20 for training and validation respectively. K-fold cross validation method is used where batch size is selected depending on the data size, and for this model, size of minibatch is 4. Trained data is visualized through a word cloud to ensure that the correct data set has been imported. Figure 3 shows a cloud visualization of the imported dataset that consists of the text from the input data.
(iii) Data tokenization and Visualization: After splitting training and validation data, tokenization is done using the required function '*tokenizedDocument*'. Then the text data is converted into lowercase, and punctuation is erased. Every sentence is tokenized and used to create a single datastore that has both documents and labels using the '*combine*' function.
(iv) Word encoding: Document is converted into sequences of numeric indices before they are input network using the *'wordencoding'* function.

5.2 *Supporting Features*

Verb Extraction: we have considered verb extraction, so the tagging is 'VB', and the verb is 'name' only. Extracted verbs are compared with Bloom's verb of each

Table 4 Identification of Verb, Bloom's Level and Noun in the questions

"'Validate' instrument used for the 'Flow' measurement in Power plant"
"Level 5" " 'Flow'" "'Validate"
"A differential pressure transmitter is calibrated 0 to 80 inches of water and transmits a 4 to 20 mA dc signal. This transmitter is placed across an orifice plate which is sized to create 80 inches of water differential 6 gallons per minute. 'Calculate' the 'flow' rate when the signal is 13 mA dc?"
"Flow" "Level 2" "Calculate'"
"Measurement of reproducibility of an instrument indicates"
"Level 1" "reproducibility" "indicates"
"Temperature sensors Used for the measurement of circulating water in Power plant"
"Level 3" "temperature" "Used"

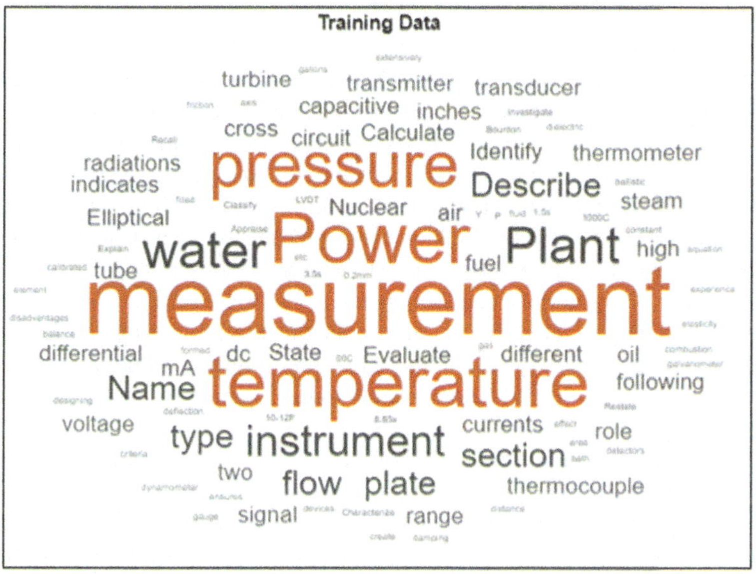

Fig. 3 Cloud visualization of trained dataset

Bloom's level. Bloom's levels are listed as Level 1 to Level 5, where Level 1 is the memorization and Level 5 is the evaluation.

WordNet Similarity: Measures the similarity between the extracted verb and Bloom's verb. The range is between 0.000 and 1.000. For the higher similarity between verbs, the value is higher and vice versa. 1.000 indicates the identical verb, and 0.000 indicates no relation. The example verb is 'name', which relates to Level 1, referring to memorization.

Universal Dependency (UD): Stanford dependency parser is used to justify the correct taxonomy level. It can produce high-quality dependency annotations to assess the accuracy of automatic dependency tools for conversion. NLP uses a dependency tree which is created using the Stanford dependency parser. In the case of two identical levels, it makes a dependency tree, where the following verb is matched with WordNet similarity. The total similarity value is calculated by adding the similarity between both scenarios, and the maximum value is selected for the classification of Bloom's questions.

Question Classification: The similarity value of the extracted verb is compared with Bloom's verb list. Bloom's value with the highest similarity denotes the level of the question. Other keywords are checked for similarity for multiple similar values, and the question is classified for the highest similarity value.

6 Construction of Bloom's Taxonomy Classifier Using Long Short-Term Memory (LSTM)

In this work, we have used LSTM (Long Short-Term Memory) for the classification of tagged dataset that can be trained with supervised learning methods. LSTM is a type of recurrent Neural Network (RNN) where computation is performed over a sequence of data, for example token sequence in text data. RNN is extensively used in NLP tasks where language modeling is prominent, and tokenization is a part of the process. RNN performs computation over sequence of tokens where computation depends on the results of previous computations. LSTM adds more power to the RNN as they can learn long -term sequence dependencies. RNN suffers from the problem of long-term dependencies as they learn by adjusting the weights by gradient descent algorithm, and for long sequences, after a certain optimized value, there is a small change in the gradient that do not contribute to the learning of model, this problem is known as 'vanishing gradient' ([4], p. 605).

LSTM is a special kind of neural network that solves the problem of long-term dependencies by using an additional gate known as 'forget gate' in the existing RNN. RNN has output and input h_t and x_t respectively, x_t is the input that takes the previous value and outputs h_t that makes a loop like structure to allow the information to pass from one step to another step of network. During the large chunks of data, this information becomes huge to be processed, and problem is resolved in LSTM where a forget gate is added to remove the extra information that is not needed. LSTM consists of a memory cell, an input gate, an output gate and a forget gate. Gates control the amount of memory to be exposed which is remembered from the previous state in the memory cells. Memory cells keep track of dependencies between elements present in an input sequence ([14], p. 215) [20]. Figure 3 shows the network structure of LSTM network. In the RNN whole dataset changes after the application of function whereas in LSTM, small modification in dataset is possible by selecting the portion of the required information.

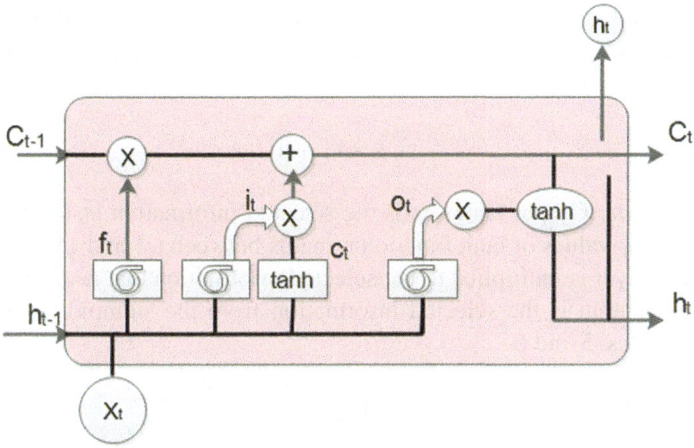

Fig. 4 LSTM network cell

Dataflow is regulated by these gates:

(i) *Forget gate*: This gate controls the amount of information from the previous state to be passed onto the next state. It is useful for removing redundant information for the program.

The network equation for the forget gate is shown in Eq. 1, initially the forget gate remains empty.

$$f_t = \sigma(W_f[h_{t-1}, x_t] + b_f \quad (1)$$

where h_{t-1}, output of previous state

x_t, current input state

W_f, b_f are weights and bias matrices for forget gate.

(ii) *Input gate*: Input gate adds the combination of present input (x_t) and previous output (h_{t-1}) that are passed through sigmoid and tanh functions. All the possible values of the vector x_t and h_{t-1} are added using tanh function.

$$i_t = \sigma(W_i[h_{t-1}, x_t] + b_i \quad (2)$$

where W_i, b_i are weights and bias matrices for input gate. Input gate in LSTM decides the information to be added into present cell from current input through the two neural network layers shown as sigmoid and tanh layers in Fig. 3. Sigmoid layer decides what value will be updated and tanh creates new values. The values of tanh values are shown by Eq. 3. C_t' is the updated values vector created by tanh.

$$c_t' = \tanh\left(W_c\left[h_{t-1}, x_t\right] + b_c\right) \quad (3)$$

New cell state c_t is created by summing the product of forget gate output and previous cell state and multiplication of both equations of input gate as shown in Eq. 4.

$$c_t = f_t * c_{t-1} + i_t * c\prime_t \tag{4}$$

(iii) *Output Gate:* Output Gate sends the selected information as output from the cell state c_t Values of tanh function remains between (-1 and 1), and output of sigmoid layer is multiplied to the selected information that needs to be passed further. Output is the selected information from the sigmoid layer which is given in Eqs. 5 and 6.

$$o_t = \sigma\left(W_o\left[h_{t-1}, x_t\right] + b_o\right) \tag{5}$$

$$h_t = o_t * \tanh(c_t) \tag{6}$$

where c_t, new cell state and h_t is the output after the output gate. h_t is the output from hidden layer whose value lies between −1 and 1 which are not normalized. There is a need for the probability distribution of total neurons that should be equal to total classes defined in the classification for a given model. To accomplish this, hidden state output is given to dense connected output layer that has SoftMax activation function for the multiclass classification with N number of classes. SoftMax activation function is defined in Eq. 7.

$$\text{softmax}(h_t) = \frac{\exp(h_t)}{\sum_{k=1}^{K} \exp(h_t)} \tag{7}$$

Softmax function is applied at the outermost layer of the network that provides the level of classification.

6.1 Designing BiLSTM Architecture

A sequence layer is added to input the sequence data in network and input size is set to 1. Then a BiLSTM layer is added where hidden layer is set to five depending on the data size, and then word embedding layer of dimension five is included, this is same as the words encoding. This model works for sequence to label classification, output mod is set to 'last'. At the end, a fully connected layer is added that is connected to SoftMax layer that classifies the output in five classes as illustrated in Fig. 5.

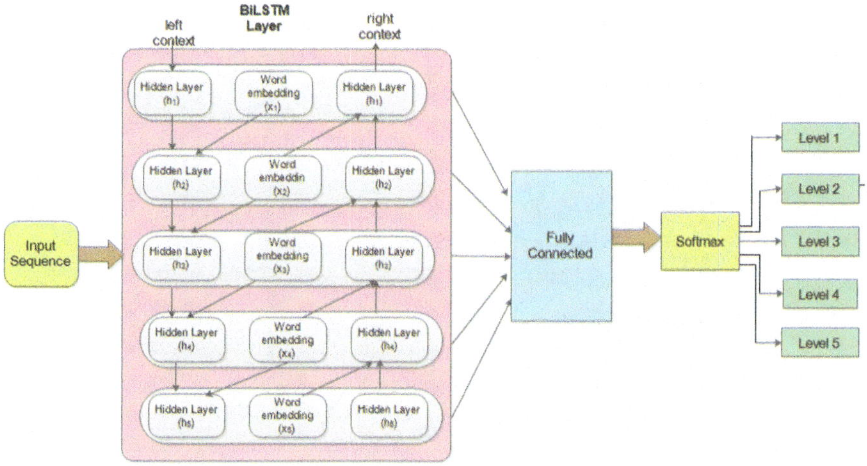

Fig. 5 BiLSTM architecture for five level classifcation

Specifications of the designed architecture are:
5 × 1 Layer array with layers:

Sequence Input	Sequence input with 1 dimension
Word Embedding Layer	Word embedding layer with 5 dimensions
BiLSTM	BiLSTM with 5 hidden units
Fully Connected	5 fully connected layer
Softmax	Softmax

Designing of architecture in the MATLAB requires the commands shown below:

```
numWords = enc.NumWords;
layers = [
   sequenceInputLayer(inputSize)
   wordEmbeddingLayer(embeddingDimension,numWords)
   bilstmLayer(numHiddenUnits,OutputMode="last")
   fullyConnectedLayer(numClasses)
  softmaxLayer]
```

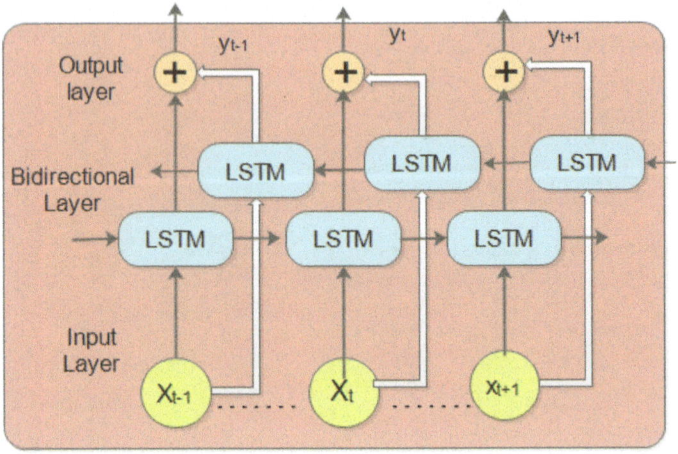

Fig. 6 Building blocks of bidirectional LSTM

6.2 Classification of Text in MATLAB Using Bidirectional LSTM (BiLSTM)

Deep Learning Toolbox ™ in the MATLAB R2023a is an effective framework for the implementing and building architecture on time-series and text data that can be classified by LSTM. In this work, classification is done using the bidirectional LSTM which can learn long-term dependencies between time series or sequence data. Figure 6 shows the basic building blocks of BiLSTM [24].

BiLSTM allows the flow of information in both directions, hence it can utilize information on both sides. In BiLSTM, an additional layer is present that allows the backward flow and output of both the layers can be combined to produce specific results. It has the advantage of combining past and present inputs so it can model the sequential dependencies between word and phrases in both direction of sequences ([10], p. 402).

In this, work BiLSTM has been used to remove the ambiguities that may arise in the presence of more than one verb or noun in the sentence.

6.3 Training the Model Using BiLSTM (RNN)

Training of the model has the following steps:

(i) Tuning of hyper parameters: Adam loss stands for Adaptive moment estimation and is a modified version of stochastic gradient descent algorithm that is used in Neural Networks. Adam Loss is suitable for LSTM network as it calculates the estimation moving average (EMA) of the gradients. The gradient decay factor is already defined and updated accordingly after each iteration. EMA is the weighted average of all the gradients, and it also reduces the effect of outliers minibatches. Adam's update rule is based on the squared gradient decay factor. Initially the model is trained by tuning hyper parameters such as selecting Adam loss function as model loss function, size of minibatch as 4, setting the initial learning rate at 0.001, and decay rate at 0.001. The gradient decay factor is 0.9 and squared gradient decay factor is 0.999. The number of epochs selected for training the model is 4. Learning rate is an important hyper parameter that decides the step size for each gradient. Optimized value of learning can reach to the convergence in fewer number of epochs.

(ii) Mini-batch Preprocessing: Function 'preprocessMiniBatch' converts the documents to sequences and one-hot encodes the labels. One hot encoding converts the categorical variables as numerical values. To accomplish this, a function is created that takes two inputs as shown below:

```
function X = preprocessMiniBatchPredictors(dataX,enc)

% Extract documents from cell and concatenate.
documents = cat(4,dataX{1:end});
% Convert documents to sequences of integers.
X = doc2sequence(enc,documents);
X = cat(1,X{:});
End
```

7 Construction of Bloom's Taxonomy Classifier Using Convolutional Neural Network (CNN)

Convolutional Neural Networks have recently found an application in text classification. Usually, CNN are good for image classification, but they can also be used for the labelled text classification. CNN consists of convolutional layer to obtain the features from text input, after the feature extraction, a pooling layer is reducing the size of feature map and select the best features present in the text. Output of the convolutional operation consists of feature map, convolved features and activation map. For text classification, text data is converted into numerical representation through 'one hot' function. This is done using arguments such as 'text' for text encoding, 'n' represents the size of vocabulary, and unwanted characters such as punctuation marks and special characters are removed through filters. All questions is converted into lower case using 'lower'[1].

Padding is done to ensure that all questions are of same size before passing to the CNN model. The maximum length of the sequence is defined, and extra zeroes are added to increase the question length and extra integers are removed from the long questions. Word embedding represents the data representation as a dense vector where words having similar semantics are grouped together [18].

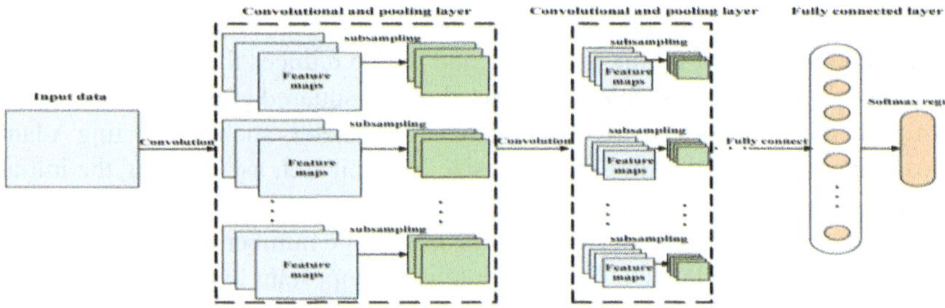

Fig. 7 Architecture of convolutional neural network [24]

7.1 Design of Network Architecture Using CNN

CNN architecture is shown is Fig. 7, network architecture is described as:

Input size is specified as one that corresponds to the channel dimension of the integer sequence input. Input is embedded using a word embedding of dimension 25. Number of successive items is termed s n-grams in Machine learning their length is specified as 2, 3,4, and 5. Blocks of layers contain convolutional layer, a batch normalization layer, ReLU layer, a dropout layer, and max pooling layer and 200 convolutional filters of size 1-by-N are specified. N depends on the size of n-gram. Input layer is connected to each block and output of each block is concatenated through concatenation layer. At the end a fully connected layer, a softmax layer and a classification layer are connected.

```
embedding Dimension = 25;
ngramLengths = [2 3 4 5];
numFilters = 200;
```

The dropout layer with the rate of frequency is aet at 0.2, and its purpose is to nullify the contribution of some neurons and leave others unmodified. After adding all the layers, the network looks like as shown in Fig. 8.

7.2 Training the Model Using CNN

A mini-batch size of 128 is trained and network is validated using the validation data. The training cycle consists of 60 epochs and 5550 iterations with the learning rate of 0.001. After training the model, it was tested for accuracy. TrainNetwork function is used to train the network in MATLAB.

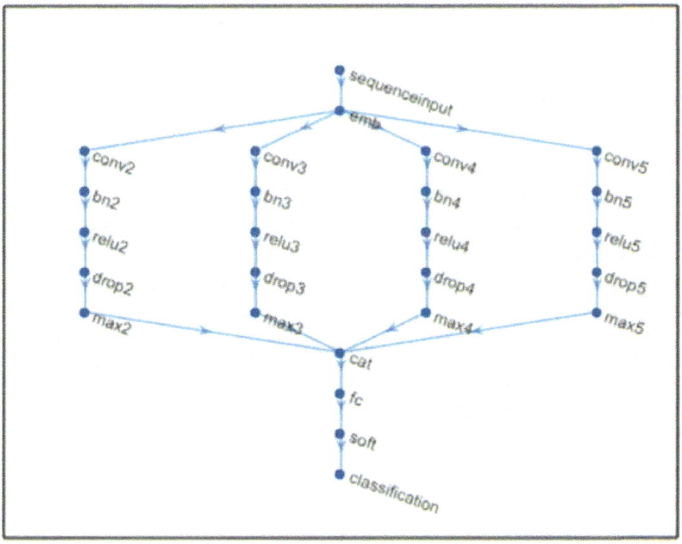

Fig. 8 Network architecture for question classification

```
options = trainingOptions("adam", ...
    MiniBatchSize=1,MaxEpochs=60, ...
    ValidationData={XValidation,TValidation}, ...
    OutputNetwork="best-validation-loss", ...
    Plots="training-progress", ...
    Verbose=true);

net = trainNetwork(XTrain,TTrain,lgraph,options);
```

8 Result and Discussion

Both the network predicted Bloom's level for new questions that are randomly input, and CNN model performed it with better accuracy. Although LSTM model is good for text classification, in this case, data was categorical and was defined in a discrete manner. This was supervised learning where data is already labeled as Bloom's verb and tagged as the level that needs to be identified. Results have shown that CNN network performed better for this system.

8.1 Network Performance in BiLSTM Model

The network is trained for the parameters discussed previously, data and loop over mini batches of data are shuffled, and after every iteration, training progress is displayed. Also, the network is validated after every epoch using validation data. There is a huge gap between training loss and validation loss. It shows that RNN is not providing meaningful results for the classification of the given dataset (Fig. 9).

As discussed above, this method is proposed to verify the Bloom's level where each question is redesigned specifically for the computerized adaptive assessment. Figure 10 shows the performance of the LSTM when run for 120 epochs and 2640 iterations. Accuracy is 70% for the given dataset. Learning rate is 0.001.

Figure 11 shows the loss in the same model. Network performance was declined when iterations were increased beyond 2640 iterations.

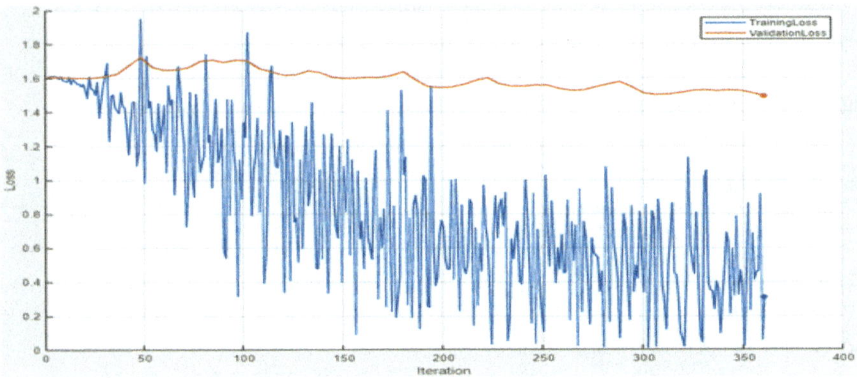

Fig. 9 Training loss and validation loss

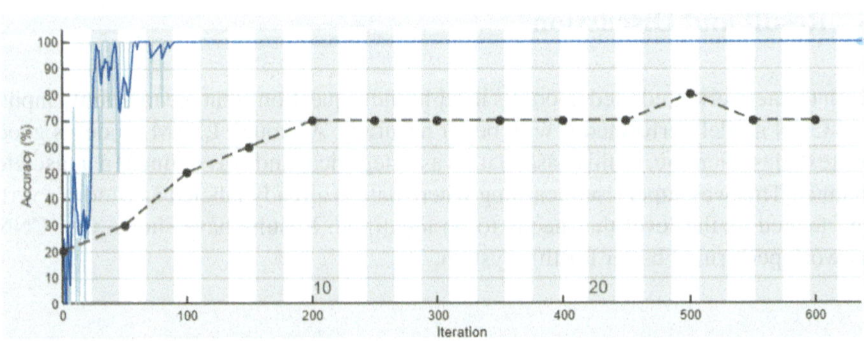

Fig. 10 Training accuracy in BiLSTM

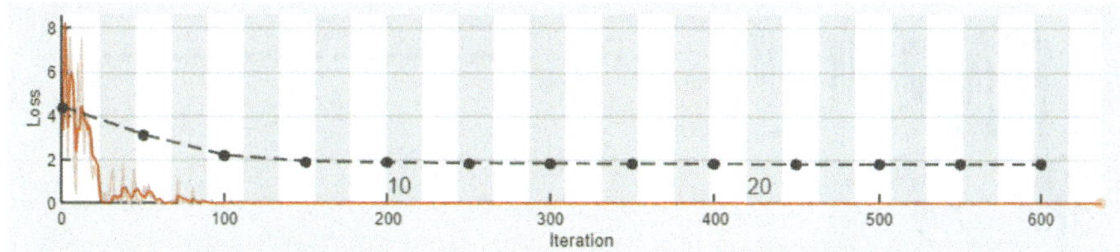

Fig. 11 Loss in the BiLSTM Model

8.2 Network Performance in CNN Model

Network is trained for the input parameters as shown in Fig. 12. Data and loop over mini batches of data are shuffled, and after every iteration, training progress is displayed. Also, the network is validated after every epoch using validation data. It is evident that validation accuracy is better in CNN than the RNN model. Also, their validation loss and training loss are following each other, thus strengthening the suitability of CNN model for the input dataset where questions are classified based on the verb defined in the questions.

Validation accuracy of 85.71% is achieved as shown in Fig. 13. Training dataset ran for 60 epochs and 5580 iterations with 93 iterations per epochs. The learning rate was set to 0. 001. There is a less difference in validation accuracy and training accuracy in CNN model, hence it is more suitable for the classification task for the input dataset. However, since the dataset is small the CNN model tends to overfit as seen in Fig. 13.

Figure 14 shows the loss in CNN model, it shows that after 480 iterations, model became stable making it robust for the classification task.

In this section, performance of RNN model and CNN model has been compared and results have shown the suitability of the CNN model over RNN model for the

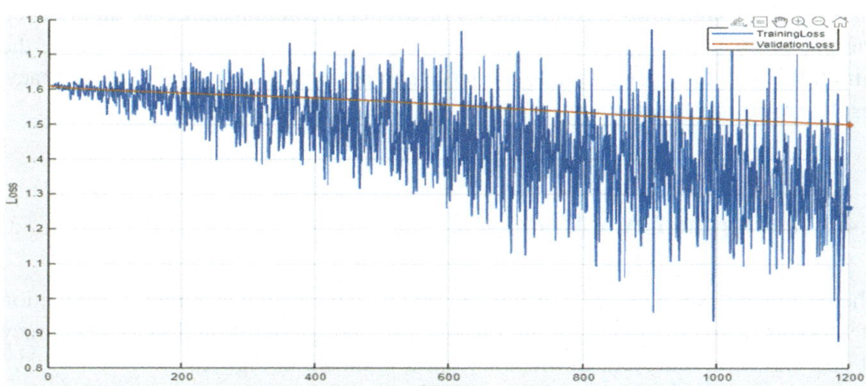

Fig. 12 Training loss and validation loss

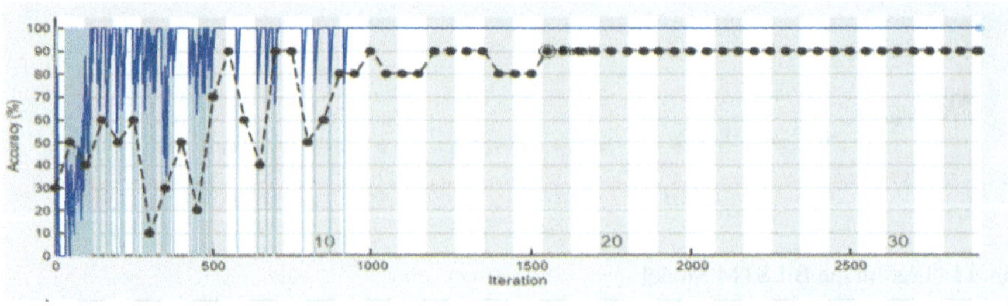

Fig. 13 Validation accuracy in CNN

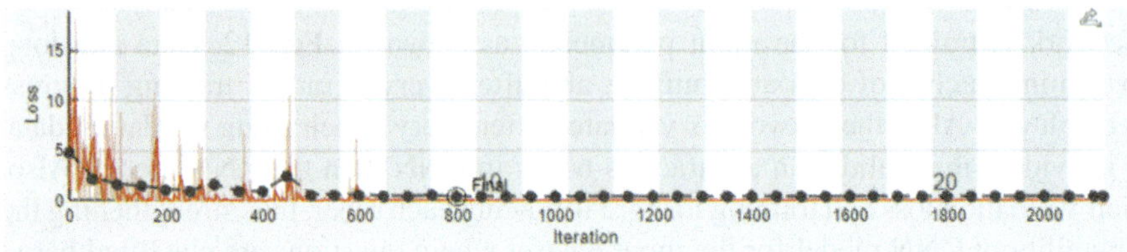

Fig. 14 Loss in CNN model

small dataset. Also, text dataset was well labeled that makes it more convenient to be used with CNN.

8.3 Testing the LSTM Model

Classification accuracy is assessed by comparing the prediction on the test set with true labels. One hot encoded label is converted into categorical and then they are transposed through proper commands. For Model prediction, function inputs deep learning array, a mini batch queue and outputs the predicted values by iterating the mini batches in the queue. Due to the large gap in training and validation accuracy, results are not good hence confusion matrix is not plotted.

8.4 Testing the CNN Model

The designed CNN network is evaluated on test data for the accurate classification of Bloom's taxonomy based on the variable provided in question. In CNN accuracy of 90% has been achieved. Confusion Matrix is shown in Fig. 15.

Accurate prediction is crucial for identification of correct questions in the adaptive assessment. Result of the LSTM and CNN based models shows that NLP based methods are suitable for text classification and can be used in the CMM.

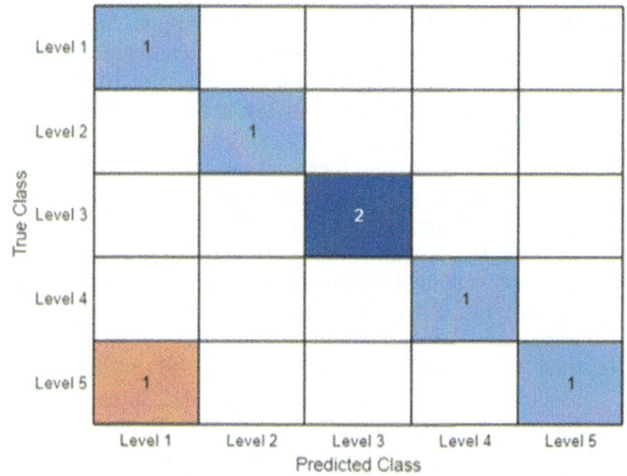

Fig. 15 Confusion matrix in CNN

9 Conclusion

This work has provided an efficient method for designing Computerized automated adaptive assessment where questions have been classified based on Bloom's learning stages. The system discussed in this chapter proposes the selection of keywords based on Bloom's taxonomy using NLP. Keywords extraction is rule-based where each word is tokenized and tagged based on verbs. Questions have been classified using deep learning models using Deep learning toolbox of MATLAB 2023a. CNN has provided more accurate results with the validation accuracy of 85.71% which is better than LSTM based model where validation accuracy is 70%. After the classification, text extraction can be used for Concept mapping needed to create adaptive path for questions generation for the assessment. The future work involves integrating the selection of nouns along with the bloom's verbs.

References

1. Amin MZ, Nadeem N (2019) Convolutional neural network: text classification model for open domain question answering system. arXiv:1809.02479v2
2. Ana P, Tawo BP (2013) Design and implementation of online examination administration system for universities. Global J Math Sci 12(1):39–51. https://doi.org/10.4314/gjmas.v12i1.16
3. Asthana P, Tanwar S, Kumar A, Mishra S (2022) Students' assessment for quantitative measurement of course learning outcomes in online class of power plant instrumentation. In: International conference for advancement in technology (ICONAT), India, pp 1–5. https://doi.org/10.1109/ICONAT53423.2022.9725978.
4. Borovkova S, Tsiama I (2019) An ensemble of LSTM neural networks for high-frequency stock market classification. J Forecasting 38(6):600–619. https://doi.org/10.1002/for.2585
5. Chang WC, Chung MS (2009) Automatic applying bloom's taxonomy to classify and analysis the cognition level of English question items. In: Joint conferences on pervasive computing (JCPC), Tamsui, Taiwan, pp 727–734. https://doi.org/10.1109/JCPC.2009.5420087

6. Dien TT, Loc BH, Thai-Nghe N (2019) Article classification using natural language processing and machine learning. In: International conference on advanced computing and applications (ACOMP), Vietnam, pp 78–84. https://doi.org/10.1109/ACOMP.2019.00019
7. Haladyana TM (1997) Writing test items to evaluate higher order thinking. Allyn and Bacon, Boston
8. Hirose H et al (2016) Questions and answers database construction for adaptive online IRT testing systems: analysis course and linear algebra course. In: 5th IIAI international congress on advanced applied informatics (IIAI-AAI), Kumamoto, Japan, pp 433–438. https://doi.org/10.1109/IIAI-AAI.2016.48
9. Huang Z, et al (2017) Question difficulty prediction for reading problems in standard tests. In: Proceedings. AAAI conference on artificial intelligence (AAAI'17). AAAI Press, pp 1352–1359
10. lfaik H, Nfaoui E (2021) Deep bidirectional LSTM network learning-based sentiment analysis for Arabic text. J Intell Syst 30(1):395–412. https://doi.org/10.1515/jisys-2020-0021
11. Jayakodi K, Bandara M, Perera P (2015) An automatic classifier for exam questions in engineering: a process for bloom's taxonomy. In: IEEE international conference on teaching assessment and learning in engineering (TALE), pp 297–303. https://doi.org/10.1109/TALE.2015.7386043
12. Kersloot MG, van Putten FJP, Abu-Hanna A et al (2020) Natural language processing algorithms for mapping clinical text fragments onto ontology concepts: a systematic review and recommendations for future studies. J Biomed Semantics 11(14):1–21. https://doi.org/10.1186/s13326-020-00231-z
13. Lokapure SS, Mahajan SA, Dhamne S, Sheikh S (2021) Automatic concept map generation and summarization from text. Int J Emerging Technol Innov Res 8(5):104–107. http://www.jetir.org/papers/JETIR2105676.pdf
14. Laddha MD, Lokare VT, Kiwelekar AW, Netak LD (2021) Classifications of the summative assessment for revised bloom's taxonomy by using deep learning. Int J Eng Trends Technol 69(3):211–218. https://doi.org/10.14445/22315381/IJETT-V69I3P232
15. Nugumanova A, et al (2015) Automatic generation of concept maps based on collection of teaching materials. In: Proceedings 4th international conference on data management, technology and applications, pp 248–254. https://doi.org/10.5220/0005554702480254
16. Prasad GNR (2021) Identification of bloom's taxonomy level for the given question paper using NLP tokenization technique. Turk J Comput Math Educ 12(13):1872–1875
17. Purohit VK, et al (2013) Design of adaptive question bank development and management system. In: Proceedings of 2nd IEEE international conference on parallel distributed and grid computing, pp 256–261. https://doi.org/10.1109/PDGC.2012.6449828
18. Rajpal D, Mishra S, Kumar A (2021) Lesion detection and classification for breast cancer diagnosis based on deep CNNS from digital mammographic data. In: Computational analysis and deep learning for medical care. Scrivener Publishing LLC. https://doi.org/10.1002/9781119785750.ch11
19. Santos VD, De Souza ÉF, Felizardo KR, Watanabe WM, Vijaykumar NL, Aluizio SM, Júnior AC (2019) Conceptual map creation from natural language processing: asystematic mapping study
20. Shaikh S, Daudpotta SM, Imran AS (2021) Bloom's learning outcomes' automatic classification using LSTM and pretrained word embeddings. IEEE Access 9(1):117887–117909. https://doi.org/10.1109/ACCESS.2021.3106443
21. Van der Linden WJ, Glas CAW, Computerized adaptive testing, theory and practice. Heidelberg, Springer
22. Von Konsky BR, Zheng L, Parkin E, Huband S, Gibson DC (2018) Parts of speech in bloom's taxonomy classification. In: Campbell M, et al (eds) Proceedings of the 35th international conference of innovation, practice and research in the use of educational technologies in tertiary education: open oceans: learning without borders, Australasian Society for Computers in Learning in Tertiary Education (ascilite), Nov 25–28 2018, pp 527–532. ASCLITE, Geelong

23. Yağci M, Ünal M (2014) Designing and implementing an adaptive online examination system. Procedia—Soc Behav Sci 116(1):3079–3083. https://doi.org/10.1016/j.sbspro.2014.01.711
24. Zhou P, Qi Z, Zheng S, Xu J, Bao H, Xu B (2016) Text classification improved by integrating bidirectional LSTM with two-dimensional max pooling. ArXiv, abs/1611.06639.2016
25. Zubrinic K, Kalpic D, Milicevic M (2012) The automatic creation of concept maps from documents written using morphologically rich languages. Expert Syst Appl 39(16):12709–12718. https://doi.org/10.1016/j.eswa.2012.04.065

Where Generative AI Fits Within and in Addition to Existing AI K12 Education Interactions: Industry and Research Perspectives

Xin Miao, Richard Brooker, and Samantha Monroe

Abstract Recent developments in Generative AI have led capital market, industry, and research institutions to explore its education applications as solutions to K12 challenges. However, there is currently a gap of analytical review of these trends. This chapter attempts to review and analyze predominant Generative AI education efforts within and in addition to existing AI education frameworks that include contributions from both industry and research institutions. Our aim is to present a holistic review of AI Education key interactions, explore the opportunities that Generative AI presents, share industry experience in implementing Generative AI in AIED products and identify future work directions.

Keywords Generative AI · K12 education · Pedagogical design · Human-AI interaction · AIED ethical concerns · 21st century skills

1 Introduction

Generative AI, a branch of Artificial Intelligence, is characterized by its ability to autonomously generate novel content, including text, images, audio, video and even 3D models. This differs from traditional AI systems that usually focus on specific tasks and domains. For this reason, Generate AI is already being applied to a wide range of industries. For instance, in software engineering, Generative AI models like ChatGPT have been used for code generation, aiding in tasks such as programming, debugging, and testing [1]. In the healthcare sector, Generative AI has been utilized

X. Miao (✉) · R. Brooker · S. Monroe
Product, Data & Research, Alef Education, Abu Dhabi, UAE
e-mail: xinmiaouae2018@gmail.com; xin@alefeducation.com

R. Brooker
e-mail: richard.brooker@alefeducation.com

S. Monroe
e-mail: 13monroes@gmail.com

M. S. Khine (ed.), *Machine Learning in Educational Sciences*,
https://doi.org/10.1007/978-981-99-9379-6_17

to create synthetic medical images for training purposes, thereby addressing data scarcity and privacy concerns [2].

The education sector also shows immense promise for the application of this technology. The integration of AI with education started in the 1970s when researchers wanted to see how computers might substitute for one-to-one human tutoring. Ever since, the application of AI in education has expanded to include learning with AI (i.e. student-facing AI, teacher-facing AI, system-facing AI), and learning about AI (i.e. AI literacy). The 2021 UNESCO AI and Education Guidance for Policymakers categorizes AI applications in education into four needs: (1) education management and delivery; (2) learning and assessment; (3) empowering teachers and enhancing teaching; (4) lifelong learning [3]. These applications are generally used by key stakeholders in the education system, such as students, teachers, school leaders, and policy-makers, who directly interact with them.

Another recent contribution comes from Rismanchian and Doroudi [4], who proposed the AI × Ed Framework to categorize four interactions between AI and education. In the framework, the horizontal axis represents end users who are interacting with AI (i.e. researchers on the left and learners on the right); the vertical axis represents the role that AI is playing, namely, AI as an applied tool and AI as an analogy to HI (human intelligence). The four quadrants created reflect interactions between AI and education from both research and industry.

This chapter aims to explore the role of Generative AI in K12 education, its applications, research and development, benefits and concerns. It maps key AI education applications and predominant AIED research efforts into the AI × Ed framework as a baseline. It then examines how Generative AI fits into this framework, enabling a critical analysis of its current state in terms of market tools, research, development, limitations and potential beyond the existing framework. This analysis will guide future work and provide insight into its benefits and concerns. Given the lack of research outlining the use of Generative AI in education within a holistic framework, and the fact that most Generative AI education applications originate from industry and capital markets, this paper's key contribution is to systematically map key generative applications from both industry and recent research into a framework to guide future work.

We propose the following research objectives:

1. To construct an analytical overview of AI K12 Education interactions based on previous AI × Ed Frameworks and learning science considerations;
2. To identify key Generative AI practical applications and research contributions within the above AI Education overview, to identify the unique role that Generative AI plays in K12 education, as well as its limitations and efforts needed in future work.

2 RQ1: Analytical Overview of AI K12 Education Interactions

In this section, we propose a new framework based on Rismanchian and Doroudi's AI × Ed four interactions. We revise the "learner interaction" to "Instructional Core End User Interaction", which includes students, teachers, and digital content. This change is motivated by (1) in any design of AI applications for the learner, we need to consider the role of teacher and learning content, as increases in student learning occur only as a consequence of improvements in the level of content, teachers' knowledge and skill and student engagement. If you change any single element of the instructional core, you have to change the other two" [5]. For instance, in a lot of student-centered AI systems (e.g. Intelligent Tutoring Systems), the role of the teacher is a core consideration especially when it comes to pedagogical design of the AI-learner interaction as the teaching tasks unfold key skills that learners have to have (e.g. metacognitive skills to set goals, self-monitor and self-reflect); (2) teacher-facing AI applications are a key aspect of learning with AI in K12 education; (3) Generative AI, as discussed above, has major breakthroughs in generating digital content, which is an integral part of many AI products, for instance, ITS.

The revised framework is divided into four quadrants:

(1) Learning with AI, which includes student-facing and teacher-facing applications; (2) Learning about AI, which is AI literacy curriculum for K12 students and teachers; (3) AIED research efforts driving the industry; and (4) AIED research exploring the analogy between AI and Human Intelligence. However, in this section we'll only discuss students and teachers as the Instructional Core End Users because traditional AI systems use digital content designed by human subject matter experts. As a result, we propose the updated framework to answer RQ1 in Fig. 1.

2.1 Learning with AI Applications: Student-Facing, Teacher-Facing

2.1.1 Student-Facing Applications

In this section, we analyze predominant student-facing AI tutoring systems, ITS (Intelligent Tutoring System) and its sub branch DBT (Dialog-based Tutoring System), adaptive testing and autograding systems.

ITS (Intelligent Tutoring Systems) have been the subject of research during the past 40 years, with over 60 commercial ITS products available worldwide. Examples include MATHia, Khan Academy and ALEKS from the US, Alef Education from the Middle East, Byjus from India, and Squirrel AI from China. These systems

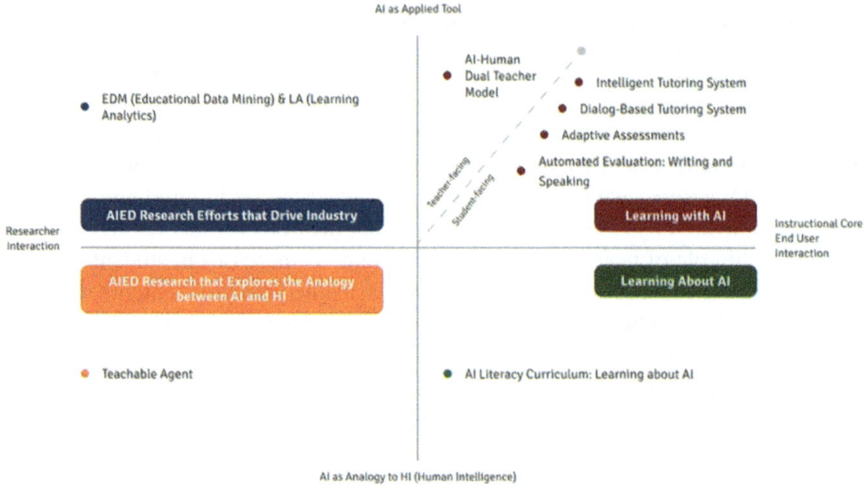

Fig. 1 AI K12 education interactions (with examples)

primarily function as cognitive tutors and have been demonstrated to effectively improve student performance [3].

ITS operate by responding to a learner's mastery of a specific knowledge domain. They identify gaps in knowledge, missing skills, and misconceptions, and then determine an optimal pathway for individual learners. This pathway includes recommended intervention materials and learning tasks, designed based on cognitive science and expert knowledge in the specific domain. This approach is sometimes incorporated into Learning Management Systems (LMS).

How intelligent tutoring systems work depends on its key components. According to (Beck et al. n.d.), there are four main components of ITS:

1. Student model: This tracks a student's performance which is used to customize instructions accordingly. This includes: (a) a record of the student's understanding of the domain to drive personalized instructions; (b) pedagogical information about the learner, such as their preference for reviewing examples before answering questions; (c) information on learner acquisition and retention (e.g. how fast students learn a new topic, how well students recall the material over time).
2. Pedagogical model uses information from the student model to determine the domain knowledge presented to the learner. This includes: (a) identifying topics the learner needs to master, potentially considering learner interest; (b) generating problems, for instance, ensuring the difficulty level of the questions is appropriate for the learner's ability; and (c) providing feedback based on learner mastery and behavior, such as identifying misconceptions and deciding how to share feedback without over-assisting and hindering the learning process.
3. Communication module: This component manages interactions with the learner. It determines the most effective way to present tutoring materials and learning

tasks to the student. A key design issue to address is learner agency. Given the procedural paradigm of ITS, learners can easily become passive recipients of information, potentially diminishing their agency. Therefore, maintaining learner agency is a crucial consideration in the design process.

4. Domain knowledge: This refers to the content and learning tasks that the tutor is teaching. A key issue is knowledge engineering, which involves representing a domain and determining how to scale it up.

Despite being the most researched domain over the last 40 years, Intelligent Tutoring Systems still have room for improvement, as observed from our industry work experience. Cultural and regional specifics necessitate further testing, emphasizing the need for testing and mixed research approaches. These include qualitative research and controlled experiments or A/B testing, which are crucial for validating the assumptions made by product managers and designers. For example, this validation process is particularly important when designing nuanced features such as interactive hints and scaffolds that drives learner agency, metacognitive skills and effectively addresses learner misconceptions. Moreover, ITS requires a substantial amount of content, such as topics, questions, and feedback, to drive learner adaptivity, as well as extensive backend data tagging. This leads to significant concerns about content production and development costs. Additionally, as [3] highlighted, learning is a social activity, and a notable drawback of ITS is the lack of social interaction.

A Dialog-Based Tutoring System (DBT) is a type of Intelligent Tutoring System that uses cooperative dialogue to stimulate critical thinking and deeper understanding. The aim is to provide constructive feedback similar to expert one-on-one human tutoring. However, there are several challenges associated with DBTs, as shared by researchers involved in the development and deployment of the IBM Watson Tutor. These include: (1) classifying student natural language responses: Interpreting learner utterances based on context can be difficult due to language diversity and the limitations of current Natural Language Processing (NLP) and artificial intelligence techniques; (2) the need for extensive work from subject matter experts for content design, creation, authoring, and data-tagging; (3) the impact of pedagogical interactions through natural dialogue on learner engagement and learning outcomes. Implementing interventions and scaffolds to keep learners actively engaged is complex, as is providing relevant examples when a student struggles [42].

Adaptive Testing. AI-powered adaptive testing has been utilized in the industry for years. A notable example from the US is NWEA MAP Growth. This tool not only measures student achievement and growth but also identifies student skill gaps in several K-12 subjects like math, reading, and science. While this is an example of a standalone adaptive testing product, some Intelligent Tutoring Systems like Alef use adaptive diagnostic testing as the basis for their "student model". This is used to diagnose a learner's skill gap or mastery of domain knowledge.

Autograding. Writing and Speaking. Automated writing evaluation uses natural language processing and other AI techniques to provide automatic feedback. The formative approach allows students to improve their work before submission, while the summative approach facilitates automatic scoring. These tools primarily aim to reduce assessment costs and save grading time for teachers. The following are a few of the key challenges: (1) Feedback: The focus should be on providing actionable feedback rather than just a numerical score, which is more beneficial for learners; (2) Biases: The algorithms may be biased due to variations in vocabulary use and sentence structure; (3) Authenticity: It might be challenging to detect AI-generated 'deep-fake' written pieces using these automated grading systems [3].

In terms of spoken language evaluation, Liulishuo is a notable product for automatic scoring of spoken English, specifically designed for English as a Second Language (ESL) learners. Its sub- product, IELTS Liulishuo, offers users speaking test simulators, intelligent corrective feedback, detailed error reports, and improvement suggestions.

2.1.2 Teacher-Facing Applications: A Complementary Relationship Between AI and HI

In this section, we'll analyze how Artificial Intelligence complements Human Intelligence (HI) from a teacher's perspective, using the AI-Human "dual teacher" model as an example. For instance, in an ITS which runs on a LMS platform, teachers were provided with instructional content, assessments, real-time analytics dashboards and features that allow teachers to group students and assign learning materials and tasks. In this scenario, teachers and ITS work together to conduct teaching and learning. In particular, when a teacher uses real time analytics dashboards to make decisions about which student to offer help and attention, what knowledge and skills her/his class has mastered and what is not, who is engaged and which student has shown off-task behavior. In this case, the human teacher assesses AI calculative predictions reflected in an analytics dashboard, to make judgment calls on what she has to do, taking into account other contextual factors. In this scenario, "AI (e.g. real-time learning analytics) excels at calculative prediction and formative decision-making, which is called reckoning; humans (e.g. the teacher) are good at judgment, which is about deliberative thought that seeks to be judged, grounded in ethical commitment and appropriate to the situation in which it is deployed" [6]. In the teaching example above, we see a complementary partnership in decision-making, between AI and HI. So, it is not a matter of AI replacing HI, but more of an exploration on how AI complements HI in specific contexts. In the teacher example above, the teacher does need to update skills to use AI (e.g. learning analytics) for instructional purposes. Therefore, as AI replaces certain specific human tasks, humans do need to update skills to work with AI.

2.2 *Learning About AI: AI Literacy for Students and Teachers*

This section reviews the AI literacy curriculum for K12 students and teachers, its significance, key stakeholders, design principles, and assessments.

> According to Williams et al. [7], students growing up in the era of AI must be equipped with the skills to build and work with AI to meet the needs of a shifting workforce. This means not only teaching students about what AI is but also enabling them to create with it, think critically about its impact, and advocate for responsible use.

According to UNESCO's K12 AI Curricula, 11 countries have developed, endorsed and implemented AI curricula, for instance, UAE, Kuwait and Qatar from the Middle East, Austria, Belgium and Portugal from Europe, China India and Republic of Korea from Asia. A few non- governmental organizations also developed AI literacy curricula, for instance, MIT RAISE (Responsible AI for Social Empowerment and Education), developed Daily Curriculum for Middle School and AI literacy curriculum for K12, Intel developed Global AI Readiness Program, Microsoft developed AI Youth Skills [8].

A few initiatives attempted to create AI Curriculum frameworks for grades K-12. For example, AI4k12 defined the Five Big Ideas for AI as: Perception, Representation and Reasoning, Learning, Natural Interaction, and Social Impact. [8].

Williams et al. [7] proposed three design principles for AI curriculum: (1) active learning (i.e. students play a key role in their learning by engaging in activities then processing information through reflection, (2) embedded ethics (i.e. the pedagogical practice of teaching technical and ethical concepts in tandem), and (3) low barriers to access (i.e. centering student and teacher needs in the design. AI integrated with subjects like art, dancing and robotics to address the barrier or engagement).

Williams and Breazeal (n.d.) reviewed 63 peer-reviewed articles that assess students' AI knowledge and perspectives, due to lack of consensus on what to teach students or how to evaluate their knowledge. This recent review on AI literacy assessments identified gaps in content coverage and assessment tools, particularly in assessing students' critical thinking and emerging beliefs about AI.

2.3 *AIED Research Efforts that Drive Industry*

In this section, we present an overview of Educational Data Mining (EDM) and Learning Analytics (LA), illustrating their use in AIED research.

EDM and LA are methodologies that extract useful information from large educational data sets. Lots of EDM and LA methodologies are used in the E-learning industry, including Intelligent Tutoring Systems and adaptive learning platforms. A common application of EDM in ITS is to understand student learning mastery using the Bayesian Knowledge Tracing algorithm. Data sources for analysis include learning management systems (e.g., log data), student assessment data, multimodal sensor data, and survey data collected by education agencies. In the following section, we review three main methodologies and their application in the e-learning industry [9]:

(1) Prediction Models: These tools predict variables that are not feasible to directly collect in real time, such as learner engagement, by mining relevant log data.
(2) Structure Discovery: This attempts to find structure in data without focusing in advance on a specific variable and common approaches include, latent analysis, factor analysis and network analysis. For instance, factor analysis was used to identify factors influencing real world teacher adoption of adaptive learning platforms [10]. In [11] paper, a series of malleable predictive factors are selected under student approaches to math learning. Through the analysis using machine learning technique, XGBoost, a latent relationship between student approaches to math learning (i.e. self- system, metacognitive strategies, instructional language skills) and math diagnostic test performance is uncovered for students from Grade 5 to Grade 9 in Abu Dhabi public schools. The research findings were used to recommend key interventions on improving readability of math content given that the number 1 challenge for children's math learning is English literacy.
(3) Relationship mining is to discover relationships in a dataset with a large number of variables, which include association rule mining, sequential pattern mining, correlation mining, and causal data mining.

Visualization of data insights on learning analytics dashboards is another important area, as it should be done in a way that is easy to understand and actionable for end users. For instance, in the case of teacher dashboards, teachers should be able to understand and know how to make judgment decisions based on the data insights and other contextual factors he/she knows. In the case of a learner, dashboard design should take into account factors such as learner motivation and engagement. For instance, it will be very demotivating to have a leaderboard that always ranks student performance by a single matrix, for instance, performance score. From industry practice experience, it is always a good idea to involve the end users in the initial designs of the product to get feedback and validate assumptions.

2.4 AIED Research that Explores the Analogy Between AI and HI

In this section, we review teachable agents as an example of AIED research efforts that explore the analogy between how humans learn by building machines that can learn. The paradigm of ITS is procedural, which potentially could over-control learner actions; however, teachable agents, by design, are about learning by teaching and self-regulation.

By definition, "teachable agents are computer agents that students teach, and in the process, learn themselves". Deep learning by the learner is achieved by three critical aspects of teaching: (1) interaction that helps students develop structured knowledge. People remember better when information is connected in meaningful ways; (2) interaction that helps students take responsibility and make decisions about learning. Spoon feeding type of instruction takes away the learning agency. Open-ended, self-directed problem solving activities are better to foster learner agency and responsibility; (3) interaction that involves the development of reflection or metacognitive skills, such as monitoring the quality of one's knowledge, behavior and learning decisions. Explicit monitoring, explicit thinking help self-evaluate how ideas are understood [12].

Researchers have been experimenting with learning science techniques to implement how learning by teaching works. For instance, Michie et al. [13] developed the math concept learning system for solving linear equations. Users supplied the strategies for solving problems by entering example solution traces, and the system learned via an inductive machine learning algorithm. Hietala and Niemirepo [14] designed EduAgents that study the relation between the competency of agents designed as learning companions and student motivation. Their findings suggest that students with higher cognitive ability preferred interacting with strong agents and weaker students preferred weak agents that initially made mistakes and were not confident of their answers, but improved their performance as the students improved in their abilities. In Betty's Brain, students instruct a teachable agent which can reason based on how it is taught. Students achieve deeper learning as they make greater effort to learn for their TAs than they do for themselves. These beneficial effects were most pronounced for lower achieving children [12].

According to [12], a few issues identified with teachable agents are (1) students might focus on getting the correct answer without trying to understand the knowledge connections; (2) the interaction between the learner and agent could explore group collaborative problem-solving or competitive environment where learners can compete with each other through their teachable agent; (3) the agent could explicitly instruct self-regulated learning strategies (e.g. goal setting, monitoring, seeking assistance, reflection on feedback). These are to be explored and tested in teachable agents or in ITS.

3 RQ2: Analytical Overview of Generative AI K12 Education Interactions

In this section, we explore our second research question, focusing on the role of Generative AI in K12 education. We will assess its applications and contributions within the context of the four- quadrant framework defined earlier (Fig. 2).

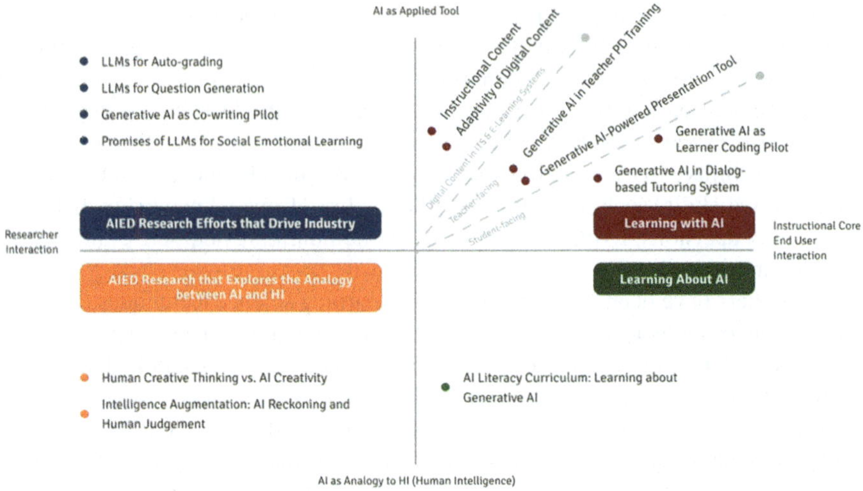

Fig. 2 Generative AI K12 education interactions (with examples)

3.1 Learning with Generative AI: Digital Content, Students, Teachers

Generative AI has been making strides in the realm of K12 education, particularly in the context of Intelligent Tutoring Systems, student-facing applications, and teacher-facing applications. In this section, we will include digital content together with student and teacher as three core Instructional Core End Users on the x-axis.

3.1.1 Generative AI in Creating Digital Content in ITS & E-Learning Systems

Creating content for Intelligent Tutoring Systems in the ed-tech and digital publishing industry is a complex, expensive process and till today, as discussed in research question one, high quality content creation is still a big challenge in the development of ITS. It requires diverse content creation, stakeholder collaboration, and careful consideration of factors like knowledge engineering, learning styles, language localization, accessibility standards, and brand consistency. The costs can be high, with a single assessment question costing up to $36, a reading passage costing between $50 and $200, and production of a video between $568 and $2200. The process is further complicated by the knowledge gap between content creators and educators, leading to costly, time-consuming revisions. For this reason, Generative AI offers potential solutions by speeding up production, reducing costs, and bridge the gap between content creation and education.

Instructional Content. Generative AI has shown potential in creating instructional content for Intelligent Tutoring Systems, offering automation that can significantly cut costs and time. For example, AI-generated synthetic videos have been found to improve learning at the same rate as to traditional instructor videos [15]. Tools like Synthesia allow easy video customization, saving time and money.

Large language models can also enhance instructional content comprehension by improving readability, diversity, cohesion, and removing bias [16, 17]. Carnegie Learning's MATHia, which uses GPT-4, effectively rewrites math problems to enhance readability, improving student learning experiences and saving time in creating personalized math problems [17].

However, the large-scale content creation in the ed-tech industry comes with its own challenges. It necessitates the development of efficient Content Management Systems (CMS) to ensure design uniformity, data security, and effective collaboration among stakeholders. Moodle, an open-source Learning Management System, has started using Generative AI for automatic image and text generation and plagiarism detection. CMS tools like Drupal and Sitecore are also beginning to incorporate Generative AI content creation and instant translations. However, to effectively utilize Generative AI, more extensive integration and the development of new interfaces, workflows, and review processes are needed.

While there are clear cost reductions with digital content created by Generative AI, its effectiveness and quality needs more comprehensive research, for instance, it is crucial to consider the long-term impact of AI generated content on K12 learners.

Adaptivity of Instructional Digital Content. In regards to digital learning content via ITS programs, Carnegie Learning's MATHia is one example in which they were able to successfully utilize LLM to improve both readability and math performance for emerging readers utilizing MATHia's math program. Word problems were rewritten by Carnegie Learning and CAST staff under the guidelines of using only recognizable content and using clear and precise language while aiming to preserve the underlying mathematical content difficulty. Across 7 weeks and 12,374 students, researchers found that rewriting word problems allowed more emerging readers to be better able to master all skills in the workspaces, do so in fewer problems, spend much less time in the workspaces, and make fewer errors [18].

Having humans create and then double check each reworded math word problem is a tedious and time consuming task. Therefore, researchers at Carnegie Learning decided to do a follow-up study in which they utilized LLM to rewrite the math word problems instead. They utilized GTP-4 to rewrite math word problems following the same guidelines as the original research, using chain- of-thought learning as a prompt engineering technique, which provides additional reasoning and scaffolding [17]. This study found that GPT-4 written math word problems improved readability, lexical diversity, and cohesion; however, it did score lower on evaluating the use of familiar words [17]. Overall, this research can be considered a success and will, in the future, allow Carnegie Learning and MATHia to save valuable time and energy in creating personalized, scaffolded, and readable math word problems for emerging readers, with the next course of action to complete a randomized trial comparing human-rewritten and GPT- rewritten problems.

3.1.2 Generative AI in Student-Facing Applications

Generative AI is advancing in student applications, particularly in dialog-based tutoring and AI- learner coding, offering personalized tutoring and code generation.

Dialog-based Tutoring System. In the past year, Large Language Models like GPT have helped address two of the three key challenges we previously identified in dialogue-based tutoring systems: (1) classifying student responses and (2) the necessity for expert involvement in content creation. These models have significantly enhanced AI's capacity to interpret student responses and develop personalized learning materials in a cost-effective way [19, 20]. Although human oversight is still necessary to avoid errors.

Several tools that utilize Generative AI have recently emerged. For instance, Khan Academy's Khanmigo, a K12 dialogue-based tutoring system, employs GPT-4 to tutor students and support teachers. Duolingo's Duolingo Max, a language learning app, uses GPT-4 for engaging students in language learning role play scenarios and for giving corrections, explanations and feedback. Other tools like Praktika and Soofy offer features such as AI speaking buddies and debate chatbots.

Despite their similarities, these tools represent a significant departure from traditional intelligent tutoring designs. They heavily depend on raw LLM outputs and less on pedagogical strategies. This shift raises concerns about over-reliance on LLMs and its impact on engagement and learning outcomes. The effectiveness of LLM-based systems is not consistent across subjects either. For example, they demonstrate potential in teaching basic mathematics but struggle with arithmetic reasoning tasks [21]. Their ability to create convincing misinformation further amplifies the risk of accuracy [7].

Generative AI as Learner Coding Pilot. Coding is another area where AI can assist students. GPTutor, developed by Chen et al. [5], uses the OpenAI ChatGPT API to provide detailed explanations of programming code. Early evaluations suggest that GPTutor is user-friendly and effective in code explanation. Several AI tools in the market aim to support students in their learning journey. For instance, Caktus.ai provides writing and coding assistance, while Writefull offers services for writing, paraphrasing, and copyediting.

3.1.3 Generative AI in Teacher-Facing Applications

Teachers in upper secondary level education spend just 43% of their time teaching, with the rest consumed by non-teaching tasks such as lesson planning and grading [22]. Generative AI offers the potential to automate these tasks, which will be detailed in the following sections.

Creating instructional presentations can be a time consuming task for educators. Kawanishi and Kawashima's [23] study suggests an automatic slide generation method using a BERT classifier. This method enables teachers to upload a document and generate slides that mirror its structure.

New tools like Alef Education's AI Toolkit are emerging as effective solutions. Alef Education's Slides Generator automates the creation of educational content, including slides, lesson plans, and quizzes, thereby easing the teacher's workload. The success of AI-generated content largely hinges on its application and guidance. Teachers, with their deep understanding of their students and subject matter, can effectively use tools like this to adapt the content to their students. This is particularly useful in context where teachers do not have access to instructional content and spend lots of time creating instructional slides.

AI in Teacher PD Training. GPTeach, an interactive chat-based tool, helps novice teachers refine their teaching strategies through interactions with simulated students, generated using LLMs [11]. The tool uses cleverly designed prompts for simulated student responses, providing a stress-free environment for teachers to improve their responses [11]. Another product, Orai, aims to enhance public speaking for professionals, including teachers. It uses speech recognition AI to give immediate feedback on presentations, pinpointing areas for improvement like filler words, pacing, and conciseness. AI tools like GPTeach and Orai aid teachers in content creation and enhancing their teaching strategies.

3.2 Learning About Generative AI: AI Literacy Curriculum for Students and Teachers

This section delves into the integration of Generative AI into AI literacy curriculum for K-12 students and teachers. It explores how this technology is being used to enhance learning experiences, foster a comprehensive understanding of AI principles, and prompt discussions around ethical implications.

Building on the AI literacy curriculum for K12 students and teachers, there's been a notable shift towards incorporating Generative AI in K-12 AI education from 2019 to 2022 [24]. Recognized for its engaging interactive potential, Generative AI allows students to immerse in AI- related activities, fostering a comprehensive understanding of core principles, highlighting AI's wide-ranging applications and prompt thought around AI's broader ethical implications [25, 26].

In line with this, Williams et al. [7] developed three middle school AI literacy curricula: Creative AI, Dancing with AI, and How to Train Your Robot. In the Creative AI curriculum, students explore tools and techniques such as neural networks and generative adversarial networks (GANs) across various forms of media. They use web-based tools to input their own text and drawings to generate new media, creating stories and illustrations using GANs. They also learn about deepfakes. This hands-on approach allows students to actively engage with generative AI, understand its potential and limitations, and reflect on its ethical implications.

Interestingly, Williams et al. [7] found that students identified more potential harms associated with GAN tools that generate anthropomorphic media, such as human faces or videos. Despite being taught techniques to identify deepfakes, students found it challenging to discern them, highlighting the convincing nature of AI-generated anthropomorphic media. This observation highlights a significant challenge within AI literacy—the anthropomorphization of AI, that extends beyond deepfakes [27]. As AI systems have become proficient at generating convincing human-like text, these systems are increasingly portrayed as intelligent in their marketing and the media; this anthropomorphization can lead students to attribute human-like characteristics to AI systems, resulting in misconceptions [28]. To address these issues, educators are advised to use more precise language when explaining AI's functionality [29]. For instance, instead of saying "AI learns" or "AI understands", they should use terms that accurately reflect the engineered nature of AI, such as "AI applications are designed to..." or "AI developers build applications that...". This careful use of language is a crucial part of AI literacy ensuring students have a clear and accurate understanding of AI's nature and capabilities [29]. Furthermore, integrating concepts from neuroscience, philosophy of science, the history of automation, and the distinctions between human intelligence and machine learning could provide a more comprehensive understanding of AI, its capabilities, and its limitations.

3.3 Aied Research Efforts that Drives Industry: Using LLMS

3.3.1 LLMs for Question Generation

As discussed in research question one, one of the key challenges for Intelligent Tutoring Systems is content quantity and quality. When it comes to assessment content, limited assessment question variety, quality of assessments and academic integrity breaches can be a hurdle to power learning analytics and adaptivity of the AI system. Generative AI, with its ability to produce a vast range of tailored practice materials, presents a potential solution. Recent studies have begun to explore the potential of large language models in generating assessment questions. For instance, [30] conducted a study using InstructGPT, a fine-tuned derivative of GPT-3, to generate educational questions catered to machine learning and biology subjects that align with Bloom's taxonomy and difficulty levels. The study found that the questions generated were substantially contextually accurate, grammatically correct, and straightforward to answer, showcasing their potential utility in classroom settings [30]. However, the model was more successful in formulating simple fact-reminder questions, but struggled to generate complex, open- ended questions that required higher-level thinking [30].

Heck and Meurers [31] propose a personalized methodology for building language learning exercises in Intelligent Tutoring Systems, employing diverse sentence structures to expose learners to various language contexts. Yet, the system's effectiveness remains unverified. Meanwhile, Belzak et al. [32] assess the fairness of AI- and human-generated test items using the Duolingo English Test (DET) as a reference. Overall, both types of tests displayed similar amounts of Differential Item Functioning (DIF), indicating comparable bias levels. The study underscores the importance of human reviewers mirroring the test-taking demographic's diversity, as they failed to identify all biases beforehand, irrespective of item source. The findings underscore human involvement as crucial in mitigating biases, despite AI's efficient test generation capabilities.

An additional noteworthy study on question generation is by Moore et al. [33]. Conducted at Carnegie Mellon University, the research team compared the performance of GPT-4 and a rule- based method in assessing multiple-choice questions. The rule-based method outperformed GPT- 4, detecting 91% of flaws, compared to 79% by GPT-4, thereby emphasizing the importance of human oversight in AI-generated content [33].

3.3.2 LLMs for Autograding

In this section we look at application of Language Learning Models for autograding tasks, particularly in grading short answers. Recent strides in Generative AI are shaping this domain, making it less taxing for educators without compromising effectiveness. LLMs have started demonstrating potential in automatically marking text-based answers and essays across a spectrum of subjects [34, 35]. For instance, Schneider and his colleagues reported an accuracy rate of about 86.5% across several languages and question types in 2022.

Despite these promising results, LLMs have encountered challenges, particularly with subjects that demand precise and unambiguous answers, like mathematics. Nonetheless, advancements in integrating LLMs with computer algebra systems show a glimmer of hope. Techniques like MathPrompter, introduced by Imani et al. [21], have significantly improved the LLMs' performance in solving arithmetic problems. Their technique employed zero-shot chain-of- thought prompting to generate multiple algebraic expressions or Python functions to answer the same math problem differently. As a result, they managed to scale the performance of a GPT- based LLM on the MultiArith dataset from 78.7 to 92.5%. However, these strides notwithstanding, Schneider et al. [35] asserted the need for human oversight in automated grading.

Thomas, Gupta, and Koedinger studied how to create user-friendly autograding for online tutoring programs using AI-based, machine learning methods [8]. The research team used two methods to study automatic short answer grading: "by using a learnersourced subset of human-graded tutor responses; and by creating a surrogate model using ChatGPT" [8]. Their study found that the learnsourced responses performed considerably better than the AI-generated responses, and were on the whole more accurate than the ChatGPT provided responses [8].

While AI-generated feedback may have limitations in their current stage that does not mean it is without use or potential: the research team noted that it may be able to currently provide comparative responses or provide similar examples to current problems students are working on. Further research completed by Li and Wu utilized BERT and multi-level semantic features to automatically grade essays [34]. Automated Essay Scoring (AES) came about as AI tools were further developed, but it is still an incomplete tool that often provides shallow or inaccurate feedback. However, by using BERT and deep neural network multi-level semantic features, the research team found that this model has the best essay scoring performance and is a marked improvement from the baseline models the study used [34].

3.3.3 Generative AI as Co-writing Pilot

A study by Stanford researchers Lee et al. [26] showcased the potential of human-AI collaborative writing. They created an interactive interface that enabled collaboration between writers and GPT-3. Utilizing an innovative tool called CoAuthor, the study captured rich interactions between 63 writers and four instances of GPT-3 across 1,445 writing sessions. All interactions were recorded at the keystroke level, including text insertion, cursor movements, and receipt of suggestions. These actions were timestamped to facilitate precise replays and multiple perspective analysis of the generated data. The researchers particularly explored ideation and linguistic capabilities of GPT-3 (e.g. grammar and vocabulary) in creative and argumentative writing tasks to define what is a good collaborative writing between human writers and AI. Two key collaborative metrics discussed in the paper are, increasing writers' productivity and increasing writers' feeling of ownership.

3.3.4 Promises of LLMs for Social Emotional Learning

Social Emotional Learning (SEL), as defined by CASEL, is "the process through which all young people and adults acquire and apply the knowledge, skills, and attitudes to develop healthy identities, manage emotions and achieve personal and collective goals, feel and show empathy for others, establish and maintain supportive relationships, and make responsible and caring decisions" (2023). With SEL considered a critical twenty-first century skill that all students (and adults) should have mastery over, this opens the door for its introduction as either instructional or auxiliary content for AI products targeted at students. As supplemental content on academic programs, rather than ITS platforms saying a simple "Good Job" or "Well Done" after a lesson, AI tools can provide targeted feedback about specific SEL skills depending on the performance of the student. Lin et al. used an LLM-based system that helps classify feedback tutors can provide to students based on student work that goes beyond the simple "Nice Work" response (2023). For example, the system utilized could help tutors who provide incorrect or insufficient responses to students: "Saying [insert Outcome] is praising students for the outcome. You should focus on praising the students for their effort and process towards learning. Do you want to try responding again?" [36]. A similar system can potentially provide further, more SEL-specific feedback; students who may not have performed their best in the lesson can receive information about stress resistance and persistence, while students who completed a lesson quickly and with minimal mistakes can receive information on creativity and self-reflection or meta-cognition [37].

Although education systems are aware of its importance, a current gap that exists in education systems is explicit instruction content for SEL; generative AI content creation processes and tools are an excellent way to bridge this gap, in which they can create explicit SEL programming based on one of the many SEL frameworks publicly available, such as the OECD, CASEL, UNICEF, or Five Cs frameworks (EASEL [27]). Such ITS programs can directly assist students in understanding various SEL skills, how they are transferable skills across various aspects of the student's life, and how improving upon and properly utilizing these skills can benefit them both in the present as well as in their future academic, work, and personal lives. Leiker et al. looked into AI-created videos to distribute content, which would be a beneficial medium for an SEL-focused e-learning content on digital learning platforms [15]. The research team found positive learning improvements from the introduction of both the experiment and control (an AI-generated video versus a traditional professor-generated video), as well as no significant difference in learner reception between the different videos [15]. This research is promising in that it could potentially allow more content to be generated and disseminated more rapidly in a video format, which is a preferred medium for certain learning and content. As mentioned above, the Kawanishi and Kawashima study highlights the benefits of automatic slide generation programs to augment teachers in the classroom [23]. This is another beneficial tool for SEL, as it allows for the quick creation of SEL-focused content with very little time or energy needed to create said content, benefiting both students and teachers.

3.4 AIED Research that Explores the Analogy Between AI and HI

In the previous section of the paper we examined teachable agents as an AIED research effort that explores the analogy between AI and HI. In this section, we include two learning science contributions: (1) AI creativity versus K12 learner creative thinking skills, (2) Intelligence augmentation between AI and HI. We believe these discussions are really needed to make judgment calls about if and how AI is going to replace humans in various workplace contexts.

3.4.1 Human Creative Thinking Versus AI Creativity

With the rise of Generative AI, particularly DALL-E, the image-generating model, there are discussions on whether AI is getting as creative as humans. In the following section, we analyze creativity by DALL-E versus creative thinking defined for K12 learners by PISA. PISA defines creative thinking as "the competence to engage productively in the generation, evaluation and improvement of ideas, that can result in original and effective solutions, advances in knowledge and impactful expressions of imagination" [37]. The competencies of creative thinking, as defined by PISA, are: generate diverse ideas, generate creative ideas, and evaluate and improve ideas (2019). In this way, creative thinking asks "students to engage productively in the cognitive processes of idea generation (the generation of diverse or creative ideas respectively) and idea evaluation and improvement. It therefore does not only look at the divergent cognitive processes of creative thinking (the ability to generate diverse or creative ideas); students are also asked to evaluate other people's ideas and develop and suggest original improvements to those ideas" [37]. This, in turn, can be evaluated across four domains: written expression, visual expression, scientific problem solving, and social problem solving [37]. In the classroom, PISA further breaks down creative thinking into three elements: individual enablers, social enablers, and achievement and progress [37]. Individual enablers are students' intrinsic resources for engaging in creative thinking, social enablers are aspect of students' school environment that either fosters or hinders creative thinking, and achievement and progress are students' individual or group displays of creative thinking that can be seen and measured, all of which come together to create a multi-dimensional model of creative thinking in the classroom.

Gordy Slack, a science writer at Stanford University, explored DALL-E as it compares to human creativity and creative thinking, doing so by speaking with four authorities on artificial intelligence, the brain, and creativity [38]. Experts told Slack that DALL-E can be considered similar to human creativity in that it is built on a neural network that is meant to mimic the brain and it works by absorbing examples to produce new work just like humans [38]. However, the differences between human and AI should be the main focus, as summed up by Isaac Kuavar and touches directly on the key PISA creative thinking competency of evaluate and improve ideas: "When I'm drawing," he says, "it's an iterative process. I usually don't know where I'm going to end up. I first just get something down and that inspires the next iteration, and that inspires the next one, and so on…DALL-E can quickly produce many variations, but it relies on a human to evaluate or modify them" [38]. Even further, Ashley Etemadi and Chris Dede of Harvard University highlight that a major component DALL-E and AI tools are lacking in comparison to humans is reflection and metacognition: "Humans have the consciousness to monitor, interpret, and regulate our own mental thinking. Research shows that metacognitive skills can be learned and homed in humans. Machines, on the other hand, have no consciousness; they cannot reflect on their designs, or functions, despite being able to "learn" from patterns in large data sets" (2021). In this regard, we cannot stop bolstering creativity in students, as human input and art is still vital; we must instead push for more reflection, metacognition, and creative thinking in the classroom, as developing these skills will ensure that students will be able to augment artificial intelligence in both work and life, rather than be replaced by it.

3.4.2 Intelligence Augmentation: AI Reckoning and Human Judgement

Researchers at Harvard University define intelligence as "the disposition to reason, plan, solve problems, think abstractly, comprehend complex ideas, learn quickly and learn from experience" [39]. This is then divided into the complementary roles of reckoning and judgment, where "reckoning refers to calculative prediction and formulaic decision-making, at which computers and AI systems already excel" and judgment is a "form of deliberative thought that seeks to be unbiased, grounded in ethical commitment, and appropriate to the situation in which it is deployed," at which humans are particularly adept [39]. This creates a complementary relationship in which AI is able to complete incredibly fast, accurate, and large-scale reckoning, which humans can use in addition to other factors to complete the necessary judgment. This partnership is the foundation of intelligence augmentation, where "AI and humans engage in a complementary partnership in which a human-and-AI team's overall performance is greater than their individual capacity" [39]. However, judgment is not a finite skill and therefore can and should be taught in schools; instructional models already exist and work to provide students with the necessary mental dispositions to improve their judgment skills. These teaching models "purposefully seek to develop intellectual character, which includes attention to curiosity, open-mindedness, and skepticism," as well as "build upon the neuroscience and cognitive science of learning that offer insight into how the human mind learns best and how brains learn better than machines" [39]. As AI takes on more and more reckoning roles and tasks, we must now focus on educating and preparing students to successfully take on the judgment role and related soft skills. This can include SEL skills, as highlighted above, as well as AI literacy, which can be understood as a "set of skills that enable a solid understanding of AI through three priority axes: learning about AI, learning about how AI works, and learning for life with AI" [40]. Through this, we can work to ensure that students are better prepared for school, work, and life, having the necessary knowledge and skills needed to find success in a world with AI.

4 Conclusion

This book chapter attempts to review predominant industry and research contributions within the AI education four-interaction framework. Research question one sets a foundation by capturing key student-facing learning with AI applications that have been researched and in industry over the past decades (e.g. ITS, dialog-based ITS, adaptive assessments and autograding) and the reviews highlighted existing challenges, or areas of improvement for ITS and dialog-based ITS. For teacher-facing applications, this section introduced AI-teacher dual partnership as an example of intelligence augmentation between AI and HI (human intelligence). From the research interaction, this section chose educational data mining (EDM) learning analytics (LA), and teachable agents among other research contributions for the following reasons: (1) EDM and LA have been used to drive core components of ITS, for instance, knowledge tracing, student mastery in adaptive learning systems, learning analytics dashboards for learners, teachers, and school leadership, prediction models that drives education policy-making; (2) teachable agent, different than the procedural paradigm in ITS, contributes to self-regulated learning, exploratory problem-solving and deep learning through teaching an AI agent, which presents analogy between AI and Hi, as well as innovations that could potentially improve pedagogical interactions between AI and learner to improve learner agency. Last but not least, AI literacy is reviewed as a key effort from governmental and non-governmental organizations to equip K12 students and teachers with knowledge and skills about AI.

Research question two introduced Generative AI contributions from industry and research within the four quadrant interactions. Given the nature of Generative AI, most of the industry's Generative AI education applications focus on automating digital content generation in the following areas:

(1) instructional and assessment content in ITS and wider e-learning platforms, which function mainly as learning inputs; (2) adaptivity and personalization of digital content, for instance, MATHia have utilized large language models (LLMs) to enhance content readability for learners who have literacy challenges for math, and this approach presents opportunities to personalize content with a lower cost in ITS industry; (3) natural language interaction in dialog-based ITS (e.g. Khanmigo from Khan Academy), brings more possibilities for human AI natural language conversation, compared with IBM watson tutor that used rule-based AI in 2018; (4) teacher instructional content creation tools allow teachers to create their own instructional content when needed, this is especially helpful in contexts where teachers do not have access to instructional content (e.g. teachers teaching Chinese language in foreign context with no instructional materials available). The above possibilities brought by Generative AI mainly offer cost reduction and time efficiencies through automation of content creation; however, there are considerable concerns including maintaining the quality of AI-generated content, ethical issues like copyright violation, plagiarism, and misinformation, along with potential loss of human input and expertise in pedagogical design. Furthermore, the adoption of Generative AI may prompt shifts in the economics surrounding the production of educational content.

From the research side, this book chapter reviewed papers that utilized LLMs for question generation, autograding, which still contributes to various intelligent tutoring systems, because ITS is powered by lots of content to be adaptive, and this is very costly and time consuming to build. LLMs potentially could reduce cost, yet the quality of various types of content generated by LLM is still a question.

Additionally, Generative AI brings new possibilities, for instance, researchers start to experiment AI as co-writing pilot to define what is good collaboration between AI and humans, AI as coding pilot, discussions on AI creativity and learner creative thinking, LLMs generating instructional content and interactive hints for social emotional skills, LLMs potentially could create simulated PD training environment for novice teachers.

Despite all the potentials and contributions of AI and Generative AI, K12 education is about cultivating human capital, and learner skill development is still at the center. Current Generative AI products marketed towards students are primarily built for cognitive skills and are designed to assist students and teachers in an academic context. However, there remains a distinct gap in products aimed at twenty-first century skills, such as SEL, creative thinking, metacognition, self- regulation, problem-solving and AI literacy. As these skills will help students augment AI reckoning skills and improve judgment skills needed in future work and life.

5 Future Work and Limitations

Based on the above reviews, we recommend the following aspects as future work directions:

1. To enhance the effectiveness of Intelligent Tutoring Systems, we should also prioritize developing AI-learner interactions that boost learner agency, problem-solving abilities, and metacognitive skills. The main objective of these cognitive tutors is to assist learners in recognizing and addressing their knowledge gaps or misunderstandings. This requires a strategic teaching approach where the design and implementation of supportive interventions are crucial. For example, scaffolds of a math misconception could be presented to the learner in various ways (e.g. worked example with scaffolds completely presented to the learner as information inputs, or interactive scaffolds that involve the learner in the process to contribute). However, it's not just about creating supportive structures, but also about asking the right questions: Are we incorporating research best practices in our designs? Are we deliberately fostering problem-solving skills? Can the tutoring system provide feedback on the learner's mastery of the missing skill? Is it practical to implement these supportive structures, and how do we identify the best methodology? It's also vital to take into account the learner's abilities and learning style in these designs. For instance, students with high cognitive and self-regulated learning skills might benefit from certain types of interactions, while those with lower cognitive skills and less self-regulation might require more interactive interventions.
2. Research, collaboration, and validation are key to solving these challenges. Initially, a cross-functional team, including researchers, pedagogical experts, UX designers, product managers, and engineers, should collaborate to clearly understand the problems and consider solutions based on research best practices

and feasibility. Subsequently, the research team should employ a multimethodology research approach to validate their assumptions, which should encompass A/B testing, qualitative research, and ideally, well- designed controlled trials.
3. The collaboration between industry and academic research is pivotal in comprehending the efficacy and impact of AI-generated content. Industry can significantly benefit from incorporating academic research, as academia is a treasure trove of innovation and new ideas. Many topics and concepts that the industry grapples with have already been extensively researched and discussed in academic circles. By tapping into this wealth of knowledge, the industry can accelerate its innovation process and avoid reinventing the wheel. Furthermore, understanding and using academic definitions and concepts can enhance communication and collaboration. Precise and accurate language can prevent misunderstandings and ensure that all parties are on the same page. For fruitful collaboration, certain practices are essential. These include clear agreements, robust data privacy and security, data standardization, and consistent communication. By adhering to these, the industry can deepen its understanding of AI-generated content and foster a more productive partnership with academia.
4. Attention and efforts are needed from the Ed-tech sector to advance twenty-first century skills (e.g. social emotional skills, creative thinking in written expression, visual expression, scientific problem solving and social problem solving, etc.). Many Ed-tech products in the market claim to cultivate twenty-first century skills, yet these products mainly focus on cognitive skills. And with the advancement of AI and Generative AI, we can see how these twenty-first century skills are the key to have AI and HI augmentation. With Generative AI trends, we can see the power and speed of the capital market and industry. Hence, there is a lot of space and opportunities to build e-learning products for these skills.
5. AI K12 literacy curriculum needs to reach a wider audience. So far only a small number of countries and institutions have started to implement AI literacy curriculum for K12 students and teachers.
6. In many contexts, such as instructional content creation, the quality of output is more important than the quantity. However, the current focus of most generative AI systems is on automation, time-saving, cost reduction, and minimizing human input. There is a clear need for a shift in research focus towards systems that prioritize quality over speed and facilitate high-skilled work. Reinforcement learning for human feedback or systems like AlphaCode could serve as inspiration for these systems.

Although this book chapter made all possible attempts to be as thorough in the review as possible, our scope was still limited and each quadrant only contains the most predominant and relevant examples. More research must be done to obtain a more comprehensive understanding of the implications now and in the future for generative AI in education.

To close this chapter, we'd like to quote Chris Dede, "Don't let anything or anyone—including me—do your thinking for you" [41].

References

1. Poldrack RA, Lu T, Beguš G (2023) AI-assisted coding: experiments with GPT-4. arXiv preprint arXiv:2304.13187 [cs.AI]. https://doi.org/10.48550/arXiv.2304.13187
2. City EA, Elmore RF, Fiarman SE, Teitel L (2009) A network approach to improving teaching and learning. Harvard Education Press, Cambridge
3. Duolingo (n.d.) Duolingo max: a language learning experience powered by GPT-4. Duolingo Blog. https://blog.duolingo.com/duolingo-max/
4. Rismanchian S, Doroudi S (2023) Four interactions between AI and education: broadening our perspective on what AI can offer education. In: Wang N, Rebolledo-Mendez G, Dimitrova V, Matsuda N, Santos OC (eds) Artificial intelligence in education. Posters and late breaking results, workshops and tutorials, industry and innovation tracks, practitioners, doctoral consortium and blue sky. AIED 2023. Communications in computer and information science, vol 1831. Springer, Cham. https://doi.org/10.1007/978-3-031-36336-8_1
5. Chen E, Huang R, Chen H-S, Tseng Y-H, Li L-Y (2023) GPTutor: a ChatGPT- powered programming tool for code explanation. In: International conference on artificial intelligence in education. https://doi.org/10.48550/arXiv.2305.01863
6. Etemadi A, Dede C (2021) Preparing students to augment artificial intelligence rather than to be replaced by machine learning. Harvard University Next Level Lab. https://nextlevellab.gse.harvard.edu/publications/
7. Williams R, Ali S, Devasia N, DiPaola D, Hong J, Kaputsos SP, Jordan B, Breazeal C (2023) AI+ ethics curricula for middle school youth: lessons learned from three project-based curricula. Int J Artif Intell Educ 33(2):325–383. https://doi.org/10.1007/s40593-022-00298-y
8. Thomas DR, Gupta S, Koedinger KR (2023) Comparative analysis of learnersourced human-graded and AI-generated responses for autograding online tutor lessons. Artificial intelligence in education. Posters and late breaking results, workshops and tutorials, industry and innovation tracks, practitioners, doctoral consortium and blue sky, pp 714–719. https://doi.org/10.1007/978-3-031-36336-8_110
9. Baker R, Siemens G (2014) Educational data mining and learning analytics. https://learninganalytics.upenn.edu/ryanbaker/Chapter12BakerSiemensv3.pdf
10. Cukurova M, Miao X, Brooker R (2023) Adoption of artificial intelligence in schools: unveiling factors influencing teachers' engagement. In: Wang N, Rebolledo-Mendez G, Matsuda N, Santos OC, Dimitrova V (eds) Artificial intelligence in education. AIED 2023. Lecture Notes in Computer Science, vol 13916. Springer, Cham. https://doi.org/10.1007/978-3-031-36272-9_13
11. Markel JM, Opferman SG, Landay JA, Piech C (2023) GPTeach: interactive TA training with GPT-based students. EdArXiv. https://doi.org/10.1145/3273051.3593393
12. Biswas G, Leelawong K, Schwartz D, Vye N, The Teachable Agents Group at Vanderbilt (2005) Learning by teaching: a new agent paradigm for educational software. Appl Artif Intell 19:3–4, 363–392.https://doi.org/10.1080/08839510590910200
13. Michie D, Paterson A, Hayes JE (1989) Learning by teaching. In: Proceedings of the second Scandi-navian conference on artificial intelligence (SCAI). Tampere, Finland. IOS Press, pp 413–436
14. Hietala P, Niemirepo T (1998) The competence of learning companion agents. Int J Arti Intell Educ 9:178–192
15. Leiker D, Gyllen AR, Eldesouky I, Cukurova M (2023) Generative AI for learning: investigating the potential of learning videos with synthetic virtual instructors. Artificial intelligence in education. Posters and late breaking results, workshops and tutorials, industry and innovation tracks, practitioners, doctoral consortium and blue sky, pp 523–529. https://doi.org/10.1007/978-3-031-36336-8_81
16. Barker C, Kazakov D (2023) ChatGPT as a text simplification tool to remove bias. arXiv:2305.06166 [cs.CL]. https://doi.org/10.48550/arXiv.2305.06166

17. Norberg K, Almoubayyed H, Fancsali SE, De Ley L, Weldon K, Murphy A, Ritter S (2023) Rewriting math word problems with large language models. AIEd23: artificial intelligence in education, empowering education with LLMs workshop
18. Almoubayyed H, Bastoni R, Berman SR, Galasso S, Jensen M, Lester L, Murphy A, Swartz M, Weldon K, Fancsali SE, Gropen J, Ritter S (2023) Rewriting math word problems to improve learning outcomes for emerging readers: a randomized field trial in Carnegie Learning's Mathia. Artificial intelligence in education. posters and late breaking results, workshops and tutorials, industry and innovation tracks, practitioners, doctoral consortium and blue sky, pp 200–205. https://doi.org/10.1007/978-3-031-36336-8_30
19. Macina J, Daheim N, Wang L, Sinha T, Kapur M, Gurevych I, Sachan M (2023) Opportunities and challenges in neural dialog tutoring. arXiv:2301.09919
20. Yadav G, Tseng Y-J, Ni X (2023) Contextualizing problems to student interests at scale in intelligent tutoring system using large language models. arXiv preprint arXiv:2306.00190 [cs.HC]. https://doi.org/10.48550/arXiv.2306.00190
21. Imani S, Du L, Shrivastava H (2023) MathPrompter: mathematical reasoning using large language models. arXiv:2303.05398 [cs.CL]. https://doi.org/10.48550/arXiv.2303.05398
22. OECD (2022) Education at a glance 2022: OECD indicators. https://doi.org/10.1787/3197152b-en
23. Kawanishi T, Kawashima H (2023) Automatic slide generation using discourse relations. In: Wang N, Rebolledo-Mendez G, Dimitrova V, Matsuda N, Santos OC (eds) Artificial intelligence in education. Posters and late breaking results, workshops and tutorials, industry and innovation tracks, practitioners, doctoral consortium and blue sky. Springer Nature Switzerland, Cham, pp 390–395. https://doi.org/10.1007/978-3- 031–36336–8_61
24. Rizvi S, Waite J, Sentance S (2023) Artificial Intelligence teaching and learning in K-12 from 2019 to 2022: a systematic literature review. Comput Educ: Artif Intell 4:100145. https://doi.org/10.1016/j.caeai.2023.100145
25. Ali S, DiPaola D, Lee I, Sindato V, Kim G, Blumofe R, Breazeal C (2021) Children as creators, thinkers and citizens in an AI-driven future. Comput Educ: Artif Intell 2:100040. https://doi.org/10.1016/j.caeai.2021.100040
26. Lee I, Ali S, Zhang H, Dipaola D, Breazeal C (2021) Developing middle school students' AI literacy. In: Association for computing machinery, Inc., pp 191–197. https://doi.org/10.1145/3408877.3432513
27. Deshpande A, Rajpurohit T, Narasimhan K, Kalyan A (2023) Anthropomorphization of AI: opportunities and risks. arXiv:2305.14784 [cs.AI]. https://doi.org/10.48550/arXiv.2305.14784
28. UNESCO (2021) AI and education. Guidance for policy-makers. https://unesdoc.unesco.org/ark:/48223/pf0000376709
29. Garside B (2023) How anthropomorphism hinders AI education. Raspberry Pi Blog. https://www.raspberrypi.org/blog/ai-education-anthropomorphism/
30. EASEL Lab (2023) Explore SEL. http://exploresel.gse.harvard.edu/frameworks/
31. Heck T, Meurers D (2023) Exercise generation supporting adaptivity in intelligent tutoring systems. In: Wang N, Rebolledo-Mendez G, Dimitrova V, Matsuda N, Santos OC (eds) Artificial intelligence in education. Posters and late breaking results, workshops and tutorials, industry and innovation tracks, practitioners, doctoral consortium and blue sky. Springer Nature Switzerland, Cham, pp 659–665. https://doi.org/10.1007/978-3-031-36336-8_102
32. Belzak WCM, Naismith B, Burstein J (2023) Ensuring fairness of human- and AI- generated test items. In: Wang N, Rebolledo-Mendez G, Dimitrova V, Matsuda N, Santos OC (eds) Artificial intelligence in education. posters and late breaking results, workshops and tutorials, industry and innovation tracks, practitioners, doctoral consortium and blue sky. AIED 2023. Communications in computer and information science, vol 1831. Springer, Cham, p 108. https://doi.org/10.1007/978-3-031-36336-8_108
33. Moore S, Nguyen HA, Chen T, Stamper J (2023) Assessing the quality of multiple- choice questions using GPT-4 and rule-based methods. arXiv preprint arXiv:2307.08161 [cs.CL]. https://arxiv.org/abs/2307.08161

34. Li J, Wu J (2023) Automated essay scoring incorporating multi-level semantic features. Artificial intelligence in education. Posters and late breaking results, workshops and tutorials, industry and innovation tracks, practitioners, doctoral consortium and blue sky, pp 206–211. https://doi.org/10.1007/978-3-031-36336-8_31
35. Schneider J, Richner R, Riser M (2022) Towards trustworthy autograding of short, multilingual, multi-type answers. Int J Artif Intell Educ. Retrieved from arXiv:2201.03425 [cs.CL]
36. Liang W, Yuksekgonul M, Mao Y, Wu E, Zou J (2023) GPT detectors are biased against non-native English writers. arXiv:2304.02819 [cs.CL]
37. OECD (2019) (rep.) PISA 2021 creative thinking framework (Third Draft
38. Slack G (2023) What DALL-E reveals about human creativity. Stanford HAI. https://hai.stanford.edu/news/what-dall-e-reveals-about-human-creativity
39. Dede C, Etemadi A, Forshaw T (2021) Intelligence augmentation: upskilling humans to complement AI. https://pz.harvard.edu/sites/default/files/Intelligence%20Augmentation-%20Upskilling%20Humans%20to%20Complement%20AI.pdf
40. Casal-Otero L, Catala A, Fernández-Morante C, Taboada M, Cebreiro B, Barro S (2023) AI literacy in K-12: a systematic literature review. Int J STEM Educ 10(1):29. https://doi.org/10.1186/s40594-023-00418-7
41. Chris Dede (2023) What is academic integrity in the era of generative artificial intelligence? Silver lining for learning. https://silverliningforlearning.org/what-is-academic-integrity-in-the-era-of-generative-artificial-intelligence/
42. Afzal S, Dhamecha T.I., Mukhi N., Sindhgatta R., Marvaniya S., Ventura M., Yarbro J., (2019) Development and deployment of a large-scale dialog-based intelligent tutoring system. In: Proceedings of the 2019 conference of the north American chapter of the association for computational linguistics: human language technologies, vol 2 (Industry Papers). Minneapolis, Minnesota. Association for Computational Linguistics, pp 114–121
43. Alef Education (n.d.) Education redefined where every learner fulfills their unique potential. Alef Education. Retrieved August 5, 2023, from https://www.alefeducation.com/
44. Beck J, Stern M, Haugsjaa E (n.d.) Applications of AI in education. https://dl.acm.org/doi/fullHtml/https://doi.org/10.1145/332148.332153
45. Bendle S (2023) AI for drupal sites: to the next level and beyond. Promet Source. https://www.prometsource.com/blog/AI-for-Drupal
46. Brand J, Israeli A, Ngwe D (2023) Using GPT for market research (Working Paper 23- 062). Harvard Business School. https://www.hbs.edu/ris/Publication%20Files/23-062_b8fbedcd-ade4-49d6-8bb7-d216650ff3bd.pdf
47. Brooks R (2023) Just calm down about GPT-4 already and stop confusing performance with competence. IEEE Spectrum. https://spectrum.ieee.org/gpt-4-calm-down
48. CASEL (2023) Fundamentals of SEL. CASEL. https://casel.org/fundamentals-of-self/
49. Chaka C (2023) Detecting AI content in responses generated by ChatGPT, YouChat, and Chatsonic: the case of five AI content detection tools. J Appl Language Teach 6(2):12. https://doi.org/10.37074/jalt.2023.6.2.12
50. Coyner AS, Chen JS, Chang K, Singh P, Ostmo S, Chan RVP, Chiang MF, Kalpathy-Cramer J, Campbell JP, Imaging and Informatics in Retinopathy of Prematurity Consortium (2022) Synthetic medical images for robust, privacy-preserving training of artificial intelligence: application to retinopathy of prematurity diagnosis. J Ophthalmic Photogr 4(1):100126. https://doi.org/10.1016/j.xops.2022.100126
51. Elkins S, Kochmar E, Serban I, Cheung JCK (2023) How useful are educational questions generated by large language models? In Wang N, Rebolledo-Mendez G, Dimitrova V, Matsuda N, Santos OC (eds) Artificial intelligence in education. posters and late breaking results, workshops and tutorials, industry and innovation tracks, practitioners, doctoral consortium and blue sky. Springer Nature Switzerland, Cham, pp 536–542
52. Feng Y, Vanam S, Cherukupally M, Zheng W, Qiu M, Chen H (2023) Investigating code generation performance of ChatGPT with crowdsourcing social data. University of North Texas
53. Helms S, Krieser J (2023) Copyright chaos: legal implications of generative AI. Bloomberg Law. https://www.bloomberglaw.com/external/document/XDDQ1PNK000000/copyrights-professional-perspective-copyright-chaos-legal-implic

54. Hermann E (2022) Anthropomorphized artificial intelligence, attachment, and consumer behavior. Mark Lett 33:157–162. https://doi.org/10.1007/s11002-021-09587-3
55. Kasneci E, Seßler K, Küchemann S, Bannert M, Dementieva D, Fischer F, Gasser U, Groh G, Günnemann S, Hüllermeier E, Krusche S, Kutyniok G, Michaeli T, Nerdel C, Pfeffer J, Poquet O, Sailer M, Schmidt A, Seidel T, Stadler M, Weller J, Kuhn J (2023) ChatGPT for good? on opportunities and challenges of large language models for education. Comput Educ 183:107010. https://doi.org/10.1016/j.compedu.2023.107010
56. Khalil M, Er E (2023) Will ChatGPT get you caught? Rethinking of plagiarism detection. arXiv preprint arXiv:2302.04335. https://arxiv.org/abs/2302.04335
57. Li Z, Yang Z, Wang M (2023) Reinforcement learning with human feedback: learning dynamic choices via pessimism. arXiv preprint arXiv:2305.18438v3 [cs.LG]. https://doi.org/10.48550/arXiv.2305.18438
58. Lin J, Thomas DR, Han F, Gupta S, Tan W, Nguyen ND, Koedinger KR (2023) Using large language models to provide explanatory feedback to human tutors. https://arxiv.org/pdf/2306.15498v1.pdf
59. Liu Z, Yu X, Zhang L, Wu Z, Cao C, Dai H, Zhao L, Liu W, Shen D, Li Q, Liu T, Zhu D, Li X (2023) DeID-GPT: zero-shot medical text de-identification by GPT-4. arXiv preprint arXiv:2303.11032 [cs.CL]. https://doi.org/10.48550/arXiv.2303.11032
60. Lucchi N (2023) ChatGPT: a case study on copyright challenges for generative AI systems. SSRN. https://ssrn.com/abstract=4483390
61. Miao X, Mishra PK, Nadaf A (2021) Evidence and promises of AI predictions to understand student approaches to math learning in Abu Dhabi K12 public schools. Gulf Educ Soc Policy Rev (GESPR) 1(2):109–134. https://doi.org/10.18502/gespr.v1i2.8458
62. Molnár G, József C (2022) AI based plagiarism checking: ease of use and applicable system for teachers to find similarities in students' assessments. In: 2022 IEEE 5th international conference and workshop Óbuda on electrical and power engineering (CANDO-EPE). Budapest, Hungary, pp 000187–000192. https://doi.org/10.1109/CANDO-EPE57516.2022.10046379
63. Neumann M, Rauschenberger M, Schon¨ E-M (2023) "We need to talk about ChatGPT": the future of ai and higher education. university of applied sciences and arts hannover. https://serwiss.bib.hs-hannover.de/frontdoor/deliver/index/docld/2467/file/neumann_etal2023-chatGPT_AI_higher_ed.pdf
64. OECD (2023) Explore SEL. http://exploresel.gse.harvard.edu/frameworks/43
65. Okado Y, Nye BD, Aguirre A, Swartout W (2023) Can virtual agents scale up mentoring?: insights from college students' experiences using the CareerFair.ai platform at an American Hispanic-Serving Institution. In: Holzinger A et al (eds) Artificial intelligence in education. Posters and late breaking results, workshops and tutorials, industry and innovation tracks, practitioners, doctoral consortium and blue sky. Springer, pp 150–155. https://doi.org/10.1007/978-3-031-36272-9_16
66. Panda S, Kaur N (2023) Exploring the viability of ChatGPT as an alternative to traditional chatbot systems in library and information centers. Library Hi Tech News 40(3):22–25. https://doi.org/10.1108/LHTN-02-2023-0032
67. Saharia C, Chan W, Saxena S, Li L, Whang J, Denton E, Ghasemipour SKS, Ayan K, Mahdavi SS, Lopes RG, Salimans T, Ho J, Fleet DJ, Norouzi M (2023) Photorealistic text-to-image diffusion models with deep language understanding. arXiv preprint arXiv:2205.11487 [cs.CV]. https://doi.org/10.48550/arXiv.2205.11487
68. UNESCO (2022) K12-AI curricula, a mapping of government-endorsed AI curricula. https://unesdoc.unesco.org/ark:/48223/pf0000380602
69. Watson D (2019) The rhetoric and reality of anthropomorphism in artificial intelligence. Mind Mach 29(3):417–440. https://doi.org/10.1007/s11023-019-09506-6
70. Williams R, Breazeal C (n.d.) Assessment for K12 AI literacy: a comprehensive review. MIT Media Lab, working paper
71. Zhou J, Zhang Y, Luo Q, Parker AG, De Choudhury M (2023) Synthetic lies: understanding AI-generated misinformation and evaluating algorithmic and human solutions. In: Proceedings of the 2023 CHI conference on human factors in computing systems, pp 1–20. https://doi.org/10.1145/3544548.3581318

SPRINGER NATURE

GPSR Compliance

The European Union's (EU) General Product Safety Regulation (GPSR) is a set of rules that requires consumer products to be safe and our obligations to ensure this.

If you have any concerns about our products, you can contact us on ProductSafety@springernature.com

In case Publisher is established outside the EU, the EU authorized representative is:

Springer Nature Customer Service Center GmbH
Europaplatz 3
69115 Heidelberg, Germany

The manufacturer's authorised representative in the EU is Springer Nature Customer Service Centre GmbH, Europaplatz 3, 69115 Heidelberg, Germany. If you have any concerns regarding our products, please contact ProductSafety@springernature.com

Printed and bound by CPI Group (UK) Ltd, Croydon, CR0 4YY
30/09/2026
02225319-0015